T0155668

THE SECOND LAW OF ECONOMICS

THE FRONTIERS COLLECTION

Series Editors:
A.C. Elitzur L. Mersini-Houghton M.A. Schlosshauer M.P. Silverman
J.A. Tuszynski R. Vaas H.D. Zeh

The books in this collection are devoted to challenging and open problems at the forefront of modern science, including related philosophical debates. In contrast to typical research monographs, however, they strive to present their topics in a manner accessible also to scientifically literate non-specialists wishing to gain insight into the deeper implications and fascinating questions involved. Taken as a whole, the series reflects the need for a fundamental and interdisciplinary approach to modern science. Furthermore, it is intended to encourage active scientists in all areas to ponder over important and perhaps controversial issues beyond their own speciality. Extending from quantum physics and relativity to entropy, consciousness and complex systems – the Frontiers Collection will inspire readers to push back the frontiers of their own knowledge.

For further volumes:
http://www.springer.com/series/5342

Reiner Kümmel

THE SECOND LAW OF ECONOMICS

Energy, Entropy, and the Origins of Wealth

 Springer

Reiner Kümmel
Institute for Theoretical Physics and Astrophysics
University of Würzburg
Am Hubland
97074 Würzburg
Germany
kuemmel@physik.uni-wuerzburg.de

Series Editors:

Avshalom C. Elitzur
Bar-Ilan University, Unit of Interdisciplinary Studies, 52900 Ramat-Gan, Israel
email: avshalom.elitzur@weizmann.ac.il

Laura Mersini-Houghton
Dept. Physics, University of North Carolina, Chapel Hill, NC 27599-3255, USA
email: mersini@physics.unc.edu

Maximilian A. Schlosshauer
Niels Bohr Institute, Blegdamsvej 17, 2100 Copenhagen, Denmark
email: schlosshauer@nbi.dk

Mark P. Silverman
Trinity College, Dept. Physics, Hartford CT 06106, USA
email: mark.silverman@trincoll.edu

Jack A. Tuszynski
University of Alberta, Dept. Physics, Edmonton AB T6G 1Z2, Canada
email: jtus@phys.ualberta.ca

Rüdiger Vaas
University of Giessen, Center for Philosophy and Foundations of Science, 35394 Giessen, Germany
email: ruediger.vaas@t-online.de

H. Dieter Zeh
Gaiberger Straße 38, 69151 Waldhilsbach, Germany
email: zeh@uni-heidelberg.de

ISSN 1612-3018
ISBN 978-1-4614-2919-7   ISBN 978-1-4419-9365-6 (eBook)
DOI 10.1007/978-1-4419-9365-6
Springer New York Dordrecht Heidelberg London

Cover design: KuenkelLopka GmbH, Heidelberg

Printed on acid-free paper

Springer is part of Springer Science+Business Media (www.springer.com)

And it was the amount of energy a single human could produce that dictated military potential, standard of living, happiness, and all besides.
Isaac Asimov, *The Naked Sun, 1956*

To
   Christa and Stephan,
   Lukas, Florian, Franziska, and Jakob

Foreword

Physicists contribute in various fields of sciences for a better understanding of our world. This is linked to the type of education they have to undergo. During their studies they are trained to solve problems, rather difficult and tricky problems in many cases. They like to do that, otherwise they would have chosen the wrong field for themselves. With this attitude in mind, they have made numerous significant contributions to such different fields as chemistry, biology, medicine, and what is of particular interest here, economics. Since physicists are used to working with equations, which represent their ideas in a transparent, logical, and consistent frame; they will naturally carry over that attitude to other fields in which they are working. This distinguishes them from those scientists who express their ideas solely in words and descriptions.

The author of this volume, Reiner Kümmel, is a physicist in the best sense. He received excellent training at world-renowned universities and despite considerable scientific success in a special field of physics, namely, superconductivity, he remained a generalist. What started as a hobby, namely, the study of the laws and driving forces of economics, rather soon became a serious occupation and a new branch of interdisciplinary work. His personal experiences, which he gained in different parts of the world, made him realize what former President Bill Clinton used as the 1992 campaign catch phrase: "It's the economy, stupid!"

After his studies in Germany Reiner Kümmel spent his postdoctoral years in Urbana (Illinois) as an assistant of John Bardeen, one of the true giants of science. In the USA he experienced the prosperity and wealth of the world's leading economic power. A few years later, he became acquainted with the life and struggle, of the people in Colombia, where he served for three years building up a master's program at the Universidad del Valle in Cali in the spirit of Kennedy's Peace Corps. The huge difference in living conditions between the USA and Germany, on the one hand, and Colombia, on the other, together with his determination to do something to improve the lives of people, provided the background for his increasing engagement in economics. He realized rather soon that as a well-trained physicist he could make an important contribution to that field.

This monograph is the result of more than three decades of research in economics. His main contribution and message is a serious study of the consequences of incorporating energy and entropy into the models of economic development. I am sure that entropy is a subject that still deserves more attention than it has hitherto received in economics, simply because it is a concept one needs to get used to. By the way, the well-known Max Planck thought lifelong about it. So it is natural that concepts of that type take a long time before they penetrate into such a different field like economics.

I am sure that this book will fill a blank space and I hope that it will stimulate students and researchers alike. In any case, it will widen our views on economics and contribute to the development of this important science for the benefit of mankind.

Dresden, October 2010   Peter Fulde

Preface

Thermodynamics was a subject I thoroughly disliked when I was a student. I just could not understand the physics behind the rattling of exact, inexact, and partial differentials. And my interest in economics matched Thomas Carlyle's characterization of the field as the "dismal science," although at that time I did not appreciate the Malthusian basis of Carlyle's description.

These attitudes changed a lot as I grew older. Fascinated by many-body quantum mechanics, I got a first inkling of the practical usefulness of thermodynamics when John Bardeen of the University of Illinois at Champaign-Urbana gave me a problem for which I had to derive the time-dependent equations of motion for quasiparticles in inhomogeneous superconductors at finite temperatures. Minimization of free energy provided the important quasiparticle distribution function, and more. Later, my colleagues in the Physics Department of Universidad del Valle in Cali, Colombia, asked me to teach thermodynamics in their newly established master program. When I objected that this was the field I was least familiar with, they recommended Frederick Reif's *Fundamentals of Statistical and Thermal Physics* as the best book to improve my state of knowledge. They were right. Reif's combined statistical and phenomenological descriptions of interacting many-body systems pulled the veil from my eyes that had prevented me from seeing the beauty and power of thermodynamics. Finally, I understood entropy.

Then came the shock of the 1972 publication *The Limits to Growth*. I realized how naive I had been when I went to Colombia to join the efforts to industrialize this beautiful, tortured country by teaching physics to its gifted students. If industrialization, done the European and American way, were to spread to the developing countries, entropy production would create problems mankind had never faced in history. The next – oil price – shock, and the concomitant economic recession in 1973–1975, showed the vulnerability of industrial economies to reductions of energy conversion. These two shocks introduced thermodynamics and economics as the third theme, besides superconductor and semiconductor physics, of my teaching and research at the University of Würzburg since 1974.

I owe a lot to people who taught me more about economics and thermodynamics. First, there is the late Wilhelm Dreier, economist and theologian at the University of Würzburg. In joint interdisciplinary seminars on economic growth and its problems, I learned that in economic theory there is practically no room for energy as a factor of production beside capital, labor, and land. I found this hard to believe and asked Wilhelm for a good introduction to economics. He recommended Paul A. Samuelson's textbook *Economics*. This book educated me as much in economics as Reif's book did in thermodynamics. After the publication of my first article in an economics journal, the dean of theoretical physics in Würzburg, the late Helmut Steinwedel, established contact with Wolfgang Eichhorn from the Institute of Economic Theory and Operations Research at the Technical University of Karlsruhe. Working together with Wolfgang during the last 30 years has, hopefully, prevented me from falling into the interdisciplinary traps that await people who venture from their field into other disciplines. At one of the international conferences on economic theory of natural resources organized by Wolfgang, I met the late Willem (Pim) van Gool from the Energy Science Project in the Department of Inorganic Chemistry of the State University of Utrecht. Pim introduced me to energy, cost, and emission optimization in industrial systems, and to all that matters in exergy and enthalpy. Interaction with colleagues from the Working Group on Energy (AKE) and the econophysics community of the German Physical Society (DPG) has also fostered research in energy science and econophysics. During the first of a series of workshops entitled "Advances in Energy Studies," organized in 1998 by Sergio Ulgiati, then at the University of Siena, Charles A. Hall of SUNY at Syracuse, New York, Robert U. Ayres of INSEAD at Fontaienbleau, France, and I discovered our common interest in heterodox economics. Since then I have benefitted greatly from our cooperation and exchange of ideas. Personal encounters with the late Gerard K. O'Neill of Princeton University's Physics Department, and participation in three "Princeton Conferences on Space Manufacturing Facilities," inspired my hope that the collision with the limits to growth on Earth might be mitigated by a timely rediscovery of O'Neill's bold vision of *The High Frontier*.

Students are the heart of research. They work out the difficult details of an idea their advisor suggests and often carry on far beyond that. I was lucky that good students took the risk of doing interdisciplinary research, despite my advising them to be rather on the safe side with theses in semiconductor or superconductor physics. This book has benefitted in one way or another from my former students (in chronological order) Klaus Walter, Bruno Handwerker, Helmuth-M. Groscurth, Uwe Schüssler, Thomas Bruckner, Volker Napp, Alexander Kunkel, Hubert Schwab, Dietmar Lindenberger, Julian Henn, Jörg Schmid, and Robert Stresing. Dietmar Lindenberger, presently at the Institute of Energy Economics of the University of Cologne, is still an active partner in ongoing research. Arne Jacobs from my superconductivity group and Andreas Vetter helped with all sorts of IT problems.

During the last few years drafts of this book have served as a text for my course on thermodynamics and economics, and the feedback from the students who took the course has been very helpful. In that course and this book, I try to summarize the basic facts on energy and entropy, which are taught in Würzburg during the first five semesters of physics studies. This material is supplemented by information on fossil, nuclear, and renewable energy sources, the technological options of using them, and the possibilities of emission mitigation. Of course, it is only possible to discuss a subjective selection from the huge amount of research on these topics. The chapter on economic evolution is quite different from the preceding two chapters. It presents methods and results of research in energy and economic growth since 1980. These things have been published in peer-reviewed journals. The results are not in line with mainstream economic thinking. There are also people in the growing field of heterodox economics who agree with the results but dislike the mathematical methods used in their derivation. For them, the methods are too similar to those of neoclassical economics. The mathematics of orthodox economics, borrowed from classical physics, is attractive to a physicist. The idea has been to incorporate energy, entropy, and technological constraints into the orthodox mathematical machinery and see how the picture of economic evolution changes. The reader may judge for himself or herself whether the new picture, with the dominant role of energy conversion in economic growth and the threat from entropy production to future growth, is convincing or not. The time travel prologue with its qualitative description of natural, technical, and social evolution may facilitate the understanding of energy conversion as the driver of change without any mathematics. Ethical problems concerning economic development, and hope that proper action will be taken, are indicated in the epilogue. Some considerations are repeated in different parts of the book so that the chapters are self-contained and can be read independently of each other.

I am grateful to my colleagues in the Faculty of Physics and Astronomy of the University of Würzburg for not only tolerating my going partly astray from the path of monodisciplinary physics, but also for being helpful in many ways.

Last but not least, I thank my wife Rita for detecting inappropriate wording and lots of typographical errors, bearing with the physicists' priorities, and all encouragement.

Würzburg, October 2010   Reiner Kümmel

Acknowledgements

Sincere thanks go to

- EOLLS Publishers Co. Ltd., for granting permission to reproduce a passage in the epilogue that was originally published in my contribution "Energy, creativity and sustainable growth" to *Our Fragile World—Challenges and Opportunities for Sustainable Development*, forerunner to the *Encyclopedia of Life Support Systems*, M.K. Tolba (ed.) Vol. 1, pp. 409–427, UNESCO and EOLSS, Oxford (2001).
- Dr. W. Kroy, Ludwig Bölkow Stiftung, for granting permission to reproduce Figs. 2.7, 2.9, and 2.10 from a talk he presented at the founding symposium of the Denkwerk Zukunft in October 2008.
- Professor C.-D. Schönwiese, Universität Frankfurt, for providing the source files from whose modifications Figs. 3.6 and 3.9 were produced.
- Professor R.U. Ayres and Drs. L. Ayres and B. Warr, INSEAD, Fontainebleau, for providing the figures from which Figs. 4.17, 4.18, 4.19, and 4.20 were produced.
- Jan Bühler and my son Stephan for kind help in the drawing of several figures.
- Professor C. Hall, SUNY at Syracuse, New York, for establishing contact with Dr. D. Packer of Springer Science+Business Media, who accepted my lecture notes on thermodynamics and economics for publication, proposed the book title, and gave valuable advice for the presentation of the material.
- Professor P. Fulde, Max-Planck-Institut für Physik komplexer Systeme, Dresden. I was his assistant when he held a chair in theoretical physics at Johann-Wolfgang-Goethe Universität in Frankfurt. He supported my going to Colombia and staying in physics after my return to Germany. Without that support, this book would not have been written.

Contents

Chapter 1
Prologue: Time Travel with Abel

Imagine an observer in the state beyond space and time from which one can watch the universe and human history unfold from the beginning to the end. Let us call him Abel. He takes us on a voyage through time and lends us his eyes and insight. This is what we see.

1.1 From the Big Bang to the Sun

A singularity out of nothing blows up in a glaring white. "This is *ENERGY* – the cosmic building stuff," Abel whispers as we watch in awe. "You see the beginning of space–time in the Big Bang 14 billion years ago. Right now the primordial content of the universe has a temperature of 10^{32} degrees."[1]

Space and time expand. The quark soup condenses out of the glittering radiation. Then quarks form protons and neutrons. These fuse into the first light elements: deuterium, helium, and lithium. "Now the universe is 100 seconds old, and its temperature is down to some billion degrees," Abel comments. The cosmos expands further. After 400,000 years matter and radiation decouple; space is filled by a multicolored glow: the cosmic background radiation and its fluctuations. A dark age follows for the next 600,000 years, when the first stars form and fuse the elements heavier than iron, such as copper, silver, and gold. Then stars and galaxies become visible. They proliferate and fill the universe with their shining glory, while it expands to size of over 100 billion light years. The cosmic background radiation has cooled down to a temperature just 2.725 degrees above absolute zero. Abel summarizes what we have seen:

---

[1] The Celsius (°C) temperature scale has its zero point at 273.15 degrees above the zero point of the absolute Kelvin (K) temperature scale.

R. Kümmel, *The Second Law of Economics: Energy, Entropy, and the Origins of Wealth*,
The Frontiers Collection, DOI 10.1007/978-1-4419-9365-6_1,
© Springer Science+Business Media, LLC 2011

"All matter has condensed out of energy, all changes are driven by energy conversion, and all structures originate from energy fluctuations, such as the ones you note in the slightly warmer and colder regions of the background radiation."

Before we rejoice about having the full cosmic vision, Abel cautions us: "You have just seen 5% of what the universe contains. The rest is 20% dark matter and 75% dark energy." He refuses to reveal more about dark matter and dark energy, stating that he is only allowed to show what is already part of human knowledge. When we ask him "What is human knowledge about energy?" he replies, "I'll just give you the grand tour. Details you may look up in the treatise I'll hand over to you at the end of our voyage."

Our vision zooms in on an average star at the fringe of a galactic spiral arm. Protuberances flicker on its surface, and flares of gleaming hot gases shoot up into the darkness of space. A distant blue planet encircles the radiating sphere. "The fountain of life," Abel comments, and recites

"Splendid are you in the heavenly mountain of light,
Living Sun, living since the Beginning,
filling all the lands with your beauty.
Great are you, shining in every country,
embracing all the earth with your life-giving rays.

This is how the Egyptian pharaoh Amenophis IV, who calls himself Echnaton, greets the Sun."

We dash toward the Sun. At its surface Abel announces: "The temperature is 5777 K. Let's go to the center. It is just 696,000 km away." Our space–time elevator speeds down past huge, swaying tubes in which gleaming hot gases are driven up by convection. After 200,000 km the tubes disappear, and there is just a glow. Then, farther down, we are surrounded by glorious gold. Abel tells us: "We are in the wedding saloon of the Sun's particles. It is the solar core with a radius of 140,000 km. Here, every second 600 million tons of hydrogen are fused into helium. The mass difference between the hydrogen and the fusion product helium is about four million tons. It is all converted into energy, at 15 million degrees. Watch out for protons and neutrons. They show up as red and black balls. You will also see red dots, the positrons. Photons, the quanta of light, will flash, and neutrinos will appear and vanish chimerically. Here we go."

Space teems with protons. Occasionally, two protons fuse into a black-red compound. "That's deuterium," Abel informs. "And did you see the positron and the neutrino escape?" Deuterium catches another proton. "Now we have helium-3." This happens many times. Each helium-3 compound chases after a partner of the same kind, and in most cases the two merge into a two proton–two neutron nucleus, emitting two protons. At each particle wedding, photons flash up. Like the neutrinos, the photons would like to dash away at the velocity of light. But they are absorbed immediately in the proton–neutron throng, then they are reemitted, reabsorbed, and in this catch-and-let-go game they diffuse away at a crawling pace.

"This has been going on for more than four billion years and should continue at least that long into the future. Four hydrogen nuclei – that's what the protons are – fuse into one helium-4 nucleus. In so doing, they generate two positrons, two neutrinos, and two photons. The photons are almost trapped in the extreme density of matter in the solar core, which is about 150 times the density of water. Therefore, they still need about a million years until they get out of the Sun and provide the Earth with light and warmth."

"People know the Sun's importance for life," Abel adds, "but only few realize that they are also children of long-gone stars."

Sensing our question, he explains: "In the Sun's atmosphere there are traces of heavy elements. These elements, quite common on Earth, can only be generated in fusion processes at temperatures much higher than those in the core of the Sun. Temperatures above 10^8 and 10^9 degrees occur in contracting stars, which have burned up all their hydrogen and fuse higher elements. These fusion processes produce energy, up to iron, ^{56}Fe. The fusion of elements heavier than iron consumes energy. Such elements are cobalt, nickel, copper, tin, silver, gold, lead, and uranium, the heaviest natural element in the periodic table. This energy may have been provided by novae and supernovae. Thus, the Sun, the Earth, and everything on Earth itself have been processed through the inside of at least one star."

We digest the feeling that most components of our bodies have been parts of dying, exploding stars. Then we move back in time by four billion years.

1.2   Light on Earth

The Sun is fainter than the one we know. The Earth is wrapped in a uniform gray layer of clouds. While we wonder how cold it may be down there, Abel tunes in: "Earth's surface temperature is about 85°C. It is so hot because of the greenhouse effect in an atmosphere that consists mainly of nitrogen, methane, water vapor, and up to 1,000 times more carbon dioxide than in the atmosphere you know." He explains that during the next 3.5 billion years most of this carbon dioxide will become dissolved in the oceans, where bacteria and algae will produce oxygen from it. The weathering of silicate rocks on the continents, followed by the deposition of carbonate sediments on the sea floor, will also remove carbon dioxide from the atmosphere/ocean reservoir. "This drastic decrease of carbon dioxide has reduced the greenhouse effect to a very convenient level. Between the two revolutions that decisively shape human history – the Neolithic revolution, with its beginning of farming and cattle breeding, and the Industrial Revolution, with its invention of the heat engine – it keeps the average surface temperature of the Earth at a comfortable +15°C. Without it you would have a deadly −18°C. And now lets move to the Cambrian, with its explosion of life forms, 530 million years before your time. Since then you can observe the forces that drive evolution."

The Earth has become the blue planet. Oceans surround land masses. Clouds sail through the thin shell of the atmosphere. In a first quick dash we ride the arrow of time through the ages of Earth. They are marked by the trilobites, the first fish and insects, the conquest of the land by plants and reptiles, the forests of giant ferns and shave-grass, the saurians, the conifer and deciduous forests, and the mammals. And during all that time the only inputs into the Earth system are energy, emitted by the Sun, cosmic radiation, and once in a while some rocks from outer space.

Solar energy activates life and fosters its growth. This also becomes dramatically patent by the mass extinctions of species we observe during periods when volcanic eruptions or dust, stirred up by the impact energy of huge meteorites, block much of the sunlight. The catastrophic disappearance of the dinosaurs 60 million years ago makes room for the mammals, which until then had barely survived in ecological niches. We also see how ionizing particles from solar or cosmic radiation, or terrestrial radioactive material, transfer energy to the genes of the living cell. This causes mutations that occasionally result in new species. "Got it?" Abel checks our understanding. "Energy conversion and genetic information processing drive the evolution of species."

Contemplating the Sun and the Earth, we understand more deeply why the Sun has been revered as sacred throughout the ages. Abel quotes from Shakespeare's Sonnet VII:

> *Lo! in the orient when the gracious light*
> *Lifts up his burning head, each under eye*
> *Doth homage to his new-appearing sight,*
> *Serving with looks his sacred majesty;*
> *And having climbed the steep-up heavenly hill,*
> *Resembling strong youth in his middle age,*
> *Yet mortal looks adore his beauty still,*
> *Attending on his golden pilgrimage.*

1.2.1 As Life Goes

Our vision zooms in on the nanoworld of the living cell. We enter the interior of an algal cell. "Watch the process of photosynthesis," Abel recommends.

We see the pulsating green compound of chlorophyll in the center of the cell. Flashes of incident photons dance over its surface. The compound pumps currents of yellow electrons along conducting chains. Red hydrogen and blue oxygen atoms flow out of the watery envelops of the chains. Brown adenosine triphosphate boxes, bearing the letters ATP, are also emitted and move into a dark reaction chamber. A gray gas of carbon dioxide molecules flows into this chamber and mixes with hydrogen. Varying its color several times, the mixture reaches the ATP boxes and reacts with them seethingly. White sugar ribbons emanate and slide toward the border of the cell. There, new cells separate and float away through blue oxygen molecules that bubble out of the wall of the cell.

Our guide comments: "In nature's sugar plant, chlorophyll converts the energy of the photons into work performed by electric currents that flow along molecular chains and produce adenosine triphosphate. ATP serves all living species as *the* universal energy currency. It is transported to places where work has to be performed. There, ATP gives off the energy from the Sun stored in it and produces sugar and new cells. Summing this up quantitatively, we note that, via the chlorophyll of the living cell, sunlight converts six water and six carbon dioxide molecules into six oxygen molecules and one sugar molecule. This breeds new cells. And now observe the complementary part of the life cycle: the conversion of sugar into work. It's called respiration."

We see an Amano shrimp devouring algae. Inside its translucent body, algae fragments merge with blue oxygen balls, which enter from the surrounding water. Brown ATP boxes emanate from the merger zone, accompanied by an undulating glimmer. The ATP boxes dissolve, their energy is transferred to the legs, which begin to move, and the shrimp crawls away, emitting gray carbon dioxide and red-blue water molecules.

Abel continues: "Here you see how the sugar of the devoured algal cell is burned with oxygen so that the moving shrimp's legs can do work. Again the whole process operates via the conversion of the solar-generated chemical energy of sugar into the chemical energy of adenosine triphosphate. This ATP acts as a sort of battery. As in photosynthesis, this battery delivers energy to those parts of the cell where work must be performed by discharging itself. To be more precise: during the combustion of food, one molecule of sugar combines with six molecules of oxygen to become six molecules of water plus six molecules of carbon dioxide plus adenosine triphosphate. The undulating glimmer you have noticed is caused by waves of waste heat into which, unfortunately, a certain part of valuable energy must always be converted. The same processes occur in the predators that feed on the shrimps, and in all other plants and animals."

Abel illustrates the cycle of life by a picture [1]: "The controlled process of the biological energy cycle can be depicted by the running of a series of water mills driving generators which charge batteries.... When photosynthesis is compared to a solar-driven pump used to bring 'water' to an elevated level, respiration can be represented as the stepwise downfall of the 'water' which drives the 'water mills' charging the 'ATP batteries.' The batteries then can be transported to sites where work has to be done; when properly connected, they can be 'discharged' by the hydrolysis reaction when work is performed."

Then we are shown how the giant stores of fossil fuels are formed from the products of photosynthesis.

In the Carboniferous and the Permian, about 300 million years before the present, huge forests grow in warm, swampy freshwater regions. When the trees in these forests die, they fall onto the swampy ground and are buried by the debris of the following years. Many generations of plants form layers of dead vegetation, which, in turn, are overlaid by sediments of nonorganic material washed down into the low-lying swamps from surrounding higher ground. Thus, the dead biomass, sealed off from the oxygen of the air, cannot rot away, and a good part of the energy stored in it is conserved, when it is squeezed and transformed into peat. As more layers of sediment pile up upon the organic deposits and these sink further down, coming under increased heat and pressure, they are further transformed, first into lignite (brown coal), then (hard) coal, and finally anthracite. Later, in the Tertiary era, which lasts between 64 million years and one million years before the present, we note the second peak of coal formation, when the large deposits of lignite are formed. We also observe the production of oil and natural gas from the remains of plants and animals, especially plankton. These remains are laid down mainly in coastal regions near or under salt water and are eventually sealed off by sediments that build up to form new layers of rock. Over millions of years, in reduction reactions with hydrogen sulfide (H_2S) and with bacterial support, they undergo chemical changes similar to those that produce coal, and become the liquid and gaseous stores of solar energy [2].

1.2.2   Fire and Grain

Abel takes us to the Quaternary, less than a million years before the present. Huge ice masses spread from the north and south poles over the northern and southern parts of the continents, and glaciers creep from high mountains into plains. When it gets warmer, the ice recedes and the land greens, then it gets colder again, and the ice comes back. The average surface temperature of Earth varies rapidly by several degrees Celsius.

In this harsh environment the first humans roam the fields and forests as collectors of plants and their fruit and live on a daily energy budget of about 2 kWh. Then they take up hunting. Although physically much weaker than their prey, such as mammoths, and competing predators, such as bears and tigers, they prevail thanks to the use of tools made from stone and wood.

A huge leap forward in the art of survival is made by the taming of fire, roughly half a million years before the present. We watch the bold leader of a horde grab the fire with a dry branch from the flames that engulf a tree ignited by lightning. The members of his horde begin to guard and nourish the fire. Quickly its domination spreads to other hordes. People learn more and more how to use the energy liberated by the oxidation of carbon and hydrogen in wood for warming their caves, defense against wild animals, cooking plants, roasting meat, and the preparation of weapons such as fire-hardened yew-tree spears. By then, the average energy consumption is 6 kWh per person per day.

Abel reminds us of Greek mythology: "Prometheus stole the fire from Olympus, the residence of the gods, and brought it to the humans on Earth. Zeus, the king of the gods, punished him cruelly for this deed, which gave humans so much power and saved them from doom." To be sure that we really understand the paramount achievement of prehistoric man, he adds Goethe's reference to Prometheus:

Kindle the Fire! Fire's on top.
Greatest the deed of stealing it.
He who lightened it,
he who made friends with it
hammered and rounded crowns for Man's head,

and quotes from Schiller's "Song of the Bell":

Power of fire, how beneficial
if carefully guarded and harnessed by man.
Whatever he forms, and what he creates,
he owes it to you, o gift of the gods.

We arrive at the dawn of human civilization in the Fertile Crescent between the Euphrates-Tigris and the Nile. Twelve thousand to 10,000 years before the present, the average temperature of Earth rises by more than 4°C and stays nearly constant after that. Still, small fluctuations occur but hardly exceed more than 1°C. After the much stronger fluctuations of the ice ages, advanced humans live for the first time in a nearly stable, warm climate. Photosynthetic biomass production occurs in bountiful, predictable cycles. In this new environment, which feels like paradise compared with the living conditions of the preceding ice age, *Homo sapiens* triggers the "Neolithic revolution": Men and women invent farming and cattle breeding. Instead of just collecting and hunting what grows and lives in grasslands, forests, and waters, humans expand their harvesting of solar energy systematically, and to an extent that grows with the area of the agriculturally utilized land.

"Look at Eve, how she did it," Abel suggests.

We see a woman who collects the seeds of grass. She separates out especially big grains and stores them for times of drought. After a number of fertile years, when the grain store overflows, she throws out the grains from the oldest harvest into the backyard of her house on the bank of the broad river. The next spring, grass plants with bigger-than-average grains of seeds grow in the backyard. The woman gets an inkling of a totally new opportunity of food provision. She sows more of the big-grain seeds and selects again the biggest grains from the blades that grow out of them. After a number of cycles of sowing, harvesting, and selecting, the woman has a field close to her house from which she gets more grain food than from the huge area of the savannah she used to roam when collecting ordinary grass. Meanwhile, her male companion continues hunting, watching her efforts with quite some suspicion. When she finally asks him to help her dig up some more ground in order to expand the area of big-seed cultivation, he first protests full of indignation. After all, he is a hunter and grain care is women's business. But his wife, seductively beautiful in her enthusiasm about her discovery, convinces him to do what he had always considered as something out of question. He joins her in digging and planting and harvesting the new fruit of knowledge. Together they cultivate the special seeds into what finally becomes wheat.

"But Adam is never quite happy with having traded free hunting for tilling the soil. He thinks that he has lost paradise," Abel concludes this vision.

Other people pick up the art of farming. Subsequent generations learn how to domesticate animals. In our privileged view provided by Abel, we see and understand the geographic advantage in food production enjoyed by the inhabitants of the Eurasian land mass and northern Africa over the humans who live in sub-Saharan Africa and the Americas: In Eurasia there are many more domesticable wild plants and animals than on other continents. Domesticated mammals such as sheep, goats, pigs, oxen, cows, donkeys, horses, and camels provide meat, milk, leather, and manure, and they also provide muscle power for plowing the fields, the transportation of goods, and rapid military attack. These animals and domesticated birds such as chicken, geese, ducks, and turkeys convert the chemical energy of plants into high-quality food and physical work for the benefit of man. Furthermore, agricultural innovations diffuse much more easily along Eurasia's east–west axis than along Africa's and America's north–south axes, occupied by geographic and climatic obstacles. "Around those axes turned the fortunes of history" [3].

Our guide adds: "Whereas the food energy harvested per hectare per year by hunters and gatherers is only about 1 kWh, it amounts to more than 3,000 kWh for Indian wheat farmers, and nearly 80,000 kWh in Chinese intensive farming [4]. The energetic yields of agricultural technologies are the foundation of the preindustrial high civilizations around the Mediterranean, in Asia, northern Europe, and southern America. A time-compressed view of the energetics of these civilizations is the next part of the tour I have to offer."

1.3   Ancient Empires

The early agricultural societies unfold. Seven thousand years before the present they produce food surpluses that can satisfy an energy demand of 14 kWh per person per day. This liberates some of their members for specialization in crafts such as pottery, and the working of wood, stone, and metal. Craftsmen join the peasants. On these pillars rest the first agrarian high civilizations that rise about 5,000 years before the present. They develop an urban business sector, pronounced social strata, trade, art, and writing. Thus, farmers and craftsmen provide the energetic and technological means that empower the ancient empires of East and Southwest Asia, Egypt, Greece, and Rome.

In the agrarian societies economic and political power is with the land owners, because they are the ones who control the energy derived from the direct and indirect products of photosynthesis. In Latin, the original expression for cattle property, *pecunia*, assumes the meaning of "money" and "wealth." The land-owning nobility accumulates far-reaching political power. Feudalism becomes the dominating political system of the agrarian societies. It gains strength with the increasing energy demand of these societies, as they advance technologically, commercially, and militarily. In medieval western Europe, about AD 1400, the energy demand per person per day is 30 kWh. Despite their impressive cultural achievements, the agrarian civilizations are handicapped in their development by the limitations of the forces that can be derived from muscle power and by the low efficiencies of energy conversion in humans and animals. Inclined planes, pulley blocks, windmills, and water mills give some, but only limited help to surmount the biological barriers.

Abel explains: "The tractive power of a horse is about 14% of its body weight and amounts to about 80 kiloponds.[2] For deep plowing one needs 120–170 kiloponds, and for mowing 80–100 kiloponds. The average performance of a horse is 600–700 W, and a donkey provides 400 W. Thus, a winch, normally powered by four donkeys in order to provide mechanical work, has a performance of less than 2,000 W. A horse can perform work of 3–6 kWh/day, and for this it needs fodder with an energy content of roughly 30 kWh. Thus, its energetic efficiency is between 10% and 20%. The energetic limits of cross-country transportation are fixed by the need of a horse to eat one cartload of fodder per week. Therefore, it does not make sense to use a horse and wagon for the transportation of feed for more than a week. The energetic efficiencies of man and horse are similar. However, the average performance of man is only between 50 and 100 W, at most one seventh of the horse performance" [4].

With this information we understand the sad fate of peasants and slaves we observe during the 5,000 years between the first Sumerian, Babylonian, and Egyptian empires and the nineteenth century. Whenever huge armies invade a country in campaigns that last much longer than a week, they have to confiscate the food for soldiers, horses, and draft animals from the peasants of that country. Thus, in times of war, peasants are often robbed of all they have. The alternative to starvation is for the peasants themselves to join the armies.

Although humans are physically much weaker than oxen and horses, the combination of their muscle power with the skills of the human hand and the creativity of the brain is indispensable for all the sophisticated tasks involved in the construction of the pyramids, palaces, temples, and castles that inspire awe in many generations. Furthermore, the members of the nobility feel entitled to a lifestyle that corresponds to the splendor of the buildings they populate. Since it is energetically impossible for the few members of the nobility to provide the means for their luxurious lives themselves, they need huge armies of slaves, serfs, and bondsmen, deprived of rights, who labor for them in quarries, on construction sites, and most important of all, in the cultivation of land. When the apostle Paul writes his letter to Philemon on behalf of the slave Onesimus, about 25% of the population of the Roman Empire are slaves.

Slavery, and its modification socage, was the prerequisite of the impressive cultural achievements of agrarian societies. The glory of the few rose from the misery of the many.

Our vision zooms in on a narrow strip of land between the Lebanon mountains in the north, the Red Sea in the south, the Mediterranean in the west, and the desert

---

[2]The technical force unit "one kilopond" is the force (weight) exerted on a mass of 1 kg by the gravitational field of Earth. It is equal to 9.81 N.

in the east. Abel comments: "The Bible's first book of Kings describes the painful transition from a primitive society of free farmers and cattle breeders to feudalism with its magnificence and compulsory labor."

We see Solomon, the first real king of Israel. His two predecessors, David and Saul, had laid the ground for kingship. But warring against each other and external enemies, they could not yet unfold court life. But Solomon does that lavishly. He builds a temple for the Lord and a palace for himself from huge, valuable stones and cedar wood. To get these materials he establishes socage in all of Israel. Many thousands of men have to take turns laboring in the quarries and the Lebanon mountains. Then Solomon dies after a life of splendor and power remembered throughout the ages. His son Rehoboam is to succeed him. All of Israel comes to Rehoboam and asks him to alleviate their statute labor. He refuses and even promises more of it. The people rebel. Only the tribe of Judah stays with Rehoboam. The rest of Israel elects another king. The empire of Solomon breaks apart into a northern and a southern kingdom, Israel and Judah. Weakened by disunity they are destroyed: first Israel by the Assyrians and then Judah by the Babylonians.

Abel takes us to the Louvre in Paris and guides us to a pillar, covered by cuneiform characters: "When more and more economic and political power accumulated in the hands of those who owned the land, people realized that societies must control power by laws and justice. Wise emperors ordered and ardent prophets demanded protection of the weak. Look at the translation of the inscription on Hammurabi's Stele, excavated in 1901–1902 in Susa." We read the solemn words in which the Babylonian emperor Hammurabi (1793–1750 BC) pronounces the reign of the law: "That I make justice visible in the country, that I exterminate the villains and wicked ones, so that the strong one does not deprive the weak one of his rights, and the land be bright." The extensive body of laws concludes with: "To provide justice for widows and orphans I wrote my delicious words into this monument." Then Abel recites words from the first chapter of the Jewish prophet Isaiah, written 1,000 years later: "So speaks the Lord: I am fed up with your burned offerings. Don't sacrifice meaningless gifts anymore. I do not listen to your prayers anymore, because your hands are stained with blood. Support the suppressed, help orphans to their right, carry on the lawsuit of the widow."

We move in time to ancient Greece. There, as reported by Homer's *Iliad*, one slave is worth four oxen. Slaves of that worth are treated well and help sustain the glory of Greek culture until the Romans take over. The Roman wars of conquest during the second and first centuries BC produce huge armies of prisoners of war that are thrown on the slave market. The price of slaves decreases and their treatment deteriorates. Slave insurrections such as those of Spartacus (73–71 BC) are the result. In new large estates of agrobusiness great masses of slaves produce food at much lower cost than the free peasants with small or medium-sized land property can do. These peasants' civic virtues, productivity, and military strength are the roots from which the Roman Republic rose to power. But small peasants cannot compete, neither with mass production by slaves nor with the imported cereals that inundate the Roman market thanks to the globalization of trade in the Roman Empire. Ruined economically, they migrate to the city and become the Roman proletariat.

The noble big land owners and merchants manipulate them by *panem et circensis*. This sometimes works, and sometimes it does not. Civil wars between the troops of big agrobusiness and the forces of the poor masses shake society. The Roman Republic perishes, torn apart by its slave-generating victories. The Roman emperors take over and rule the world.

When the Roman Empire finally breaks down under the onslaught of barbarian Germanic tribes, civilizing them nevertheless during its agony, new feudal societies evolve in northern Europe. Again, initially free peasants become more and more dependent on big land owners. They suffer in bondage, until European feudalism is swept away by the French Revolution and the Industrial Revolution. Before that the Europeans reestablish slavery on a grand scale, to satisfy the luxury needs of the noble few, and provide wealth for the many who lack land in Europe and acquire it overseas.

Nearly ten million enslaved Africans are brought to the Americas between 1520 and 1850. This by far exceeds the number of European immigrants. Shortly before the outbreak of the American War of Independence, 192 British ships with a total carrying capacity of 47,000 humans participate in slave trade. Slavery in European–American civilization ends with the American War of Secession in 1861–1865. In this first modern war, the industrialized, abolitionist North States defeat the agrarian, feudal, slave-dependent southern Confederate States. Abel explains: "The northern unionist states can afford abolition economically, because they have a new army of slaves at their service: energy slaves."

While we wonder what is meant by "energy slaves," Abel suggests postponing a clarification until we have looked into preindustrial energy services *not* derived from muscle power.

1.4   Wind Power, Gunpowder, and Wood

We see sails on the sea. They belong to the ships of Tarshish which bring "gold, silver, ivory, apes, and guinea-fowls" to Solomon. They belong to Greek, Carthaginian, and Roman fleets that cruise the Mediterranean. They belong to Chinese and Arab vessels navigating along the shores of the Pacific Ocean and the Indian Ocean. Their sails catch the wind and transform its kinetic energy into the kinetic energy of the boats. Often galley-slaves provide additional power.

The sailing ships serve trade and war. Armed with guns, they become the instrument of Europe's rise to power.

Europe's global dominance between the sixteenth and the twentieth centuries is due to the ever more efficient maritime use of wind power by oceangoing sailing ships and the military use of gunpowder energy in firearms – and a Chinese folly in the fifteenth century.

The Chinese are the first to discover gunpowder. From the ninth century they use it in fireworks. But they do not succeed in casting guns that can withstand the sudden release of energy in a gunpowder explosion. About 500 years later, in the fourteenth century, Roger Bacon in England and Berthold Schwarz in Germany discover gunpowder too. Metallurgical progress in bronze and iron casting enables the Europeans to build guns that resist the high pressure of the explosion gases. Guns are the first weapons whose destructive impact does not depend on muscle power. Rather, they transform the chemical energy of gunpowder into the kinetic energy of bullets, cannonballs, and grenades. From the end of the fifteenth century, Portuguese, Spanish, English, Dutch, and French sailing ships with cannons and fire-armed soldiers carry the European conquerors to the Americas, Africa, Asia, and Australia. Much of the world is divided up into European colonies.

Abel takes us to the imperial court of medieval China: "In the early fifteenth century China is the world leader in technology. See how it misses its chance of beating Europe in colonizing the globe" [3].

A big, fat dignitary bows before the emperor. Then he straightens up and pronounces solemnly: "Son of the Heavens, let me summarize the proud achievements of your empire. Their glory shines even more brilliantly against the primitive technological background of the barbarian tribes in the realm of the sinking sun, with whom we have come in contact recently. The long list of our major technological firsts includes cast iron, the compass, gunpowder, paper, printing, and many others. And, may I add, the present administration of your faithful and humble civil servants has elevated China to world leadership in political power, navigation, and control of the seas. During the last few decades we have sent treasure fleets, each consisting of hundreds of ships up to 400 feet long and with total crews of up to 28,000, across the ocean as far as the coast of the black continent the barbarians call Africa. We trade the treasures of our ingenuity with the natural resources of distant countries. As a result, no one matches your majesty's wealth."

Abel adds: "And the Chinese did this decades before Colombus's three small ships crossed the narrow Atlantic Ocean to the Americas' east coast. Now you may be wondering: Why didn't Chinese ships proceed around Africa's southern cape westward and colonize Europe, before Vasco da Gama's three ships rounded the Cape of Good Hope eastward and launched Europe's colonization of East Asia? Why didn't Chinese ships cross the Pacific to colonize the Americas' west coast? Why, in brief, did China lose its technological lead to formerly so backward Europe? Watch a turning point in Chinese history."

A slim, muscular man jumps up and shouts: "Wicked, corrupt eunuchs. You have wasted the empire's resources. You have shielded the Son of the Heavens from the people. You don't care about what is going on in the country. The treasure fleets, captained by the likes of you, serve most of all your personal wealth. This must not go on any longer. Your Highness," he bows before the Emperor, "Your truly faithful servants will now save you and the country from these eunuch parasites." At the shout of the last word the gates of the throne room burst open, warriors storm in and arrest the eunuch administration. The Emperor cannot help but hand power over to the eunuchs' enemies.

Abel extends our vision: "That's the end of China's treasure fleets. Seven of those fleets had sailed from China between 1405 and 1433. Now they are suspended as a result of the power struggle between the two rivaling political factions at the Chinese court you just have witnessed. The eunuchs have been identified with sending and captaining the fleets. Their opponents, having gained the upper hand, reverse the eunuchs' maritime politics: the fleets are detained in the harbors, the shipyards are dismantled, and oceangoing shipping is forbidden altogether in imperial China. That one temporary decision becomes irreversible, because no shipyards remain to turn out ships that would prove the folly of that temporary decision, and to serve as a focus for rebuilding other shipyards. This sort of typical aberration of local politics can happen anywhere in the world and block social, political, or technological progress."

While we wonder how this fits with the general notion that technological progress will prevail one way or the other, Abel adds: "A lasting loss by a folly is much more likely in an isolated society than in societies interacting with each other. This is shown by another folly, this time the abandonment of guns by China's neighbor across the Yellow Sea. Here, cultural tradition finally abhors muscle power being beaten in warfare by the chemical energy of gunpowder. Let's go to Japan and watch its military development since the year 1543, when firearms first reach that country."

A Chinese cargo ship drops anchor in a Japanese harbor. Two Portuguese adventurers armed with harquebuses go ashore. They make big shows with their primitive guns. The Japanese are so impressed by the new weapon that they commence indigenous gun production, greatly improving gun technology. By 1600 they own more and better guns than any other country.

But there are factors working against the acceptance of firearms in Japan. The country has a numerous warrior class, the samurai, for whom swords rate as class symbols, and works of art, and means for subjugating the lower classes. Japanese warfare has previously involved single combats between samurai swordsmen. They stand in the open, make ritual speeches, and then take pride in fighting gracefully. Such behavior becomes lethal in the presence of peasant soldiers ungracefully blasting away with guns. In addition, guns are a foreign invention and grow to be despised, as do other things foreign in Japan after 1600. The samurai-controlled government begins by restricting gun production to a few cities, then introduces the requirement of a government license for producing a gun, then issues licenses only for guns produced for the use of the government, and finally reduces government orders for guns, until Japan is almost without functional guns. Only because Japan is a populous, isolated island can it get away with its rejection of the powerful new military technology. Its safety in isolation comes to an end in 1853, when the visit of Commodore Perry's US fleet bristling with cannons convinces Japan of its need to resume gun manufacture [3].

We look across the seas and are fascinated by the proud, powerful ships under sail that race the oceans, transport goods and people, battle in fierce maritime combats between rivaling sea-faring nations, and tighten Europe's grip on the world.

Abel puts Turner's painting *The Fighting Temeraire* before our eyes (Fig. 1.1). "This painting comprises the end of the sailing ships' glory and the beginning of the greatest leap to power ever made by a civilization."

Fig. 1.1 *The Fighting Temeraire* by William Turner (1775–1851) [5]

The painting depicts a huge, pale, three-mast battleship with tied-up sails. A small, black tugboat, its chimney belching flames and smoke, tows her on the River Thames to her last berth to be broken up.

We know that the tugboat is powered by a coal-burning steam engine and understand that Turner's painting from 1839 heralds the dawn of the age of the fossil-fuel-powered heat engine and the end of the million-year period when humanity only thrived on the daily influx of solar energy and its storage in the living biomass.

"We should observe the use of fire more thoroughly than we have done so far," Abel suggests. "Before we watch its power unfold in the furnaces and heat engines of industry, let us see how fire and wood shape the technology and living conditions of preindustrial societies."

Scanning history, we note the following. Agrarian society obtains heat essentially from wood. Peat and coal are burned to a much smaller extent. There is no technology to transform this heat into mechanical work. Rotatory motions are generated only by wind, water, and animal and human labor – energy forms not permanently available and not very reliable. The corresponding machines such as winches, windmills and water mills are limited in size and power by the small pressure resistance of the universal raw material wood and its intense abrasion. Greater use of iron is frustrated by the high energy cost of iron ore smelting. Iron is very expensive because of technological and resource constraints, which limit the size of blast furnaces. The larger the blast furnace, the larger the quantities of wood required per unit time, and thus the larger the distances over which wood from sustainable forestry has to be transported. Except for the case of timber rafting, long-distance transportation results in negative energy balances as soon as the draft animals consume more scarce biomass energy than is contained in the wood they move. Thus, it is more advantageous to produce iron in small, decentralized production units scattered around in the woods, always close to their energy source. The iron output of a small, charcoal-fired blast furnace in 1 year is less than that of a twentieth century steel mill in 1 day. From an economical point of view, this method of iron production is extremely expensive, although it has ecological advantages such as small emission quantities, which can be absorbed by (and damage only) the local ecosystems. Furthermore, the energy source is renewable – if used sparingly. But often the woods as the main source of energy and raw materials are overused and destroyed.

Around the Mediterranean the once lavish forests are cut down during antiquity to satisfy the timber demand of ship and home construction. North of the Alps, wood becomes scarce first regionally, e.g., close to the German cities of the fourteenth century, and then everywhere in western Europe during the eighteenth century. These wood and energy crises lead to the medieval administrative regulations of forest utilization close to the cities, and in the eighteenth century to the reforestation of wastelands and the development of the science of forestry with its principle of sustainability.

Deforestation is proceeding at an alarming pace in the developing countries of the twentieth and twenty-first centuries. These countries still satisfy about 10% of their energy needs by the burning of wood. However, wood is burned not only for energy purposes, but also, and quite often, just to get rid of the forests which contain the largest variety of living species on Earth. Estimates of carbon release from all forests range from 2 to 10 gigatons (Gt; 1 Gt is one billion tons) annually. This is a nonneglegible fraction of gross annual terrestrial carbon production of 120 Gt in the annual vegetational cycle from carbon to carbon dioxide to carbon, where the total carbon content in all terrestrial biomass is 560 Gt, with 80% in the trees [6]. Most of the tropical rain forests are destroyed to clear land for small-scale farming colonization and accumulate pasture for large agroindustrial cattle ranches; in addition, there is the demand of international timber dealers.

In the northern countries the importance of wood as fuel declines. In 1850 there are 23.2 million US citizens, who, by the burning of wood, consume 76 kWh per capita per day; after peaking at 52 kWh per capita per day for 50.1 million citizens in 1880, US wood energy consumption steadily declines to 3.6 kWh per capita per day for 179 million citizens, until in 1960 it starts to rise again [7, 8]. During the same period, the share of wood energy in total US energy consumption declines from 90% in 1850 to about 50% in 1890, when it is surpassed by the share of coal, and reaches less than 5% by the middle of the twentieth century [9], when oil, coal, and gas have become the principal energy carriers. In Great Britain and continental Europe the switch from wood fuel to fossil fuel occurs much earlier than in the USA. From 1850 to 1950 coal commands a bigger than 90% share of British energy use, use of wood having almost completely disappeared by 1870 [9]. The cause of coal's rise to dominance is the Industrial Revolution.

1.5   Industrial Revolution

"1776 is a magic year. It is one of those years, when things mature, when creativity, hard work, and crises lead to breakthroughs that stir society on a new course," Abel states. "See what happens in 1776:
The first steam engines of James Watt are installed and are working in commercial enterprises. They trigger the Industrial Revolution.
The Wealth of Nations is published by Adam Smith. This book lays the foundation for economic science.
The Declaration of Independence is approved on July 4 by the Second Continental Congress in Philadelphia. It starts the history of the USA with the noble words:
'We hold these truths to be self-evident, that all men are created equal, that they are endowed, by their Creator, with certain unalienable Rights, that among these are Life, Liberty, and the pursuit of Happiness.'
    The first German translation of the Declaration of Independence is published 4 days later by the printing press of Steiner & Cist of Philadelphia.

> The human rights, as proclaimed by the Declaration of Independence, and market economics, as established by *The Wealth of Nations*, would not have become ruling principles of free societies had not steam engines and more advanced heat engines provided the energy services that create the preconditions for freedom from toil.

One may quantify these energy services by the number of 'energy slaves' in an economy. This number is given by the average amount of energy fed per day into the energy conversion devices of the economy divided by the human daily work-calorie requirement of 2,500 kcal (equivalent to 2.9 kWh) for a very heavy workload. In this sense, an energy slave, via an energy-conversion device, does physical work that is numerically equivalent to that of a hard-laboring human. Dividing the number of energy slaves by the number of people in the economy yields the number of energy slaves per capita. The number of energy slaves at the service of a person has increased in time from one, 100,000 years ago, to roughly ten in medieval western Europe, to between 40 and 100 in modern Europe and North America. And, of course, modern energy slaves work much more efficiently than medieval ones. It is also interesting that Jefferson's original draft of the Declaration of Independence included a denunciation of the slave trade, which was later edited out by Congress. Only after industrialization had provided enough energy slaves could the noble words of the Declaration of Independence be finally put into practice – albeit not without the sufferings of the Civil War."
    With this comment on energy and society Abel takes us to Glasgow in the year 1765. We see a 29-year-old Scottish instrument maker wandering across Glasgow Green. His name is James Watt. He is in deep thought: "Two years before I was given

the job of repairing a model Newcomen steam pump for the University of Glasgow and realized the great inefficiency of this engine. But compared with muscle-powered pumps, Thomas Newcomen's engine represents tremendous technical progress. Even the first example from 1711 was able to replace a team of 500 horses that had powered a wheel to pump out a mine. In over 50 years few detailed changes have been made to the basic design. Seventy-five of these engines can now be found at mines all over England. Coal is dearly needed, because wood has become scarce. But following the seams of coal down into the Earth, the mines get flooded by water. Getting that water out of the mines is vital for exploiting the only available substitute for wood fuel. But Newcomen's pump consumes too much of the coal it unearths." Suddenly he takes a leap: "A separate condensing chamber for the steam could save much fuel! That's it!" This idea gets James Watt started on the scheme for an improved steam pump. Soon he has a working model.

Watt meets Mathew Boulton, a dynamic entrepreneur, and tells him about his ideas for improving Newcomen's engine. Boulton agrees to fund development of a test engine and becomes his business partner. For a while, progress is frustratingly slow and Watt repeatedly almost gives up on the project. But Boulton always succeeds in convincing him to continue.

Watt finally gets access to some of the best iron workers in the world. The difficulty of manufacturing a large cylinder with a tightly fitting piston is solved by John Wilkinson, who had developed precision boring techniques for cannon making. Finally, in 1776, the first engines are installed and working in commercial enterprises. These first engines are used as pumps and produce only reciprocating motion. Orders begin to pour in and for the next 5 years Watt is very busy installing more engines, mostly in Cornwall for pumping water out of mines.

The improvement of the steam pump is dramatic: Watt's design uses about 75% less fuel than a similar Newcomen engine. Since the changes are fairly limited, Boulton and Watt license the idea to existing Newcomen engine owners, taking a share of the fuel cost saved by their improvement.

Smoothing the movement of the piston by injecting steam alternately on the two sides of the piston, and adding the sun and planet gear system to transform reciprocating motion in rotary power, Watt develops the pumping steam engine into the multipurpose steam engine. This engine becomes the main driver of the Industrial Revolution.

But there are also social forces that drive the revolutionary change of production and living. The traditional social fabric of the village ruptures, people abandon the countryside en masse and flock to the cities for a living. In the beginning city life is miserable for the newcomers. Manual labor in the small craft shops and factories with primitive machines is extremely hard. The working day has 14–16 hours, and people die early. But the huge surplus of cheap labor continues to exist and becomes another factor for the beginning industrial expansion. Furthermore, England comes under severe economic and political strain. First, because of the loss of its American colonies and, later, because of the continental blockade during the Napoleonic wars. Crises contribute to stimulating innovations.

Abel displays the acceleration of innovations [10]: Watt's multipurpose steam engine becomes operational for industrial applications in 1786, together with Cartwright's loom. An improved weaving machine follows in 1803, and in 1825 the automated Selfaktor spinning machine enters production. The mechanized textile industry grows rapidly, in conjunction with coal and iron industries. Blast furnaces burn no longer charcoal, but coke. This opens up new ways of using coal, e.g., in the form of coal gas, and the chemistry of iron is developed. English coal production, which in 1780 is a mere 6.4 million tons, grows to 21 million tons in 1826 and 44 million tons in 1846. The increasing demand for fuel results not only from the expansion of the textile industry and other manufacturing enterprises, but also to a large extent from the rapidly growing railroad system, which revolutionizes the early industrial era. In 1803 the first locomotive is built for coal mines and in 1829 the first train pulled by steam engines runs between Manchester and Liverpool.

Iron and more iron is needed. The puddle process (1784) makes possible the production of pig iron, which can be forged and rolled. English iron production grows from 68,000 tons in 1788 to 500,000 tons in 1825 and then jumps to one million tons in 1835, two million tons in 1848, and three million tons in 1855. The many branches of mechanical engineering emerge. New building materials are found: Roman cement in 1796, Portland cement in 1824. Coal and iron are the catalysts of the industrial transformation of England.

The appearance of cities changes drastically. Factory chimneys rise above church towers and belch huge, dark clouds of smoke. Charles Dickens writes in *A Christmas Carol* of 1843: "...candles were flaring in the windows ... like ruddy smears upon the palpable brown air." Soot covers buildings and plants. Air pollution causes pneumonia and heart diseases. Occasional smog catastrophes claim many lives in London.

From England the Industrial Revolution leaps to the continent. There, it first catches on in France. In French industry the number of steam engines increases from 625 in 1830 to 5,200 in 1848 and 26,146 in 1870. The railroad system grows slowly between 1832 and 1851 to 3,541 km and then expands rapidly to nearly 18,000 km by 1870. Compared with industrial growth in England, industrial growth in France is restrained during the first half of the nineteenth century, despite France having rich coal and iron reserves. One reason is that high customs barriers protect French heavy industry from international competition. As a result, French blast furnaces are fired by wood and charcoal until the middle of the century. The wood for this backward technology is provided at good profit by the big landowners who have recaptured their prerevolutionary forest properties and are in a strong political alliance with the masters of heavy industry. "This folly is similar to the Chinese abandoning of seafaring and the Japanese banning of firearms," Abel comments. "But since France is anything but isolated, the folly lasts for a much shorter time."

Germany is a latecomer. The Thirty Years' War (1618–1648) and its aftermath had shattered the country into more than 300 states, ruled by absolute monarchs and bishops, who often wastefully and sometimes ridiculously tried to imitate the French *Roi Soleil*. The transformation of the feudal agrarian economy into a system of capitalistic dependencies is way behind that in England. Feudal structures survive into the twentieth century. When this country with its broken national identity finally starts to organize its economic and scientific powers after the formation of the German customs union in 1830 and the recuperation of political unity in 1871, and belatedly but vigorously joins her neighbors on the path of industrial expansion, it shows the typical behavior of the insecure latecomer, which causes Churchill to remark: "The Germans are either at your feet or at your throat."

Germany's way into the industrial era is facilitated by a nearly inexhaustible supply of coal and rich reserves of iron ore. However, initially the growth of the coal and iron economy is tardy, and the accompanying industrial dynamics only slowly gains momentum. In the state of Prussia, coal output increases gradually from 1.1 million to hardly four million tons between 1825 and 1848, whereas England produces 44 million tons in 1846, and the number of steam engines in manufacturing grows from 419 in 1837 to 1,444 in 1849. Machine and vehicle production have not yet conquered the internal market. There are 245 locomotives operating in 1842, of which only 38 have been built in German factories, the rest are imported. German production of pig iron grows from 100,000 tons in 1837 to just less than 230,000 tons in 1847. But after this slow start the industrial economy begins to boom in the second half of the century: coal production leaps from 11.3 million tons in 1857 to 21.8 million tons in 1865 and climbs to 109 million tons in 1900, whereas the output of iron increases to 500,000 tons in 1860 and 1.8 million tons in 1876. The railroad network expands from 549 km in 1840 to 6,044 km in 1850 and 19,575 km in 1870; it reaches a length of 61,148 km in 1910. The Industrial Revolution accelerates with technical progress, which brings improved iron and steel production processes (Bessemer converter, 1856, Siemens–Martin process, 1864), electrical engineering (in 1866 Werner von Siemens develops the electricity generator and the electric motor), telecommunications, chemical industry, and individual mobility (in 1888 Carl Benz builds the first automobile powered by a petrol engine, and in 1897 Rudolf Diesel has the first engine running that bears his name). The Industrial Revolution is accompanied by a rapid population increase due to gradually improving living conditions and medical progress: Ignaz Semmelweis's discovery of the septic and contagious nature of puerperal fever (1847), Joseph Lister's antiseptic wound treatment, and vaccinations against the principal contagious diseases initiate and accelerate the drastic reduction of mortality in industrialized countries. The concomitant decline of the birth rate prevents an even greater population increase.

The Industrial Revolution sweeps over to North America, where the USA grows to become the dominating industrial giant of the twentieth century, and to Japan. Since then, all countries have striven for industrialization.

Abel summarizes the fundamental energetic steps in human history:

"Universal history can be subdivided into three parts. Each part is characterized by a certain energy system. This energy system establishes the general framework, within which the structures of society, economy, and culture form.

> Thus, energy is not just one factor acting among many. Rather, it is possible, in principle, to determine the formal basic structures of a society from the pertaining energetic system conditions" [4].

Becoming a bit more specific he adds: "Humans lived as hunters and gatherers on the solar energy stored in naturally growing biomass for 90% of the time of their existence, that is until the Neolithic revolution. The social structure is that of the horde. For 98% of the time of civilized life, that is between the Neolithic revolution and the Industrial Revolution, people continued to live on the daily influx of solar energy, using naturally growing *and* cultivated biomass, wind power, and water power. Peasants and craftsmen were the pillars of society, whose structure changed from tribal to feudal. The last 2% of civilized life, the years since the Industrial Revolution, has been determined by the combustion of fossil fuels in heat engines. Since then, in addition to the daily influx of solar energy, people have to their avail the giant stores of fossil fuels accumulated on Earth by the Sun over more than 200 million years. They have already used one third of that in 200 years. Now they are turning to the nuclear fuels on Earth and in the Sun. The social structures that have emerged for industrial societies are either democratic or authoritarian. In the battle between democratic and authoritarian societies, democracy seems to emerge as the winner. About 35 countries are considered as full (or only somewhat flawed) democracies by some standards.[3] Among them are the seven industrial countries Canada, France, Germany, Italy, Japan, the UK, and the USA, which have about 11% of the world population and generated more than 64% of global domestic product in 2006."

1.6 Golden Age

We are in 1945. Soviet tanks roar toward Berlin. Allied bomber fleets unload their cargo on Dresden and Würzburg. Some Messerschmitt Me 262 jet fighters take off from remote German air fields and attack Flying Fortresses. The last U-boats launch their torpedoes at oil tankers and are sunk. Nazi Germany collapses. Trucks and trains transport concentration camp survivors and prisoners of war across Europe. Ships bring food from the Americas to fight hunger in devastated Europe. World War II continues in the Pacific theater. On August 6 a lone US aircraft, the B-29 Superfortress *Enola Gay*, drops the first atomic bomb on Hiroshima, destroying 80% of the city and killing 140,000 people. Four days later, Japan surrenders.

The vehicles of war and peace are propelled by fossil-fuel-powered heat engines: tanks, trucks, and submarines by diesel engines, aircraft by gasoline engines,

---

[3]The Economist Intelligence Unit's index of democracy, 2007.

jet fighters by gas turbines, ships by steam turbines, and locomotives by steam engines. Without heat engines the European wars of the twentieth century could not have gone global.

Modern heat engines serve peace even better than war. They put nature's resources and forces at the service of humans in quantities we realize by just looking at areas. The 800-MW steam turbine of a fossil-fuel-burning electric power station occupies an area of $44 \times 14 = 616 \, \text{m}^2$. A horse, which provides muscle power of about $0.7 \, \text{kW}$, needs roughly $10,000 \, \text{m}^2$ of pasture. In total, 1.143 million horses provide the same power as the steam turbine. The total area of pasture they need is 11.43 billion square meters. This area is more than 18 million times larger than the area occupied by the steam turbine. Even if one includes the area required for mining and transportation of the steam turbine's fossil fuel, the power per area of the fossil energy system tremendously exceeds that of the horse system.

After 1945 a Golden Age begins for the industrial democracies. Peace reigns in Europe for a longer time than ever since the end of the Thirty Years' War – except for conflicts on the Balkans. With generous support from the USA, western Europe is rebuilt – first slowly, then at an accelerating pace. In 1951–1952 Belgium, the Federal Republic of Germany, France, Italy, Luxembourg, and The Netherlands form the *European Coal and Steel Community*. This grows into the *European Union* of 27 nations in 2007. Although coal is the energy carrier whose jointly administered power initially drives political unity, its importance decreases when cheap, abundant oil begins to flow from the wells in the Middle East and the Americas and drives the production of wealth in North America, western Europe, and Japan. Progress is slower in the eastern parts of Europe under the inefficient regime of "socialist" planned economies.

The communist party claims to know the course of history and economic evolution. Even at universities and in intellectual circles of western democracies, Marxists teach the inevitability of the collapse of capitalism and predict the global victory of communism. Their religious zeal is based on the belief that all wealth is created by labor, so the accumulation of private property is only possible by the exploitation and pauperization of the working masses. These masses will eventually revolt, establish the dictatorship of the proletariat, and then create the classless society where everybody receives according to his abilities and needs. But experience proves Marxism wrong.

In 1989 the Berlin Wall falls. The Iron Curtain comes down. People who had to live behind it accept the often painful transition from planned to market economies. Most of the former East European satellites of the Soviet Union eventually become members of the European Union.

Abel comments on the failure of the Marxist prophecy: "Marxism failed to realize that private wealth can be created by the exploitation of energy sources instead of the exploitation of people. Industrial democracies have demonstrated that so convincingly that it came to this other magic year, 1989. And thanks to the threat from the devastating energies stored in nuclear weapons the Cold War between the communist and the capitalist camps never grew hot.

An example for the big increase of material well-being in industrial democracies in one generation is the growth of the buying power of the work minute. This is the average working time of an industrial employee the remuneration of which can buy a given quantity of goods. For instance, in the Federal Republic of Germany the number of required work minutes decreased between the years 1958 and 2005 by factors of 2 for bread, 10 for butter, 6 for sugar, 4 for milk and beef, 2 for potatoes, 5 for beer, and 3 for gasoline. Thus, for most of the basic goods of everyday life, industrial workers have to work much less at the beginning of the twenty-first century than they had to during the middle of the twentieth century. They owe this to the growing support from energy slaves. In 1960 each West German had about 20 energy slaves and each US citizen had 60 energy slaves at his service, and these numbers grew to more than 45 and 90 by the end of the twentieth century. The energy slaves toil in furnaces, heat engines, and information processors. They take over hard and boring jobs so that people have to work less. Their services are much cheaper than those of human labor. Thus, goods become cheaper and can be afforded more easily by the average citizen. The situation is, of course, quite different in developing countries, where, on average, a person only commands six energy slaves."

Abel grabs behind himself and then stretches out his hands. Each hand carries a device. One is a triangle balancing on a point contact with a thin plate. Three metallic wires are connected to its upper base. The other device is a slab with three stripes on its top. One metallic wire contacts the center stripe, covered by a thin metal layer, and two other wires are connected to the two ditches between the three stripes. "Look at the transistor. The very first specimen of its kind was the point-contact transistor. Its all-important successor, however, is the field-effect transistor. The transistor has revolutionized information processing. In combination with heat engines it propels automation and liberates people from hard and boring work. Let's watch its birth, which triggered the second industrial revolution."

We move back to October 22, 1945, and enter the Bell Telephone Laboratories at Murray Hill, New Jersey. John Bardeen, who just has joined Bell Labs, Walter Brattain, and William Shockley begin the work that leads to the transistor [11].

A foreseeable energy crisis is one reason why some of the best brains in solid-state physics are dedicated to research into a substitute for the electronic vacuum tube. The vacuum tube can act as an amplifier of signals and as a valve. As an amplifier it is at the very heart of telecommunications. As a valve, which blocks electric currents or lets them pass, it is the device that can represent the basic elements of computing: 0 and 1. Unfortunately, the vacuum tube consumes so much energy that in the 1940s people were speculating when the rapidly growing number of radios and television sets in the USA would consume all electricity generated by the power stations in that country. Furthermore, the vacuum tube, consisting of metallic filament, grids and a plate within a glass mantle, is bulky, fragile, and expensive. The idea is to replace it by suitably tailored semiconductor compounds, where the flow of electrons can be modulated by electric fields in a similar way as in the vacuum tube.

Shockley asks Bardeen to look over a design that he had sketched in his notebook 6 months earlier for a silicon "field-effect" amplifier. An electric field is applied perpendicular to a thin slab of silicon; the field draws charges in the slab to its surface. In a thin sample, Shockley argues, the field should cause a substantial change in the number of available charge carriers. In this design the field would play the role the grid plays in a vacuum tube. But the design does not work in practice. By March 1946, Bardeen theorizes that a substantial number of the electrons close to the surface may be trapped in surface states and thus cannot contribute to conductivity. An experiment proves the existence of the surface states and also shows that charge carriers in thin films are less mobile than in bulk material. Now the group knows why Shockley's field-effect design failed. But during the next 18 months little progress is made toward a semiconductor device that can act as an electronic amplifier and valve. The "magic month" that culminates in the transistor begins in the middle of November 1947. Bardeen and Brattain work closely together on a new design, based on point contacts between metal tips and a germanium plate. The breakthrough comes when Brattain, at the suggestion of Bardeen, wraps a gold foil around the apex of a polystyrene triangle and slits it carefully open. With use of a spring, the triangle is pushed down on the germanium and by "wiggling it just right," it makes contact with both points of the two metallic lines, separated by about 0.004 cm. On December 16, 1947, this design is tried for the first time. Immediately it achieves substantial power and voltage amplification. The (point-contact) transistor is born. Later Bardeen jokes that his and Brattain's invention of the point-contact transistor slowed the development of the field-effect transistor by several years [11].

The field-effect transistor, especially the silicon-based metal oxide semiconductor field-effect transistor becomes the workhorse of semiconductor electronics. The transistor first expels the vacuum tube from radios and television sets. In fact, "portable radio" and "transistor" become synonymous. Computers shrink in size and tremendously increase their computing power as transistors replace vacuum tubes. After the first landing on the Moon, one of the Apollo 11 crew says to Bardeen: "Without you I wouldn't have been there" [12].

The semiconductor industry becomes one of the principal driving forces of technological progress. Transistors decrease in size to micrometers and nanometers. Their density on a microchip approximately doubles every 18 months. This progress in integrated circuitry is given the name "Moore's law." The early information processors such as vacuum tubes and electromagnetic switches were too voluminous, and too energy- and material-consuming, to be integrated in very large numbers in the capital stock. But the tiny, energetically much more efficient transistor and progress in the miniaturization of information processors accelerate the pace of automation.

International trade and travel boom. Tariffs are lowered, or abolished altogether, by international agreements. After 1989 ideological barriers are no obstacle anymore. Highly automated container ships and wide-body jets with small crew, moved by cheap fuel and embedded in a system of sophisticated, computerized logistics, transport goods over long distances from cheap-labor production sites to high-wage consumer countries. Electromagnetic waves, propagating through cables or beamed from ground-based and satellite-based antennas, carry information services around the globe. Globalization, the worldwide division of labor, is established. Average-income citizens of the industrialized countries enjoy the beauty of the Blue Planet as tourists. At home they are offered a rich variety of goods from all over the world. Never in history have so many lived so well.

1.7   Outlook

Abel turns stern: "And this very success embodies the germ of self-destruction." Noting our bewilderment, he adds: "Industrial evolution has been driven by human creativity that has intensified energy conversion in furnaces and heat engines and combined it with information processing in machines. But all wealth production by energy slaves must face the problem that, unfortunately, energy slaves do not only consume and convert energy. They also excrete life-threatening substances such as sulfur dioxide (SO_2), nitrogen oxides (NO_x), carbon monoxide (CO), hydro carbons (C_mH_n), dust, radioactive waste, and climate-destabilizing carbon dioxide (CO_2). Global warming by emissions of infrared-active greenhouse gases, especially CO_2 from the combustion of fossil fuels, is likely to become a very serious problem. People in industrial countries have put a big burden on the fragile biosphere of Earth during their Golden Age.

You will need lots of creativity to relieve that burden, because, like all slaves, energy slaves make life comfortable and threaten the ones they serve. An irrefutable law of physics is behind all that. It says that energy conversion is coupled to entropy production, and entropy production is coupled to emissions of particles and heat. Large quantities of emissions change the molecular composition of and the energy flows through the biosphere. If these changes are so big and occur so rapidly that adaptation deficits of the living species and their societies develop, they are perceived as environmental pollution."

After having let us ponder energy conversion and entropy production for a while, Abel sums up:

> "Global society faces a threefold challenge: provide sufficient energy for the future, observe the biosphere's limited capacity of absorbing pollution, and prevent the growth of social tensions."

"Social tensions arise from the loss of decently paid jobs for common people. Energy slaves, toiling in heat engines and transistors, take away these jobs in the rich countries. In the poor countries, many of their citizens will find it increasingly difficult to escape poverty. Dwindling oil and gas resources and environmental constraints may prevent the developing countries from catching up with the rich countries in per capita income. Whether the battle for survival can be restricted to migration movements without armed conflicts is uncertain.

Geologists and petroleum engineers expect that sometime in the second decade of the twenty-first century global oil production will start to fall – for ever. "Peak Oil" is the name of this turning point. Already, for every barrel of oil discovered, three barrels are consumed [13]. The energy slaves are just too thirsty. Natural gas will not last much longer than oil. There is still plenty of dirty coal. But so far only the pollutants SO_2, NO_x, and dust are being removed from the flue gases of coal-fired power plants in the countries that can afford the technology and the additional energy. Theoretically, it also seems possible to remove the greenhouse gas CO_2 and store it underground. But the financial and energetic costs are quite high, and no one knows whether carbon capture and storage can ever be done on a significant scale.

To sum it up: If industrial evolution is restricted to the surface of Earth, the first half of the twenty-first century will become an unstable time of painful transitions. You'll need a lot of creativity to avoid a Dark Age after the Golden Age."

Abel pauses, sensing our bewilderment. Then he goes into economics: "Problems have begun to accumulate since 1989, this other year of freedom's victory. When capitalism lost its competitor 'socialism,' the protagonists and champions of capitalism also lost interest in sharing the wealth, produced by the energy slaves, with the general population. Gaining the upper hand in more and more market economies, especially in the new ones, they implemented the rules of the game established by the most powerful player, the USA. These rules are shaped by the belief that a market economy works best if only a minimum of regulations controls human greed in its strife for profit – never mind occasional market crashes. Thus, facilitated by automation, a growing share of production has been transferred to the shareholders and managers of the capital stock. They are the masters of the energy slaves toiling in the machines of the capital stock. As such they are endowed with economic and political power comparable to the power of the aristocratic landowners when the Roman Republic began its decline.

Look at the economics of automation and globalization. In the course of industrial progress, energy has become cheaper – except for occasional oil price shocks – and labor has become more expensive. Therefore, automation continues to take over routine jobs, even rather sophisticated ones in banking and trading. In addition, globalization exports jobs to countries were labor is cheap and taxes are low. Net income for the lower and middle classes stagnates or even decreases, whereas the income of the richest 10% of households increases steadily. Everywhere in the world the gap between rich and poor widens. The question is how long free, democratic societies can sustain that."

When we ask whether anything can be done about that, Abel answers: "There is a way to distribute the wealth created by the energy slaves more evenly and avoid disruptive social tensions: decrease the taxes and levies on labor substantially and increase the taxes on energy correspondingly. Such a tax system will stimulate employment and, in addition, energy conservation. Of course, to minimize problems in the competitiveness of energy-intensive industries, it should ideally be introduced in an internationally harmonized and gradual manner."

We remember the idea of an "ecological tax reform" and that this idea has been torpedoed so far by powerful special interest groups. We wonder,whether a stronger emphasis on the productive power of energy and the distribution of wealth would make it more acceptable to politicians and their electorate. But what about the problems of future energy scarcity and pollution?

Sensing our question, Abel makes his last remark: "Earth receives about 1.2×10^{17} W of solar radiation, which is the same as 33×10^9 kWh/s. This energy influx exceeds present world energy demand by a factor of roughly 10,000. You can collect solar power on a grand scale if you are willing to pay for the necessary investments. You can even overcome the limits to growth, which exist in all finite systems, by expanding the production system into space with the help of solar power satellites and space manufacturing facilities [14]. There is also the option of nuclear fission, which liberates the energy stored in thorium and uranium. There are inherently safe nuclear reactors, whose core cannot suffer a melt down. But you must solve the problem of radioactive waste disposal. And, eventually, the attempts to reproduce the solar fire in terrestrial fusion reactors may succeed.

Choosing your energy system you'll choose your road map to the future. Your choice will very much depend on how you assess financial and environmental risks. Risks concern the future, and since I am not entitled to show you the future, my tour ends here. I hope that you have got a feeling for the long way people had to go until they understood that heat, food, work, and light can be summed up by the concept of energy, and that motion, fire, wind, flowing water, coal, oil, gas, cereals, sunshine, and the masses of the lightest and the heaviest elements can provide energy services.

Understanding energy and its ugly sister entropy will be all important for future well-being and stability. This understanding must especially prevail in economic theory, because economists are in modern industrial societies what priests and theologians were in antiquity and during the Middle Ages. You can read more about this in the treatise I herewith hand over to you. Good luck and good bye."

References

1. Sybesma, C.: Biophysics. Kluwer, Dordrecht (1989)
2. Gribbin, J.: The Strangest Star. Athlone Press, London (1980)
3. Diamond, J.: Guns, Germs, and Steel. W.W. Norton & Co., New York, London (2000). This great book provides inspiration far beyond that of the passages quoted.
4. Sieferle, R. P.: Das vorindustrielle Solarenergiesystem. In: Brauch, H. G. (ed.) Energiepolitik, pp. 27–46. Springer, Berlin (1997)
5. National Gallery London
6. Hall, D.O., Rosillo-Calle, F.: CO_2 cycling by biomass: Global bioproductivity and problems of devegetation and afforestation. In: Lingeman, E.W.A. (ed.) Balances in the Atmosphere and the Energy Problem, pp.137–179. European Physical Society, Geneva (1990)
7. Statistical Abstracts of the United States, 1990

8. Hewett, C.E., High, C.J., Marshall, N., Wildermuth, R.: Wood energy in the United States. In: Hollander, J.M., Simmons, M.K., Wood, D. E. (eds.) Annual Review of Energy 6, pp. 139–170. Annual Reviews, Palo Alto (1981)

9. McMullan, J. T., Morgan, R., Murray, R. B.: Energy Resources, 2nd Ed., Arnold, London (1983)

10. Gurland, A.R. L.: Wirtschaft und Gesellschaft im Übergang zum Zeitalter der Industrie. In: Mann, G. (ed.) Propyläen Weltgeschichte 8, pp. 280–336. Propyläen Verlag, Berlin (1991)

11. Hoddeson, L., Daitch, V.: True Genius. The Life and Science of John Bardeen. Joseph Henry Press, Washington (2002). This excellent biography tells the details of how the transistor was invented and gives a fascinating insight into the life and science of the only winner of two Nobel prizes in physics.

12. Bardeen, J.: private communication, 1972

13. Strahan, D.: The Last Oil Shock. John Murray, London (2007)

14. O'Neill, G. K.: The High Frontier. William Morrow & Co, New York (1977)

Chapter 2
Energy

Fire is the basis of all craft.
Without it, Man could not persist.

Hesiod, 700 BC

2.1 Understanding the Prime Mover

Energy is the capacity to cause changes in the world; it is stored in matter and force fields. Energy conversion provides the work that drives the processes of life and the production of goods and services.

Energy lets the Sun shine, the rain fall, and the wind blow. It surrounds us as light and heat. It penetrates us and keeps us alive as food. If hard work drains us of too much energy, we collapse. Energy moves masses in construction, when dredgers dig and cranes heave, and deals blows of destruction when storms rage, lightning strikes, and the Earth quakes. It transmits information imprinted upon sound and electromagnetic waves. Liberated by the combustion of chemical fuels, and in nuclear reactions, it moves vehicles, ships, planes, and rockets and performs more physical work in the machines of our factories than all humans on Earth could supply. And yet, science got a clear concept of energy only in the nineteenth century. And energy's significance is hidden to economic theory even these days.

2.1.1 How the Energy Concept Evolved

The problem with understanding energy is that we do not perceive it in a unique way. Rather, it stimulates our senses and contacts with the outside world in many different and seemingly incoherent ways. Energy is present in so many different forms, such as light, fire, flowing water, wind, wood, wheat, meat, gunpowder, coal, mineral oil, and natural gas that for a long time people did not realize what they have in common. Sometimes energy carriers also serve different purposes, which also

R. Kümmel, *The Second Law of Economics: Energy, Entropy, and the Origins of Wealth*, 29
The Frontiers Collection, DOI 10.1007/978-1-4419-9365-6_2,
© Springer Science+Business Media, LLC 2011

creates confusion. For instance, many economic production processes are powered by the burning of mineral oil. Besides, a small share of oil also serves as a lubricant. Thus, people often call oil "the lubricant of the economy," camouflaging its real role as the still most important *fuel* of the economy.

The following short tale of energetics is based on the treatise *Die Energie* [1], written by the 1909 Nobel laureate in chemistry *Wilhelm Ostwald* (1853–1932). It shows how in the course of history some of the brightest minds have struggled to penetrate the essence of what we now call energy, and that they finally succeeded with the help of experiments and mathematical analysis.

The Greek philosopher and scientist *Aristotle* (384–322 BC), starting from equilibrium considerations for balances, thought about levers and the motion of their ends when different weights act on lever arms of different lengths. He used the Greek word *energeia*, but the meaning remained vague.

The situation stayed this way for nearly 2,000 years until, at the end of the Middle Ages, inquisitive minds turned their attention away from philosophical and theological speculations to asking nature experimental questions.

Galileo Galilei (1564–1642) showed that there is equilibrium on an inclined plane, and in other simple machines of that time, when the center of mass of all movable masses neither goes up nor goes down. His pupil Torricelli added the remark that in stable equilibrium the center of all masses assumes the lowest position. Investigating the free fall of bodies, Galileo found that their velocity is independent of the body mass and increases linearly with time. Experimental observations were the basis of all his theoretical reasoning. This made Galileo the founder of modern science. He stated that the book of nature is written in mathematical symbols.

Isaac Newton (1646–1727) published his *Philosophiae naturalis principia mathematica* in 1687. In this work Keppler's laws for the motion of the planets are derived from Newton's law of gravitation. This breakthrough, Newton's axioms of mechanics, and calculus (invented independently by Newton and Leibniz) became the foundations of classical theoretical physics. But energy was not yet an issue.

In 1717 *Jean Bernoulli* (1667–1748) wrote in a letter to Verignan: "En tout équilibre de forces quelconques, en quelque manière qu'elles soient appliquées, et suivant quelques directions qu'elles agissent les uns sur les autres, ou médiatement, ou immediatément, la somme de énergies affirmatives sera égale à la somme des énergies négatives, prises affirmativement." ("In every equilibrium of whatsoever forces, no matter how they are applied and under what angles they act directly or indirectly relative to each other, the sum of the positive energies is equal to the positively taken sum of the negative energies.") "Energy" is explicitly defined as the product of force times the (virtual) way in its direction. Further elaborations on the "principle of virtual works (or displacements)" resulted in the basic equations of statics. They are all based on the experience that it has never been possible to build a perpetual motion machine, i.e., a machine that creates work out of nothing, where work is the sum of all products of force times the parallel (positive) and antiparallel (negative) parts of way elements along which the force acts.

On the basis of observations such as those of a swinging pendulum, *Gottfried Wilhelm Leibniz* (1646–1716) reasoned that there should be an invariant whenever work is consumed in a machine, in order to produce velocities of moving masses. This invariant was recognized as the sum of all work performed and of "all living forces," as Leibniz called what later was given the name "kinetic energy" by *William Rankine*.

After this discovery of what is now known as the conservation law of mechanical energy, the concept of "energy" was used as a popular name for the "living force" by *Thomas Young* (1773–1829) at the beginning of the nineteenth century. *William Thomson* (1824–1907) later Lord Kelvin, firmly introduced the energy concept in science and emphasized that every state of a body is characterized by a well-defined energy value, its eigenenergy.

The decisive steps that extended the law of energy conservation beyond the realm of mechanics, and revealed energy as the fundamental constituency of the world besides matter, were taken by the physician *Julius Robert Mayer* (1814–1878) and the brewery owner and private scholar *James Prescott Joule* (1818–1889).

The story of Mayer's discovery deserves to be told in some more detail, because it shows how an outsider achieved a breakthrough and how difficult it was for him to communicate it to the scientific community.

Mayer, as a ship physician on a Dutch sailing ship that sailed to the West Indies, observed during occasional bleeding of the crew that after arrival of the ship in the tropics, the blood came out of the veins much redder than in colder latitudes. He concluded that in the warmer environment the body has to produce less heat by the physiological combustion of food in order to maintain constant temperature, so vein blood contains more unused oxygen, which causes the redder color. This conclusion was near at hand, because *Antoine Laurent de Lavoisier* (1743–1794) had shown that the warmth of human and animal bodies is indeed due to food combustion. In his youth, Mayer had tried to build a perpetual motion machine. In a more mature way, he now rethought the issue of work. The human body can perform work and produces heat. Is there a relation between the two? In trying to find the answer, he was handicapped by his lack of formal training in physics. On the other hand, he was neither handicapped by traditional scientific thinking habits nor afraid to wonder whether mechanical work and heat, two seemingly so different things, are manifestations of one and the same entity. But when he tried to put his thoughts into a mathematical-physical form, nobody understood them at first. A paper he sent to the *Annalen der Physik* in 1841, and subsequent inquiries about its fate, were simply ignored by the editor Poggendorf. Mayer was more successful with his second attempt. The article entitled "Bemerkungen über die Kräfte der unbelebten Natur" ("Remarks on the forces of inanimate nature"), written at the beginning of 1842, was soon published by the *Annalen der Pharmacie und Chemie*. One of the editors of this journal, *Justus Liebig*, had thought himself about the utilization of food in animal bodies, which was the starting problem of Mayer's considerations. The article cut its way through conceptual jungles. It talked of "forces" when "energy" was meant. In the introduction Mayer said that hitherto "force" was the concept of an unknown, impenetrable, hypothetical entity and that his article tried to contribute to clarification. Observations guided the way. For instance, Mayer pointed out that often a motion stops without inducing another motion or lifting a weight. He stated that an existing "force" cannot vanish but can only change into another form and asked what other form might be taken by the "force" that is the sum of motion and the "falling force" (*Fallkraft*). He noted that by shaking water sufficiently intensely, one can enhance the water temperature from 12°C to 13°C and raised the question of where the corresponding heat comes from. Mayer recalled the inverse process in the steam engine, which "decomposes" heat into motion and the lifting of weights. Finally, he reported his discovery of the mechanical heat equivalent with the following words: "...es ergibt sich ...dass dem Herabsinken eines Gewichtsteils von einer Höhe von zirka 365 m die Erwärmung eines gleichen Gewichtsteils Wasser von 0° auf 1° entspreche." (" ...one finds ...that the sinking of one weight unit from a height of about 365 m corresponds to the heating up of the same weight unit of water from 0° to 1°.") Although Mayer's heat equivalent is by a factor of 0.85 smaller than that in later textbooks, he was the first to state clearly that heat and mechanical work are two forms of the same thing and to establish their quantitative relation nearly correctly. In 1845 he published the article entitled "Über die organische Bewegung und den Stoffwechsel" ("On organic motion and metabolism"), where he stated the law of energy conservation more generally.

Independently from Mayer and only a little later, Joule arrived at the same conclusions in a completely different way. The revenues from his Manchester brewery enabled him to follow his scientific interests. Electromagnetism fascinated him, because he expected to gain cheap work from the attractive forces exerted by an iron core encircled by electric currents. He investigated all different factors that are relevant for the operation of electric machines. (In so doing, he found a number of important physics laws. In cooperation with William Thomson he discovered the Joule–Thomson effect, which is the basis for the liquefaction of gases.) He observed that the heat developed in the current-conducting wires – this heat was later given Joule's name – was uniquely related to the consumption of chemicals in his galvanic elements. Furthermore, he found that the currents in the wires of a machine produce less heat when the machine works than when the machine stands still, all other factors being equal. His starting point was similar to Mayer's. The animal organism is only replaced by the electromagnetic apparatus and its galvanic elements. In both cases chemical reactions are the source of mechanical work and, more or less, heat. The situation was rather complicated, and Joule tried to simplify it. He asked himself: How can the simplest relation between work and heat be established? He answered: When work is changed into heat by friction. If the idea is correct that work and heat are equivalent, then a given amount of work must always produce the same amount of heat, independently of the special ways in which one may convert work to heat. Joule performed experiments in which the work of falling weights was converted to heat in many different ways and concluded that there is indeed an invariant relation between work and heat. The corresponding communication was published in 1843, just one year after Mayer's first article.

Subsequently, the physiologist and physicist *Hermann Helmholtz* (1821–1882) deepened the understanding of the principle of energy conservation mathematically and in many details. Still, he used the name *Kraft* ("force") for what is energy. When *James Clerk Maxwell* (1831–1879) developed his theory of electromagnetism between 1861 and 1864, electromagnetic energy was readily included in the general concept of energy.

With that the beautiful structure of classical physics was firmly established. Physicists believed that their deterministic laws would, in principle, allow them to predict the evolution of everything, from the beginning of the universe to its end.

At the start of the twentieth century, more than 2,300 years after Aristotle had talked about *energeia* in the context of natural phenomena, the overwhelming importance of "energy" for a proper understanding of the world had been realized by all natural scientists. For instance, Ostwald wrote in [1]: "Will sich heute ein Physiker oder Chemiker recht fortschrittlich gebärden, …so …definiert [er] die Naturwissenschaft als die Lehre von der Umwandlung der beiden unzerstörlichen Dinge, der Materie und der Energie…" ("If a physicist or chemist wants to show himself really progressive these days …he defines natural science as the science of the conversion of the two indestructible things, matter and energy …"). And he predicted: "…der Begriff der Materie [wird] als ein untergeordneter …sich herausstellen …" ("…the concept of matter will turn out to be secondary").

Between 1948 and 1978 the International Union of Pure and Applied Physics developed and recommended the International System of Units, abbreviated by *SI system*. The basic SI units and the derived SI units relevant for the measurement of energy are:

> meter (m) for distance, second (s) for time, kilogram (kg) for mass, ampere (A) for current, kelvin (K) for absolute temperature. newton (N) for force ($1\,N = 1\,m\,kg/s^2$), joule (J) for energy ($1\,J = 1\,m^2 kg/s^2 = 1\,N\,m$), and watt (W) for power ($1\,W = 1\,m^2\,kg/s^3 = 1\,J/s$).
> One energy unit that often appears on energy bills is the kilowatt-hour (kWh); $1\,kWh = 3.6 \times 10^6\,J$.

2.1.2   Energy for All and Forever

Ostwald's prediction that the concept of energy will dominate physics was proven right by the deepening understanding of the energy–matter world during the twentieth century. *Albert Einstein's* special theory of relativity showed the equivalence of energy and mass in 1905. This later explained the liberation of energy in nuclear fission and fusion. And according to the present theory of cosmic evolution, all matter has condensed out of the primordial Big Bang energy in a series of phase transitions.

Einstein discovered that energy E and mass m are equivalent, when he realized that classical Newtonian mechanics must be augmented to account for experimental observations such as those of *Albert A. Michelson*, according to which all observers measure the same speed of light in a vacuum, which is $c = 299,792$ km/s, no matter how fast they move at constant velocity relative to each other and to the source of light. The equivalence relation is[1]

$$E = mc^2. \tag{2.1}$$

Energy was also the key that opened the door to quantum mechanics, which shattered the foundations of classical physics. This happened between 1900 and 1930. Among the pioneers were the Nobel laureates in physics *Max Planck, Albert Einstein, Niels Bohr, Werner Heisenberg, Erwin Schrödinger, Paul Dirac, Wolfgang Pauli*, and *Max Born*. They discovered that experimentally observed phenomena such as the radiation emitted by black bodies, the emission of electrons by illuminated metal surfaces (photoelectric effect), and all processes in the microworld of atoms can only be understood if one assumes that energy is quantized in the sense that atoms emit electromagnetic radiation of frequency f in energy packets (photons) of size hf, where $h = 6.625 \times 10^{-34}$ W s^2 is Planck's constant. The reason is that atoms can accommodate their electrons only in discrete (quantized) energy levels, between which the electrons jump down or up if they emit or absorb energy. This contradicts a fundamental conviction of classical physics, expressed by the old saying *natura non fecit saltus* (nature does not make jumps). Quantum mechanical computation of the atomic energy levels requires knowledge of the classical energy function, called a *Hamiltonian*, for the atomic system. (Ironically, neoclassical economics, which has little room for energy, has borrowed the mathematical formalism of Hamiltonians from classical mechanics.) In quantum mechanics the Hamiltonian is usually the sum of kinetic and potential energies. An example is given by (2.27) in Appendix 1 of Chap. 2. Replacing these energies by operators that operate on quantum mechanical states, one can calculate all physical properties of atomic systems in full agreement with experiment. However, quantum mechanics allows only predictions of *probabilities* for properties and processes in the microworld of atoms. These probabilities can also affect the macroworld

---

[1] If m is the mass m_0 of a body at rest relative to an observer, this relation gives the amount of energy the observer would obtain from nuclear reactions that convert all of m_0 into energy. Similarly, if an electron and its antiparticle, the positron, meet, they annihilate and turn into photons, the quanta of electromagnetic radiation, whose energy is given by (2.1), m being the sum of the equal masses of the electron and the positron in this case. On the other hand, if a body moves with velocity $\mathbf{v}$ relative to the observer, one has $m = m_0/(1 - \mathbf{v}^2/c^2)^{1/2}$ in (2.1). A popular saying is that mass increases with velocity, although Einstein objected to talking about the mass of a moving body (letter dated June 19, 1948 to L. Barnett). Einstein preferred to describe the inertia of rapidly moving bodies by the concepts of energy, momentum and rest mass m_0 [2]. In any case, no rocket can exceed the velocity of light. Just to reach $|\mathbf{v}| = c$ would require an infinite amount of energy. (That is why science fiction on space travel invented "jumps through hyperspace.") Only massless "particles" such as photons propagate at the speed of light.

of our daily lives, for instance, when cosmic radiation interacts with a gene or a transistor and damages it with a probability between zero and one. Thus, quantum mechanics has destroyed the rigid determinism of classical physics. Fortunately, the statistical *averages* of properties and events in the macroworld, computed with quantum-mechanical probabilities, obey the laws of classical physics to an excellent approximation. Thus, our technical devices are safe, as a rule.

One principal finding of classical physics has remained unchanged by quantum mechanics. In fact, it has been confirmed even more deeply[2]:

> Energy, including the energy equivalent of mass, is a conserved quantity. It can be neither created nor destroyed.

This is the first law of thermodynamics.

2.1.3 Energy Quantity and Quality

Since energy is eternal, why do people worry about economic problems because of energy shortages? The answer is: Although the quantity of energy is conserved, the quality of energy is not. Let us look into both.

The *quantity* of energy contained in an energy carrier is measured by the heat given off to the environment in a chemical (or nuclear) reaction that occurs under precisely controlled conditions, and which takes all reaction partners from an initial state before the reaction to a final reference state. The initial state is characterized by the enthalpy H_1, the final state has the enthalpy H_2, and the heat produced is $Q_{12} = H_1 - H_2$ according to (2.42) in Appendix 1 of Chap. 2. Usually, the temperature of the reference state is chosen as $25°C \approx 298$ K. Often the pressure is that of the natural environment. The reference state of each chemical element is the most stable, naturally occurring compound of this element.

For carbon and hydrogen these compounds are carbon dioxide (CO_2) and gaseous or liquid water (H_2O). One measures the energy quantities contained in coal, oil, gas, and biomass of given mass or volume by the specific heating value in a controlled combustion process. Fuel and combustion air, both at the same initial temperature T_0, are brought into a reaction chamber (calorimeter), and the combustion products must be cooled down to precisely this initial temperature T_0. The heat that leaves the reaction chamber in this process divided by the quantity of fuel is the heating

---

[2]For instance, just to preserve the law of energy conservation for the β decay in nuclear reactions, *Wolfgang Pauli* postulated the existence of an uncharged particle with energy, spin 1/2, and vanishingly small mass in 1930. *Enrico Fermi* called it a "neutrino" in 1940. It was found experimentally in 1956.

Table 2.1 Average specific heating values of primary energy carriers; in kilojoules per kilogram (kJ/kg) and kilojoules per cubic meter (kJ/m³); 1,000 kJ = 0.278 kWh		
Oil	41,900 kJ/kg
(Hard) coal	29,300 kJ/kg
Lignite (raw)	8,200 kJ/kg
Wood	14,650 kJ/kg
Peat	12,600 kJ/kg
Oil gas	40,730 kJ/m³
Natural gas	32,230 kJ/m³

value of the fuel. There is an upper heating value and a lower heating value. They differ by the heat that has to be supplied by the environment to liquid water to evaporate it. When gaseous water condenses to liquid water, this heat is recovered by the environment as condensation enthalpy. Therefore, one measures the upper heating value of the fuel if the final state of the water in the combustion products is liquid and the lower heating value if the final state is gaseous [5]. In modern heating systems, condensing-value boilers prevent gaseous water from escaping through the chimney. Instead, water is condensed to its fluid state and adds its condensation enthalpy to heating.

Average specific heating values of primary energy carriers are given in Table 2.1. Different fossil fuels contain different concentrations of carbon and hydrogen. Oxidation of carbon according to the molar reaction equation

$$C + O_2 \rightarrow CO_2 + 394 \, kJ/mol$$

yields roughly 80%, 70%, 60%, and 50% of the energy gained by the combustion of lignite, coal, oil, and gas, respectively. Oxidation of hydrogen according to the reaction equation

$$2H_2 + O_2 \rightarrow 2H_2O + 242 \, kJ/mol$$

provides the rest of the heating value of fossil fuels.

The energy content of nuclear fuel is much higher than that of fossil fuels. According to (2.1), the complete conversion of a mass of 1 g into energy yields 25×10^6 kWh. To get the same energy quantity from the combustion of coal, one has to burn about 3,100 metric tons (t) of hard coal, emitting 8,000 t of CO_2 [6]. A 1,300-MW$_{electric}$ nuclear power station with a boiling water reactor generates 25×10^6 kWh of electricity (and about two times more heat) in 19 h; during this time it produces about 0.15 t of nuclear waste in the form of burned-out nuclear fuel rods [7]. If one considers the 25×10^6 kWh of enthalpy from coal combustion and (only) the 25×10^6 kWh of electricity from fission as "useful" energy, one may say that the mass ratio of CO_2 to nuclear waste per generated quantity of "useful" energy is more than 50,000. If one compares electricity generation and assumes an efficiency of 40% for the coal power plant, the mass ratio exceeds 100,000.

Energy *quality* came into the focus of energy science when the energy balances of industrial production processes where analyzed quantitatively. This led energy analysts to differentiate between two components of a given energy quantity: a useful one, called *exergy*, which can be converted into any type of physical work and measures the quality of an energy quantity, and a useless component, called *anergy*.[3] Entropy reduces exergy and enhances anergy, as (2.45) and (2.47) show. Examples for exergy are the energy of solar radiation, the chemical energy stored in coal, oil, and gas, nuclear energy, potential and kinetic energy, and electric energy. Saying "energy" when talking about these energy carriers is the same as saying "exergy." When combustion processes or friction convert exergy into heat, anergy is produced. A quantity Q of heat at temperature T in an environment at temperature $T_0 < T$ contains only the exergy $Q(1 - T_0/T)$. Heat at the temperature of the environment is all anergy.

Exergy drives the machines in mines and on drilling sites, in power stations, factories and office buildings, on rails, roads and farms, in the air, and on the sea. In short, it activates the wealth-creating production processes of industrial economies. Appendix 1 of Chap. 2 presents more physical details on the basic forms of energy and exergy.

In terms of exergy and anergy the first law of thermodynamics says that

$$Energy \equiv exergy + anergy = constant. \tag{2.2}$$

In all energy-conversion processes, which produce entropy, useless anergy increases at the expense of useful exergy. This is meant by "energy consumption."

2.2 Sun and Earth

The Sun maintains life on Earth. It has provided all the energy carriers that are presently used in our economies, except for uranium, and is the source of the principal renewable energies, which will become important in the future. Absorption and re-emission of the energy from the Sun controls our climate, whose stability is part of the general framework that determines further economic evolution. The basic data on the Sun show the might of the gravity-assisted fusion reactor 150 million kilometers away.

2.2.1 Energy Production in the Sun

The Sun shines because hydrogen (H) is being converted into helium (He) in the central core at a rate of 600×10^6 t/s. Fundamental to the solar fusion process is

---

[3]The concept of "anergy" has been transferred from medicine and psychology. It has not yet been accepted as widely as "exergy," where "availability" was one of the older names of the latter.

the proton–proton reaction, where a proton (p) combines with another proton in its vicinity to form deuterium (^{2}H), a positron (e^+) and a neutrino (v). (This first step is extremely slow, because it depends on proton tunneling through the Coulomb barrier and on the weak interaction.) The deuterium, which consists of a proton and a neutron, and a proton fuse into helium-3 (^{3}He), where a photon (γ) is emitted. In 85% of the cases two ^{3}He nuclei merge into a ^{4}He nucleus and two protons, and in 15% of the cases a more complicated fusion chain passes through ^{7}Be and ^{7}Li until it terminates in ^{4}He. The proton–proton chain of helium fusion can be summarized by the reaction equation

$$4p \longrightarrow {}^4He + 2e^+ + 2v + 2\gamma. \tag{2.3}$$

The kinetic energy released by this reaction is about 26 MeV, where 1 MeV $= 1.6 \times 10^{-13}$ W s. The mass of four protons is bigger by a factor of 1.007 than the mass of a ^{4}He nucleus and the two positron masses. The mass difference between the 600×10^6 t of hydrogen and the fusion product helium is

$$\Delta m = (1 - 1/1.007) \times 600 \times 10^6 \text{ t} \approx 4.2 \times 10^6 \text{ t}. \tag{2.4}$$

It is converted into energy E according to $E = \Delta m \times c^2$, Einstein's equation (2.1). Thus, the solar photoluminosity, i.e., the energy emitted per unit time by the Sun in the form of photons, is

$$L \approx 4.2 \times 10^6 \text{ t/s} \times c^2 = 3.845 \times 10^{26} \text{ W}. \tag{2.5}$$

(A small part of the power of 4.2×10^6 t/s $\times c^2$, namely, $0.023L$, is carried away by the neutrinos.) In addition to the mass loss of 4.2×10^9 kg/s because of photon and neutrino production and emission, the solar wind blows another 10^9 kg/s away. Thus, during its lifetime (from its birth until now) of about 4.5 billion years (equivalent to 1.4×10^{17} s) the Sun has lost less than 10^{27} kg. This is negligible with respect to its mass of

$$M = 1.99 \times 10^{30} \text{ kg}. \tag{2.6}$$

The solar fusion process may last for another five billion to ten billion years.

The solar mass occupies the volume of the solar sphere with a radius of

$$R = 6.96 \times 10^5 \text{ km}, \tag{2.7}$$

and the Sun's mean density is $\rho = 1.41$ g/cm^3. This is 1.41 times the density of water at 4°C. About 92% of the Sun's atoms are hydrogen, nearly 8% are helium, and all other elements make up 0.1% of the total. The mass fraction of hydrogen is 74%, of helium is 24% and of all other elements is 2%. The vast bulk of the Sun is highly ionized, and the free electrons make it an extremely good electric conductor.

Energy production by hydrogen–helium fusion is largely confined to a small volume within 1.4×10^5 km of the Sun's center. There, the temperature exceeds 10^7 K, and pressure is about 10^{10} atm. In the center itself, the temperature is 1.5×10^7 K, and the density of matter is larger than that of water by a factor of 150.

It takes time for the energy to get out of the Sun into space. The main transport process from the center up to about 500,000 km through 98% of the Sun's mass is diffusive radiation. "Radiation is emitted when hot electrons collide with atoms. The radiation travels at the speed of light (300,000 km/s) until absorbed. Although in free space radiation would travel a distance equal to the Sun's radius in little more than 2 s, absorption is so strong inside the Sun" the photon mean free path being typically a fraction of 1 cm, "that it takes 1 million to 2 million years to diffuse out. Thus, the light and warmth we receive were produced near the Sun's center over a million years ago" [8].

Convection, i.e., the transport of heat by the motion of hot gases (or fluids), takes over from about 500,000 km up to the photosphere at the surface of the Sun. There, in electron–hydrogen collisions, negative hydrogen ions are formed. These emit the vast bulk of sunlight and solar radiation. The associated effective solar surface temperature $T_{S,eff}$ is obtained from the Stefan–Boltzmann law. This law says that the energy flux density Q emitted by a body of temperature T and emissivity ε is

$$Q = \varepsilon \sigma T^4 \tag{2.8}$$

with $\sigma = 5.67032 \times 10^{-8}$ W/m²K⁴ being the Stefan–Boltzmann constant. Using the black-body emissivity $\varepsilon = 1$, the luminosity L of (2.5), and the solar radius R of (2.7), one obtains from (2.8)

$$T_{S,eff} = (L/4\pi R^2 \sigma)^{1/4} = 5777 \text{ K}. \tag{2.9}$$

The power L emitted by the Sun spreads into space as a continuous flow of spherical electromagnetic radiation. When this energy flux hits the top of Earth's atmosphere after having travelled across the distance

$$D = 149.6 \times 10^6 \text{ km} \tag{2.10}$$

it has thinned out to

$$S = L/4\pi D^2 = 1{,}367 \text{ W/m}^2. \tag{2.11}$$

This irradiance at the mean Sun–Earth distance D is also called the "solar constant." Like the luminosity L, it is the integral over the electromagnetic spectrum. It is the solar quantity primarily accessible to measurement by balloon, rocket, and satellite experiments, and from it the solar luminosity is deduced. All solar models have to give results which are compatible with these basic data.

About 30% of the solar irradiance S, the so-called albedo α, is reflected back into space. Thus, at the top of Earth's atmosphere the power flux

$$S(1 - \alpha)/4 = 239 \text{ W/m}^2 \tag{2.12}$$

is absorbed. The factor 1/4 is the quotient of the cross-sectional area and the surface area of the Earth. (Here, the Earth is approximated by a sphere of radius R, so the surface area is $A_E \approx 4\pi R^2$, and the cross section is πR^2.) The product of $A_E = 510 \times 10^6 \text{ km}^2$ with 239 W/m^2 yields the solar power P_{solar} absorbed by the Earth as

$$P_{\text{solar}} = 1.2 \times 10^{17} \text{ W}. \tag{2.13}$$

This solar power has produced the stock of fossil fuels and grows all biomass on Earth. More about the Sun can be found in [8–12].

Finally, there is the stock of energy that is not a gift of the Sun. In the Sun's atmosphere one observes traces of heavy elements. These elements, quite common to us on Earth, can only be generated in fusion processes at temperatures much higher than those in the core of the Sun. Temperatures above 10^8 and 10^9 K occur in contracting stars which have burned up all their hydrogen and fuse higher elements. Up to iron (^{56}Fe) these fusion processes produce energy. The fusion of elements heavier than iron, however, consumes energy. This energy may have been provided by novae and supernovae. Thus, the Sun, the Earth, and everything on Earth itself have been processed through the inside of at least one star. Especially uranium, the heaviest natural element in the periodic table, is most probably the product of a stellar explosion before the time of our Sun. Its fission into lighter elements with a total mass less than that of the original nucleus liberates some of the energy caught in it during a cataclysmic cosmic event.

2.2.2 The Natural Greenhouse Effect

The simplest quantitative description of the natural greenhouse effect considers the Earth in radiative equilibrium with the Sun. In this equilibrium the total solar power that arrives at the top of the atmosphere is

$$P_{\text{top}} = 1.7 \times 10^{17} \text{ W}. \tag{2.14}$$

The solar irradiance, i.e., the power flux at the top of Earth's atmosphere, is given by the solar constant S in (2.11). According to (2.12), Earth's biosphere absorbs the power flux $S(1 - \alpha)/4 = 239$ W/m^2; the spectral range of wavelengths is between 0.2 and 2 μm. This power is radiated back into space in the spectral range of the infrared between about 5 and 30 μm. The required effective temperature T_{eff} can be calculated from the Stefan–Boltzmann law (2.8). The equation

$$S(1 - \alpha)/4 = \sigma T_{\text{eff}}^4 \tag{2.15}$$

yields

$$T_{\text{eff}} = 255\,\text{K} = -18°\text{C}. \tag{2.16}$$

The solid surface of Earth, however, has an average temperature of 15°C (288 K). At this temperature the Stefan–Boltzmann law yields a radiated power of 390 W/m². The difference of 151 W/m² between the power emitted by Earth at the bottom and the top of its atmosphere is trapped by the infrared-absorbing trace gases. Trapping so much heat radiation, these gases provide the life-cradling warming blanket around the Earth. They play roughly the same role as the glass roof and glass walls in a greenhouse: Visible sunlight passes through the glass almost unhindered and is absorbed by the plants and soil inside the greenhouse, which are thus warmed up. The heat in the infrared range, radiated by the warmed plants and soil, is absorbed by the surrounding glass, and is then emitted partly to the outside and partly back to the inside. Because of this back-radiation of heat, the temperature inside the greenhouse rises to a level higher than the outside temperature.[4]

The infrared-active gases in the atmosphere, their concentrations in parts per million (ppm), and their contributions to the temperature enhancement above $-18°$C are as follows: water vapor, H_2O (from 2 ppm to 3×10^4 ppm, 20.6°C), carbon dioxide, CO_2 (preindustrial 280 ppm, 7°C), ozone, (near the ground), O_3 (0.03 ppm, 2.4°C), nitrous oxide, N_2O (0.3 ppm, 1.4°C), methane, CH_4 (1.7 ppm, 0.8°C), and others (0.6°C).

Figure 2.1 shows schematically the energy flows associated with the natural greenhouse effect. Here, the incoming solar radiation at the upper limit of the

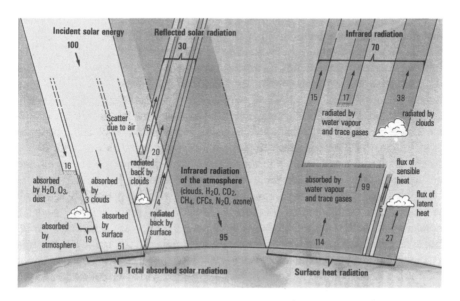

Fig. 2.1 Radiation budget of Earth's surface/atmosphere system [13]

---

[4]Inhibition of convection by the glass roof also contributes to warming.

atmosphere is given the reference value 100%. Twenty-six percent of it is reflected by the clouds and 4% by the surface – the sum makes up the albedo. The atmosphere swallows 19% (clouds 3%, water vapor, ozone, and dust 16%), and the surface absorbs 51%. Thus, Earth's net absorption of the incoming solar energy is 70%. All this has to be radiated back into space. This occurs partly directly and partly through the buffer of the greenhouse gases in the atmosphere. Let us first see how the surface gets rid of the absorbed 51%. It emits 15% as infrared radiation directly back into space and sends 32% via convection as sensible heat (5%) and latent heat (27%) into the atmosphere. The remaining 4% is the difference between an upward and a downward energy flow in the infrared radiation field caused by the greenhouse effect: The infrared-active trace gases radiate heat down to the surface. The energy of this heat is huge, namely, 95% of the reference value. In infrared-response the surface radiates back 99%. Thus, energy absorption and energy emission are balanced for the surface. The same is true for the atmosphere: 19% absorbed from the incoming solar radiation plus 32% and 4% received from the surface are emitted into space. Adding the 15% of Earth's direct infrared radiation into space yields the 70% absorbed by Earth's surface/atmosphere system as a whole, and the law of energy conservation is satisfied on all levels.

2.2.3   Solar Activity and Climate

The data given for the solar luminosity L and the effective solar surface temperature $T_{S,eff}$ are valid only as averages over the solar surface and time. A closer look, first done by Johannes Fabricius in 1610 and Galileo Galilei about the same time, reveals dark spots on the Sun, the number of which varies with an 11-year cycle. In these sunspots magnetic flux penetrates the surface of the Sun. This flux blocks the convective energy transport to the surface so that in the sunspots the temperature is only about 4,000 K. Variations in the number of sunspots are the most easily detectable sign of variations in solar activity. The more sunspots, the more active is the Sun. Increasing and decreasing with the number of radiation-blocking sunspots is the number of radiation-enhancing brighter regions, called faculae and plages, where faculae cover an area up to four times that of the associated sunspots and are about 1,000 K hotter than the surrounding photosphere [14]. Thus, all solar emissions exhibit an 11-year periodicity. Correlated with this periodicity is a small but significant modulation of the solar constant S by about 2 W/m^2, S being maximum when the sunspot number is maximum. This 11-year modulation comes from two sources of mutually counteracting effects: enhanced emissions from bright faculae during solar maximum and enhanced blocking by sunspots – obviously the former wins over the latter. In addition to the 11-year sunspot cycle and the associated 22-year magnetic cycle, one has observed long-term secular changes of sunspot cycle amplitude, the so-called Gleissberg cycle. There are also short-duration, sporadic events, such as solar flares, the sun-controlled solar wind shocks, and reversals of the interplanetary magnetic field.

Whether varying solar activity has a perceptible impact on terrestrial climate is a subject of ongoing controversial debate. The "yes" advocates point to empirical findings such as the following ones:

1. Tree rings, which are thicker in wet years and thinner in dry years, show a periodicity of 22 years, the solar magnetic period [8].
2. Carbon-14 is produced by cosmic-ray bombardment of nitrogen. During times of high solar activity these rays are deflected from the Earth by impulsive solar flares, resulting in an anticorrelation of galactic cosmic ray flux with sunspot number. Measurements of the amount of carbon-14 in tree rings has enabled the the number of sunspots to be traced back until AD 1000. This number was nearly zero during the "Little Ice Age" in the second half of the seventeenth century and during the cold period in the fifteenth century (Maunder and Spörer minima) [14].
3. All major midwinter warmings that have been investigated occurred during a special dynamic state of the stratosphere which happened only at times of solar maximum [16].

Arguments of the "no" advocates are as follows:

1. The power involved in solar variability is too small.
2. One cannot think of any mechanism responsible for solar–climate effects.
3. The time intervals for correlation studies are too short, it's all just coincidence.

In response to this it is said that the statistics are becoming better and better, and the energy argument is countered by calling upon the fact that the atmosphere–ocean–biosphere system with its large reservoirs of latent energy is a complex, highly nonlinear system, which can behave chaotically, so very small changes in energy input can trigger large global energy redistributions.

Thus, the sun–climate situation is far from being clear. There are forecasts that solar activity will increase until the year 2030 to the highest level ever recorded, but this is contradicted by expectations of a decrease of solar activity in the next few cycles [8]. The uncertainties encountered here seem to be greater than those associated with the anthropogenic greenhouse effect, to be discussed in Chap. 3. Testimony to that is the following statement: "Should another Maunder minimum begin within the next few decades, (as has been suggested by some authors...) then this would clearly offset future greenhouse gas induced climatic change to some considerable degree. However, the magnitude of Man's impact is such that it would still be the dominant factor in future climatic change" [17].

Furthermore, multiple linear regression analyses show "that within recent decades the solar signal systematically decreased whereas the greenhouse gas signal increased to become a dominant factor of climate variability" [18].

2.2.4 Photosynthesis, Respiration, and Food Production

Photosynthesis provides the chemical energy for life on Earth. Respiration converts this energy into work within living organisms.

Solar photons excite electrons in a special type of molecule (chlorophyll) in the photosynthetic reaction centers of plants and some algae. The excited electrons are passed along a chain of molecules, forming a tiny, solar-driven electric current. This current does two jobs. It breaks up water molecules into hydrogen and oxygen atoms, and it turns molecules of adenosine diphosphate (ADP) into the more energy-rich compound adenosine triphosphate (ATP). In a series of chemical reactions sugar (glucose) and oxygen are formed from hydrogen, carbon dioxide, and ATP. In the net input–output balance of the photosynthetic reaction six molecules of water and six molecules of carbon dioxide plus sunlight result in one molecule of sugar and six molecules of oxygen:

$$6H_2O + 6CO_2 + \text{ light } \longrightarrow (HCOH)_6 + 6O_2$$

The energy of the sunlight is essentially transformed into and stored as the chemical energy of sugar and oxygen.

The process of respiration completes the basic cycle of life. Plants and animals liberate the stored solar energy by the recombination of sugar and oxygen in order to perform work. This work may be mechanical work of muscle contraction, electric work when charges are transported, osmotic work when material is transported across semipermeable barriers, or chemical work when new material is synthesized. At the constant temperature prevailing in most cells, a net output of work can be obtained only when some energy is dissipated. This dissipated energy finally ends up in useless heat at the temperature of the environment.

Thus, respiration converts the exergy stored in sugar into the chemical energy of ATP in a balanced sequence of oxidation–reduction reactions:

$$(HCOH)_6 + 6O_2 \longrightarrow 6H_2O + 6CO_2 + \text{"38ATP"}$$

"38 ATP" stands for the energy of about 2,800 kJ, which is stored in 38 ATP molecules. ATP serves as the universal energy currency in all living systems. When at a given time work has to be performed, ATP is converted into ADP and inorganic phosphate and releases its energy in a hydrolysis reaction under controlled conditions. The chemical end products of respiration are carbon dioxide, emitted back into the air, and water [19].

Agricultural technologies put photosynthesis at the service of humans. Edible plants are grown systematically, and livestock grades up plants that are inedible for humans to tasty meat. The chemical energy stored in food powers the human body.

The quantity and plant composition of crops depend on two properties of the biosphere: (1) weather and climate, which are the short-term fluctuations and the long-term cycles of sunshine and rain; (2) the available land and its content of nitrogen, phosphorus, and other useful trace elements, on the one hand, and the absence of noxious salts and pests, on the other hand. Technical progress in tilling the soil, such as the transition from the wooden pick axe to the animal-drawn iron plow, and crop care, such as manuring and the introduction of rotation of crops, increase the crop yield substantially. Domesticated animals make the energy content of grass available to humans. However, the efficiency of producing animal biomass from plants is only about 20%. Thus, in times of population growth and food scarcity, changing the dedication of pasture to arable land can enhance the production of food energy fivefold. Estimates for different agricultural technologies [20] arrive at the following annual energetic yields per hectare (10,000 m²), including fallow:

- Rice with fire clearing (Iban, Borneo): 236 kWh
- Horticulture (Papua, New Guinea): 386 kWh
- Wheat (India): 3,111 kWh
- Corn (Mexico): 8,167 kWh
- Intensive farming (China): 78,056 kWh

Hunters and gatherers only get 0.2–1.7 kWh/(ha year). Thus, on the basis of Chinese intensive farming, 50,000 times more people can live on a given area than under the conditions of hunting and gathering [20].

2.3 Amplifiers of Muscles and Brain

Food powers the human body, whose hands perform work and whose brain processes information. Horses, asses, oxen, and mules also convert fodder into work. Food, fodder, and wood were the main sources of chemical energy for economic activities before the Industrial Revolution. Besides, there are the kinetic energies of wind and water. The use of all these energy forms is quantitatively constrained by the technical potential of harvesting the annual input of solar energy into the biosphere. Qualitatively, their conversion into work by muscles, and machines built mostly from wood, had been constrained by the forces organic structures can take and exert. Heat engines changed this situation drastically. They opened up the huge store of fossil fuels accumulated by the Sun on Earth in more than 200 million years. This had two revolutionary consequences. First, a positive-feedback circle was established in which fossil fuels facilitate the cheap production of metals, from which heat engines are built, which make more natural resources accessible. Second, heat engines convert the chemical energies of coal, oil, and gas into work *outside* the limitations of human and animal bodies. Transistors, powered by electricity, further reduce biological limitations. They assist the human brain in processing and storing huge quantities of information. Heat engines and transistors are amplifiers of human muscle and brain power, and as such they have tremendously enhanced wealth creation by work performance and information processing. Their technical details merit attention in order to understand the physical basis of technological changes in the economy.

2.3.1 Heat Engines

We would still be living in agrarian societies if we had no heat engines. They relieve humans from toil, and provide an ever-expanding realm of energy services. Steam engines were the first heat engines that burned fossil fuels. They triggered the Industrial Revolution, but they are nearly obsolete now. Modern heat engines are steam turbines, diesel engines, gasoline engines, and gas turbines.

2.3.1.1 Steam Engine and Modern Heat Engines

Steam engines and steam turbines are heat engines with external combustion. Gas turbines, diesel engines, and gasoline engines operate with internal combustion. In external combustion heat flows from a furnace or a reactor into a boiler and generates steam. The energy of the hot steam under high pressure is then converted into mechanical work. In internal combustion engines ignition converts the chemical energy of the fuel into heat, which spreads explosively throughout the gas. Then work is performed at the expense of internal energy.[5]

In the steam engine a piston is pushed back and forth in a cylinder by the steam fed from the boiler into the cylinder by either one or two inlet valves, so the pressure of the expanding steam operates either on one side or on both sides of the piston. After a mass of steam has given off its energy, it is ejected through outlet valves. The motion of the piston can be transmitted by a crank to wheels, e.g., in steam locomotives or old steamships, or to the moving parts of working machines. Control of the engine is facilitated by an indicator, which plots a diagram of the different phases of operation: high-pressure steam injection, expansion, ejection of the relaxed steam, and compression of the residual steam in the cylinder.

The steam turbine is the heat engine with the highest power limit, this limit being set by the corrosion of steel at steam inlet temperatures above 540°C; the outlet temperature is given by that of the environment, or the available cooling water, and is usually not below 30°C. Since the efficiency of heat engines increases with the difference between the inlet and the outlet temperatures according to (2.17), considerable efforts aimed at energy conservation go into materials research. The aim is to find coatings for the surfaces of the turbine blades that will make them less susceptible to corrosion at higher steam temperatures. In contrast to the steam engine, in the steam turbine the pressure of the steam is not used directly for work performance. Rather, it is first transformed into the high velocity of steam molecules, when the steam expands in nozzles, which direct it onto rotor blades. Hitting these blades, the steam molecules give off their kinetic energy to the blades and set the rotor in motion. This motion then drives the propellers of steamships, the pinion gear of machines, and the electricity generators of modern conventional or nuclear power plants. In these plants the steam-generating boiler is heated either by furnaces burning fossil fuels or by heat exchangers through which the coolant of a nuclear reactor runs.

---

[5]Equations (2.27), (2.28), and (2.38) in Appendix 1 of Chap. 2 indicate how internal energy, i.e., chemical energy, is calculated and measured.

In diesel and gasoline engines the combustion of an air–fuel mixture performs work within a (thermodynamic) cycle consisting of four steps:

1. Air, into which fuel will be injected, or a fuel–air mixture, is sucked into the combustion chamber through the inlet valve; subsequently, it is compressed adiabatically (i.e., without exchange of heat with the environment) by the piston.
2. In gas engines a spark plug ignites the mixture and the gas heats up at constant volume. In diesel engines, after fuel injection and self-ignition, the gas heats up at constant pressure.
3. The hot gas expands adiabatically, driving the piston down, delivering its energy via the piston to the shaft drive.
4. The gas, cooling down at constant volume, is ejected through the outlet valve and the exhaust.

Gasoline engines power cars, aircraft, and outboard motors. Diesel engines drive cars, trucks, locomotives, ships, and electricity generators .

Gas turbines deliver mechanical power either via shaft drives or in the form of jet power, using the kinetic and thermal energy of hot or combustion gases to drive turbine wheels. Normally, one or more compressors compress air and transport it into one or more combustion chambers, where fuel is injected and ignited. The energy-rich gas is carried away into one or more turbines, where it is decompressed so much that the power required by the compressor(s) can be supplied. Then the remaining energy is transformed into mechanical work by another turbine, coupled to a shaft drive; alternatively, in jet engines, the pressure energy accelerates the exhaust gases to velocities between 500 and 800 m/s at takeoff. The thrust-generating gas turbine powers more than 90% of world air transportation capacity. In combination with shaft drives, the gas turbine is used in helicopters, locomotives, ships, pumping stations, gas-turbine power stations, and combined gas-turbine and steam-turbine power stations. Because of its increasing profitability, the gas turbine is often considered as the final stage of the heat engine.

2.3.1.2   Carnot's Ideal Heat Engine

The physical principles of conversion of heat to work were analyzed by the French officer-engineer *N.L.S. Carnot*, who, in 1824, founded thermodynamics with his article entitled "Réflexions sur la puissance motrice de feu et sur les machines propres à développer cette puissance." He investigated the theoretical construction of an ideal heat engine, which now bears his name. This engine consists of a gas in a cylinder, which is closed by a movable piston. It operates between a heat source of absolute temperature T and a heat sink of lower temperature T_0 in four *reversible*, infinitely slow subprocesses $1 \rightarrow 2 \rightarrow 3 \rightarrow 4 \rightarrow 1$, which represent the Carnot cycle. In process $1 \rightarrow 2$ the gas is compressed adiabatically, i.e., in thermal isolation. Process $2 \rightarrow 3$ is isothermal decompression at constant absolute temperature T. Adiabatic decompression then follows in process $3 \rightarrow 4$.

Isothermal compression at temperature $T_0 < T$ closes the cycle in process $4 \to 1$. During isothermal expansion(process $2 \to 3$), the gas absorbs the amount Q of heat from the heat source at temperature T, and during isothermal compression (process $4 \to 1$), it rejects a smaller amount of heat $Q_{0C} < Q$ to the heat sink, e.g., the environment, at the lower temperature $T_0 < T$. The difference between received heat Q and rejected heat Q_{0C} is the work W_C the Carnot heat engine performs. In general, the energetic efficiency of a heat engine that receives the heat Q and rejects the heat Q_0 is defined as $\eta \equiv W/Q$, with $W = Q - Q_0$. Carnot showed that for the ideal heat engine this efficiency is

$$\eta_C \equiv W_C/Q = (Q - Q_{0C})/Q = 1 - T_0/T. \tag{2.17}$$

The *Carnot efficiency* η_C determines the exergy of heat, which is $\eta_C Q$. Since, in principle, fossil fuels, solar radiation, and nuclear reactions can produce heat sources of absolute temperatures far above those of the environment, i.e.,$T \gg T_0$, so $\eta_C = 1$, their enthalpy is practically all exergy.

No heat engine can have a higher efficiency than the Carnot efficiency η_C. If the heat sink is the natural environment with $T_0 \approx 290\,\text{K}$ and the heat source has a temperature T of, say, $900\,\text{K}$, then η_C is close to 68%. Real heat engines, which complete one cycle not infinitely slowly but within a fraction of a second, dissipate energy by internal friction and turbulences. The pure engine efficiency in motor cars is 20–25% for diesel engines and for gas engines is somewhat lower. Steam turbines have efficiencies up to 40%, whereas the efficiency of gas turbines is higher.

2.3.1.3 Historical Note: Newcomen's and Watt's Steam Engines

Newcomen's engine is shown in Fig. 2.2. It was used industrially for the first time in 1711, when it replaced a team of 500 horses that had powered a wheel to pump out a coal mine. It consisted of a boiler A, where the steam was generated. This was usually a haystack boiler, situated directly below the cylinder. It produced low-pressure steam, all that the current state of boiler technology could cope with. Steam at this pressure would be unable to move a piston of any size. One side of the beam was attached by a chain to the pump at the base of the mine, and the chain at the other side suspended a piston within a cylinder B. The cylinder was open at the top end to the atmosphere above the piston P. The piston had a bevelled edge, around which hemp rope, kept in place by metal weights, acted as a primitive seal. (The rope was kept wet, so that it would expand against the sides.) When the valve V was opened, the steam was admitted into the cylinder. After this valve had been closed, valve V′ was opened to allow cold water from the tank C into the cylinder, thus condensing the steam and reducing the pressure under the piston. The atmospheric pressure above then pushed the piston down in the power stroke. This raised the working parts of the pump, but their weight immediately returned the beam to its original position. Steam was then readmitted, driving the remains of the condensate out through

Fig. 2.2 Newcomen's steam
pump [21]

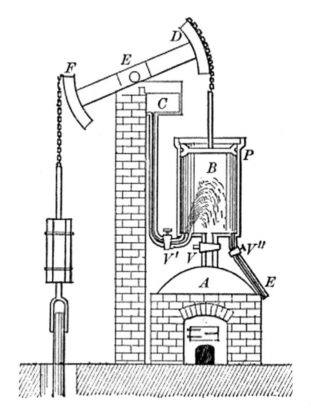

a one-way snifter valve V″ as the process started all over again. Unfortunately, the water also cooled the cylinder walls, so when the next charge of steam was introduced, a considerable part was spent simply warming the cylinder back up to boiling temperatures, condensing while this occurred. This reheating of the cylinder consumed much energy. The efficiency of Newcomen's steam engine was 0.5%.

James Watt's first improvement of Newcomen's steam engine was enhancing its energetic efficiency by a separate condensing chamber C; see Fig. 2.3. The cooling water spray was injected into this condensing chamber, attached to the main cylinder B through a valve V′. When the piston P had reached the top of the cylinder, the valve V was closed, so that no more steam from the boiler entered the cylinder B, and valve V′ was opened. External atmospheric pressure then pushed the steam and piston toward the condenser. Thus, the condenser C could be kept cold and under less than atmospheric pressure, whereas the cylinder B, connected to the boiler, remained hot.

Watt further developed his pumping steam engine into what became the multiple-purpose steam engine by substituting low-pressure steam for atmospheric pressure and providing rotary power. A double-acting engine, in which the steam acted alternately on the two sides of the piston, gave a very even movement of the beam. The reciprocating motion of the piston was transformed into rotational power

Fig. 2.3 Watt's pumping
steam engine [21]

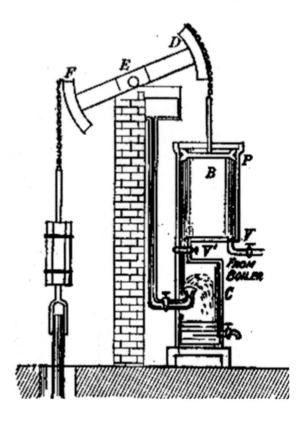

for grinding, weaving, and milling by means of the sun and planet gear system invented by Watt's employee William Murdoch. Later, the more familiar crankshaft was adopted. This allowed the steam engine to be used to replace water wheels, eliminating geographical constraints on the utilization of rotary power.

To improve reliability, Watt introduced further improvements. The operation of the condenser was assisted by an air pump, driven by an eccentric rod attached to the beam. As the pace of the operation of the machine increased, it needed to operate at a constant speed. A centrifugal governor, earlier used in windmills to automatically control the pressure between the millstones, was installed for automatically controlling steam flow to the engine and keeping it at a steady speed. Watt also introduced the manometer to measure steam pressure within the engines. This, when connected to a linkage to the position of the piston and a pencil that recorded both, could record the action of the machine throughout the cycle, producing the indicator diagram. The efficiency of Watt's steam engine was eventually 3%.

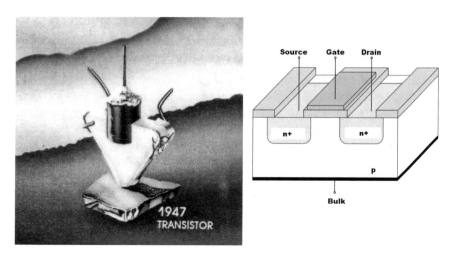

Fig. 2.4 *Left*: Model of the first transistor presented at the celebration of John Bardeen's 80th birthday at the University of Illinois at Champaign-Urbana. *Right*: An n-channel metal–oxide–semiconductor field-effect transistor

2.3.2 Transistors

The metal–oxide–semiconductor field-effect transistor (MOSFET) shown in Fig. 2.4 consists of the following elements:

1. The p-type bulk material. This is silicon doped with acceptors. Acceptors are atoms that accept and bind electrons to themselves, leaving positively charged, mobile holes in the valence band of silicon.
2. Two n+-type regions close to the surface of the p-type bulk. The n+-type regions are silicon heavily doped with donators. Donators are atoms that donate negatively charged, mobile electrons to the conduction band of silicon. Two metallic contacts, source and drain, connect the n+-type regions to the rest of the electronic circuit.
3. Three separate stripes of insulating silicon dioxide on top of the p-type and n+-type regions. The center stripe is connected to a metallic contact, the gate.
4. The most important part is the (invisible), 5–10-nm-thick n-type inversion channel between the two n+-type regions. It forms at the surface of the p-type silicon when a positive voltage is applied to the gate, depleting this region of holes. Current is conducted through it only when the positive gate potential is high enough to attract electrons from the source into the channel. When zero or negative voltage is applied between the gate and the source, the channel disappears and no current can flow between the source and the drain. Thus, by varying the gate voltage, one varies the number of electrons in the inversion channel and the conductivity of the device. This way the MOSFET can act as an amplifier of electromagnetic signals in telecommunications or as a valve for

electric currents. Digital information processing is done by blocking or letting pass a current through the inversion channel below the gate: no current flow represents the number 0, current flow represents the number 1.

Since the 1960s the transistor has superseded the vacuum tube in electronics. Substituting the transistor for the bulky, massive, and energy-consuming vacuum tube has saved enormously on materials, energy, and space. For instance, a vacuum-tube computer with the computing power of a 2008 notebook computer would have had a volume of many thousands of cubic meters. According to "Moore's law," transistor density on a microchip has doubled every 18 months during the last four decades. It may continue in that way for a while until the joule heat, produced by the currents circulating in the transistors, can no longer escape sufficiently rapidly to prevent a meltdown of the microchip.

2.4 Energy Services

The wealth of industrial nations has grown so much thanks to energy services from heat engines and transistors[6] that all other countries try to industrialize rapidly, despite the emerging collateral damage pointed out in Chap. 3. Economists interpret the energy services that heat engines and transistors provide under the control of human hands and brains as enhanced labor productivity.

2.4.1 Freedom from Toil

The liberation of humans from hard and dangerous work is arguably the most important energy service provided by machines. Tilling the soil, excavating foundations, lifting weights, drilling holes, breaking and hewing stones, processing metals, and transporting goods and people draw so much on the maximum power of about 100 W an average human can supply that agrarian societies could hardly do without complementing the work of domesticated animals by slavery and socage.

Heat engines provide industrial societies with energy services that are numerically equivalent to that of a mighty workforce of humans. To provide a quantitative idea, let us consider the economy of the Federal Republic of Germany before reunification. In 1990 it consumed 7400 PJ of final energy, approximately 40% of which entered heat engines [22]. This includes electric devices, which may be considered as extensions of the heat engines in power stations. The corresponding power input into heat engines was 8.22×10^{11} kWh/year $= 9.39 \times 10^7$ kW. The

---

[6]Switching devices such as relays and vacuum tubes are now information processors of minor importance.

average human daily work-energy requirement in the form of food for a very heavy workload is 2.9 kWh, corresponding to an average power input of 0.12 kW. Although heat engines are energetically more efficient than humans, let us assume for the sake of simplicity that their output of work per unit of energy input is the same as that of humans. Then the work performed by West German heat engines in 1990 would have been energetically (at least) equivalent to the work performed by $9.39 \times 10^7/0.12 = 780$ million humans laboring 365 days per year.

More specifically, a comparison of forces and powers illustrates the progress introduced in agriculture by farming tractors. Deep plowing requires forces of about 1,500 N. (The tractive power of a horse is about 780 N. On average, humans can pull with a force of about 570 N for short times, and with much less in continuous operation.) For plowing 1 ha of land, and assuming a plow–furrow width of 1 m, one has to apply the force of 1,500 N along a length of 10,000 m. In so doing, one has to perform total work of 1.5×10^7 J, as one can compute from (2.32) in Appendix 1 of Chap. 2. If four men could work continuously, each at his maximum power of 100 W, they would have performed an amount of work equivalent to that of plowing 1 ha after 10.5 h. Two horses, each working continuously at 700 W, would require 3 h. Diesel engines of farming tractors, running at nominal powers[7] between 20 and 200 kW, would deliver 1.5×10^7 J within a couple of minutes. (Of course, a tractor needs more time for pulling its plows through a hectare.) On the whole, the agricultural energy services of farming tractors, usually driven by diesel oil and serving a wide variety of purposes, exceed those of men and animals by several orders of magnitude. No wonder that now in industrial countries the percentage of total civilian employment in agriculture is only small, as shown in Table 4.1 in Chap. 4. In Germany it decreased from 25% in 1950 to 2.4% in 2009.

Heaving heavy loads in the construction of buildings and cargos from storage sites to transportation vessels and vice versa has been hard work throughout the ages. In our days it is being done by cranes. Consider a container that has a mass of 10 t. Its center of mass has to be craned against the pull of Earth's gravity from a harbor's loading platform to a containership's cargo area at 10 m above the platform. According to (2.19) and (2.32), the crane must perform work of 9.81×10^5 J $= 0.27$ kWh. If the crane were powered by electricity, and if its owner had to pay €0.2/kWh – the electricity price for German private households in 2009 – the energy price for heaving the container to the ship would be €0.054. By the same reasoning, one finds that an electricity quantity of 4.8 kWh, costing €0.96, would be required to lift a 100-kg mountaineer in a 100-kg seat from sea level to the top of Mount Everest at 8,848 m.

Excavating building pits and drilling holes in the ground to extract matter from the depths of Earth have also been hard and dangerous work for humans. What if fossil fuels became scarce and could no longer power diggers and drillers? Recently, two economists discussed the decline of nonrenewable energy resources. One expected big problems. The other one was more optimistic and reasoned that there

---

[7]Mc Cormick D-439: 26 kW, Fendt 824: 177 kW.

are many unemployed people in the world who could provide the physical work required for making renewable energies available, for instance, by drilling 1,800-m-deep holes into the Earth's crust to get geothermal energy. But to provide the required drilling power , one would have to chain huge masses of people to winches. The conventional rig used for drilling a 4,000-m-deep hole in preparation for drilling the world's deepest hole of 9,101 m in the project Kontinentale Tiefbohrung (Windischeschenbach, Germany) required a total power of 2,320 kW.[8] Thus, 23,200 people, permanently supplying 100 W, would have been needed to provide the power for this drilling equipment.

Smelting iron from ore in the fire from wood and charcoal, and hammering it into tools, vessels, and weapons kept many hard-working men busy in olden days. Modern blast furnaces burn fossil fuels, mined and transported by heat engines, and turn out iron in quantities unimaginable to the blacksmiths of the past.

2.4.2 Comfort, Mobility, Information

Steam, gas, water, and wind turbines generate electricity, which lightens homes and powers vacuum cleaners, washing machines, dishwashers, refrigerators, and air conditioners. In kitchen stoves wood and coal have been replaced by (joule heat from) electricity and gas for cooking and baking. Until the middle of the twentieth century, most homes in northern latitudes had just one warm room on cold days; an oven or a tile stove burned dirty coal, which had to be carried upstairs from its bunker in the basement. Nowadays, oil and gas are pumped through pipelines over thousands of kilometers from distant wells to local distribution sites, or immediately into buildings. Their burning in efficient boilers is regulated electronically by thermostats, which keep room temperatures at convenient levels everywhere.

Modern materials allow good thermal insulation of homes and reduce fuel consumption substantially. The temperature of the walls of a room determine the temperature we feel in the radiation field of that room. Since we are only comfortable within a narrow temperature range around 20°C $\approx$ 293 K, we program thermostats to correct for small temperature changes ΔT that fall out of this range. If heat losses through walls and windows have lowered the temperature from T to $T - \Delta T$, raising the temperature back to T requires an energy quantity that increases with the third power of T according to (2.25) and (2.26) in Appendix 1 of Chap. 2.[9]

---

[8]Lifting equipment 1,240 kW, scavenging pumps 720 kW, rotary table 360 kW (F. Holzförster, private communication). The power of drilling equipment units offered by Drill-Quest Engineering in Hünenberg, Switzerland, is 82.9 kW (112 hp) for a stationary diesel power unit and 555 kW (750 hp) for a mobile drilling rig.

[9]The difference between the radiation energy in a room with wall temperature T and the same room with wall temperature $T - \Delta T$ is proportional to $T^4 - (T - \Delta T)^4 \approx 4T^3 \Delta T$.

But few people bother about good thermal insulation of homes as long as they have to invest more in insulation than they can save in fuel costs in, say, 20 years.

A huge gain in traveling comfort and mobility has been obtained since heat engines entered the field of transportation. Until the beginning of the nineteenth century, common people such as journeymen traveled on foot. A daily walking distance was usually not much more than 30 km. Wealthier people, who could afford a carriage drawn by horses, traveled at speeds of about 15 km/h and suffered from poor suspension on bad roads. In our times, we enjoy comfortable seats in well-cushioned cars, buses, and trains going at 80–300 km/h.

Energy balances for a compact passenger car with a total mass of 1,500 kg are as follows. If the car moves at 120 km/h, its kinetic energy is 8.34×10^5 J $= 0.23$ kWh, according to (2.18). The velocity is achieved, and maintained against friction, by a 75-kW gas engine with a gas consumption of, say, 7 L per 100 km when driving at 120 km/h on a flat road during a calm day. (About 100 horses could also supply 75 kW and would need 10^6 m^2 of pasture.) If one could travel at a constant 120 km/h for 5 h, covering 600 km, one would have used 42 L of gasoline, less than a tankful, which cost €59 at a price of €1.40/L. Assuming an energy content of 8.8 kWh $= 31,700$ kJ/L of gasoline, one would have used 370 kWh on the trip, essentially by producing heat via the friction from air and wheels. Braking the car from 120 to 0 km/h dissipates the kinetic energy of 0.23 kWh as heat in the car's brakes. Braking by converting the kinetic energy of forward motion into the kinetic energy of a rotating flywheel within the car could conserve most of the exergy. It could be reused for subsequent acceleration. Since kinetic energy increases with the square of velocity, a car that hits a wall at 50 km/h releases four times more destructive, and often deadly, kinetic energy than the same car hitting the wall at 25 km/h.

Trucks, ships, and airplanes, burning diesel oil and kerosene, move people and freight rapidly over distances that span continents and the globe. Rockets, powered by solid boosters and engines that burn liquid fuels such as hydrogen, transport telecommunication and Earth-observation satellites to low and high Earth orbits.

Telecommunication via telegraphs, phones, radio, TV, and the Internet makes more information available to the modern individual than his or her brain could ever process and store. This is quite different from the times when letters and books where the only carriers and stores of information outside the human body. Information in telecommunication is imprinted upon electromagnetic waves that propagate through metallic wires, the air, and a vacuum. Their energy density (in a vacuum) is given by (2.22) in Appendix 1 of Chap. 2. One source[10] estimates that "power consumption of telecommunications and data networks ... (is) as much as 3% of European electricity".

Electric energy also serves information processing. Electric waves carry information to a computer's processor, which consists of transistors. Either a 1 (electric

---

[10]http://www.optoiq.com/index/photonics-technologies-applications/lfw-display/lfw-article-displ ay/articles/optoiq2/photonics-technologies/news/applications-_markets/communications-_it/2010/ 8/BIANCHO-project.html.

current) or a 0 (no current) is received per cycle. The rate at which the wave pulses enter the processor is measured in cycles per second, i.e., hertz (Hz). This determines the speed of information processing. Modern computers are so fast that this rate is measured in billions of cycles per second, i.e., gigahertz (GHz). The battery of a 2.4-GHz laptop, which stores about 0.06 kWh when new, is emptied within 1–2 h of operation. Quite a bit of this energy is consumed for ventilating the joule heat developed by the currents in the processor. The higher the pulse frequency, the more joule heat is produced. The fact that haste makes waste will be discussed further in Chap. 3 on entropy production.

2.4.3 Political Power

A nontechnical energy service flows from the ownership of energy sources. This was seen impressively after World War II, when abundant cheap oil powered rapid economic growth in North America and Europe. The party ended abruptly when the Arab members of the Organization of Petroleum Exporting Countries (OPEC)[11] implemented an oil embargo against the West in retaliation for US support of Israel during the Yom Kippur War from October 6 to October 26, 1973. Although the embargo was lifted in March 1974, the OPEC members decided to raise their real incomes from oil exports by raising world oil prices. As a consequence, the inflation-corrected price of a barrel of crude oil in 2009 US dollar prices jumped from about $\$_{2009}15$ in 1973 to $\$_{2009}50$ in 1975. This first "energy crisis" caused the first postwar economic recession. The second energy crisis hit when the price of the barrel climbed to more than $\$_{2009}90$ in 1981, after the Iranian revolution and the Iraq/Iran war had reduced oil supply. Figure 2.5 shows how the price of a barrel of crude oil in constant and nominal US dollars developed between 1861 and 2009. OPEC became a major economic and political player in the world.

The Soviet Union was not an OPEC member but, being a major oil exporter, profitted greatly from the oil price explosions. When the economies of the West suffered from the high oil prices on international markets, the Soviet Union and its allies enjoyed the "socialist" block's low oil prices set by the Soviet planners. Thus, East Germany remained practically unaffected by the two oil price shocks, whereas West Germany went into recession. This caused an East German leader to boast that the superiority of socialism had finally been proven. In 1979 the Soviet president, Leonid Brezhnev, and the politburo of the communist party felt the Soviet Union was strong enough to invade Afghanistan and start an arms race. The navy was reinforced to catch up with US sea power and SS-20 rockets were deployed against western Europe. Then came the third, negative oil price shock, that is, the

---

[11] OPEC is a cartel of 12 countries made up of Algeria, Angola, Ecuador, Iran, Iraq, Kuwait, Libya, Nigeria, Qatar, Saudi Arabia, the United Arab Emirates, and Venezuela. OPEC has maintained its headquarters in Vienna since 1965.

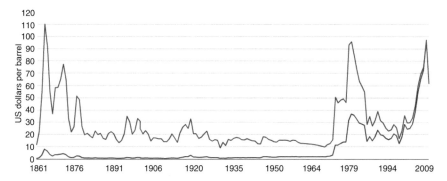

Fig. 2.5 Development of the price of one barrel of crude oil since 1861 in 2009 US dollar prices (*upper curve*) and in dollar prices of the day (*lower curve*); see also Fig. 4.7. (Source: http://www. pdviz.com/historical-crude-oil-prices-1861-to-2009)

fall of the oil price from over $_{2009}90$ per barrel in 1981 to just $_{2009}30$ per barrel in 1986. The flow of petrodollars into the Soviet Union subsided and could no longer finance the import of urgently needed consumer goods. The nation's resources had mainly gone into heavy industry, whereas the consumer industry remained weak and inefficient under the control of the planning bureaucracy. Realizing the need for change, Mikhail Gorbachev introduced the reforms of perestroika and glasnost, but it was too late. After the collapse of the Soviet Union in 1991, Russia and the other successor states introduced market economics, however without the appropriate legal framework and the institutions to enforce it. This was one cause for the economic and political decline of the former superpower during the decade from 1990 to 2000. The other cause was the low oil price in the 1990s. When the oil price started to rise again at the turn of the century, and its annual average climbed to nearly $_{2009}100$ per barrel in 2008, Vladimir Putin rose to power in Russia. He became acting president of the Russian Federation on December 31, 1999, won the presidential elections in 2000 and 2004, and would have been reelected in 2008 had the constitution not forbidden a third term in office. Instead he was nominated to be Russia's prime minister by his successor Dmitry Medvedev. Analysts attribute a good part of Putin's popularity with the Russian electorate to the increase in the standard of living during his reign, thanks to Russia's large revenues from the exportation of oil and gas.

Industrialization of populous emerging economies such as those of China, India, and Brazil, and the resulting growing demand for energy, has driven up the oil price since 1999. Its fall by one third because of the global recession that was triggered by the collapse of the US housing market and the ensuing bank crashes in 2007–2008 is most likely to be temporary. If the global economy recovers despite the huge debts accumulated by governments in their fight to avoid recession, energy prices will rise again. By how much they will rise will depend on the reserves and resources of the traditional fuels, the potential for energy conservation, and the technological options for exploiting all energy sources that are fed by the natural or technological conversion of mass into energy.[12] Economic and political power will be with those societies that meet best the challenges arising from the energy problem.

2.5 Consumption, And What Is Left

Energy services dissipate the valuable exergy of energy carriers by producing entropy. Equations (2.28) and (2.45–2.47) in Appendix 1 of Chap. 2 indicate how entropy reduces exergy. In this sense, economic activities lead to "energy consumption."

2.5.1 Consumption of Energy Carriers

Average energy consumption per person per day is a rough indicator of material well-being in an economy. However, when comparing it during different times and for different economic systems, one has to take into account differences in energy-conversion efficiencies. In general, a system with higher efficiencies of its energy-converting devices provides more material well-being than a system with lower efficiencies and the same per capita energy consumption. Despite this qualification, the evolution of per capita energy consumption during human history can still be seen as running parallel to the evolution of civilization. Material well-being in this sense can be also illustrated by the average number of energy slaves serving every person in an economy.

> The number of energy slaves in an economy is given by the average amount of energy fed per day into the energy-conversion devices of the economy, divided by the human daily work-calorie requirement of 2,500 kcal (2.9 kWh) for a very heavy work load. Dividing this by the number of people in the economy yields the number of energy slaves per capita.

The following scheme shows the evolution of average energy consumption per person and day and the resulting *per capita* energy slaves from one million years before the present (BP) until our time:

One million years BP: 2 kWh (gatherer without fire).

100,000 years BP: 6 kWh (hunter and gatherer with fire), approximately one energy slave

7,000 years BP: 14 kWh (simple peasant society), approximately four energy slaves

---

[12]Fusion in the Sun provides all renewable energies, except tidal power and those geothermal energies that result from the radioactive decay of minerals and volcanic activity.

AD 1400: 30 kWh (western Europe), approximately nine energy slaves
AD 1900: Germany 89 kWh, approximately 30 energy slaves
AD 1960: West Germany 61 kWh, approximately 21 energy slaves; USA 165 kWh, approximately 59 energy slaves
AD 1990: West Germany 117 kWh, approximately 40 energy slaves; USA 228 kWh, approximately 79 energy slaves
AD 1995: Germany 133 kWh, approximately 45 energy slaves; USA 270 kWh, approximately 92 energy slaves
World average 46 kWh, approximately 15 energy slaves; developing countries 20 kWh, approximately six energy slaves

Of course, the 21 energy slaves per West German in 1960 worked much more efficiently, i.e., provided more energy services, than the 30 energy slaves per German in 1900, thanks to technical improvements in the energy-conversion devices. Furthermore, the 40 energy slaves per capita in West Germany in 1990 include those that provide process heat and room heating, and which add to the heat-engine energy slaves – roughly 13 per person in 62 million people – calculated in Sect. 2.3.1.

The provision of energy slaves by a production system determines the number of people who can live on a given land area. Hunters and gatherers with fire, who command one energy slave per person, have an average per capita energy consumption of 6 kWh/day, or 2,190 kWh/year. Harvesting plant and animal biomass, they can get between 0.2 and 1.7 kWh/ha/year, as we saw in Sect. 2.2.4. Thus, each member of a hunter-and-gatherer society needs at least 13 km^2 to satisfy his energy needs. Germany's land area of 351,000 km^2 could therefore accommodate at most about 27,000 hunters and gatherers. In 2010, 82 million people were living in industrialized Germany.

Figure 2.6 illustrates how world energy demand grew between 1970 and 2004, and which energy sources satisfied it.

In 2004 global primary energy consumption was 427 exajoules (EJ) $= 427 \times 10^{18}$ J (approximately 10.198×10^9 t of oil equivalents, approximately 14.571×10^9 t of coal equivalents, approximately 117.483×10^{12} kWh). Oil provided 157 EJ, gas 101 EJ, coal 116 EJ, nuclear power 26 EJ, and other sources 27 EJ (see Fig. 2.6). In 1992 global primary energy consumption was 344.8 EJ, and in 1970 it was 209 EJ. Oil has the biggest share (approximately 40%) of global primary energy consumption, followed by coal and gas.

Primary energy consumption by G7 countries has been more than 40% of global consumption. Table 2.2 gives the shares of the different energy sources in satisfying the demand of these industrial nations. The USA is the biggest energy and oil consumer. Its per capita energy consumption is only slightly smaller than that of Canada, as shown in Table 2.3.

The European G7 countries and Japan are much more densely populated and have much smaller daily per capita energy consumption than Canada and the USA. Furthermore, private households and vehicles use energy more sparingly in Europe and Japan than in North America. Comparison of the data in Table 2.3 with the 1995

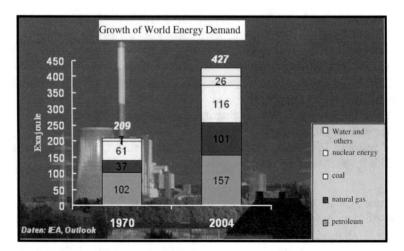

Fig. 2.6 Global energy demand in 1970 and 2004. From *top* to *bottom*: Water and others, nuclear energy, coal, natural gas, mineral oil. $1\,EJ = 10^{18}\,W\,s$; $427\,EJ = 10.2 \times 10^{9}\,t$ of oil equivalents (Source: www.weltderphysik.de)

Table 2.2 Primary energy consumption (million tons of oil equivalents) in the G7 countries Canada, France, Germany, Italy, Japan, the UK, and the USA in the years 1995 and 2003 [23]. (*Others* includes water, geothermal, and solar power, combustion of biomass and garbage, export balances of electricity and heat)

G7 countries	Coal 1995	2003	Oil 1995	2003	Gas 1995	2003	Nuclear 1995	2003	Others 1995	2003	Total (rounded) 1995	2003
Canada	25.4	30.0	78.2	91.7	67.1	79.2	25.6	19.5	36.2	40.2	232	261
France	16.1	14.4	86.6	91.0	29.6	39.4	98.3	115.0	10.9	11.6	241	271
Germany	91.0	85.1	135.7	126.5	66.4	79.2	39.9	43.0	5.9	13.3	339	347
Italy	12.3	14.9	94.5	87.4	44.6	63.3			10.1	15.4	162	181
Japan	82.6	107.7	269.6	257.0	52.0	71.0	75.9	62.6	17.0	18.9	497	517
UK	48.6	38.2	84.6	81.4	65.1	85.9	23.2	23.1	3.0	3.3	225	232
USA	475.3	531.2	804.4	921.4	508.7	519.2	186.0	205.3	114.0	103.7	2089	2281
Sum	751.3	831.5	1553.6	1656.4	833.5	937.2	449.5	468.5	197.1	206.4	3785	4090

data in the energy slave listing shows that daily per capita energy consumption in the USA decreased from 270 kWh in 1995 to 246 kWh in 2005. The US population increased from 257.6 million in 1993 to 295.7 million in 2005.

The lifestyle of industrialized countries such as the G7 counties is associated with average energy quantities required for products and activities that are indicated in Table 2.4.

A broader view of energy and oil consumption per person is presented in Fig. 2.7.

Table 2.3 Population of G7 countries in 2005 [24], (approximated) daily per capita primary energy consumption, and the corresponding per capita power consumption

G7 countries	Population (millions)	Primary energy consumption per person per day (kWh)	Power consumption per person (kW)
Canada	32.8	253	10.5
France	60.6	143	6.0
Germany	82.4	134	5.6
Italy	58.1	99	4.1
Japan	127.0	130	5.4
UK	60.4	122	5.1
USA	295.7	246	10.3

Table 2.4 Average energy requirement per product or activity [25]

Good/activity	Energy requirement (kWh)
1 kg bread	10
1 kg book	50
1 kg motor car	200
1 kg laptop	1,000
1 warm shower	5
1 h cell phone conversation	2
1 h watching TV	3
1 university classroom hour per student	20
1 h car driving	200

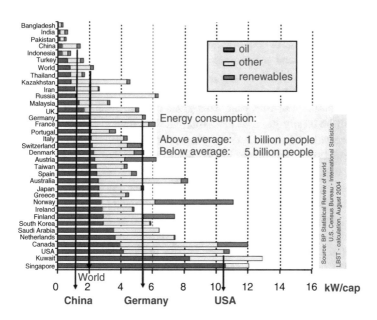

Fig. 2.7 Energy and oil consumption per person (kW/capita) in different countries [26]

Table 2.5 Global and regional reserves of coal (hard coal and lignite), mineral oil, and natural gas in 2005, and their depletion times (*DT*) in years at 2005 exploitation rates [27]. *MtCE* 10^6 t of coal equivalents (equivalent to 29.3 PJ), *MtOE* 10^6 t of oil equivalents (equivalent to 41.9 PJ; 1 t of oil equivalents is equivalent to 7.3 barrels of oil equivalents), *EJ* exajoules; 10^9 m^3 natural gas is equivalent to 32.23 PJ. North America includes Canada, the USA, and Mexico; Europe is without the Commonwealth of Independent States (*CIS*; the former USSR)

Region	Coal MtCE	EJ	DT (years)	Oil MtOE	EJ	DT (years)	Gas 10^9 m^3	EJ	DT (years)
World	656,200	19,227	161	161,600	6,771	41	179,059	5,771	63
Middle East				100,427	4,208	83	72,652	2,342	>200
Africa	47,600	1,395	232	15090	633	32	14,082	454	86
North America	210,400	6,165	213	6,840	286	11	7,737	249	10
South America	15,000	440	246	14,209	595	41	7,174	231	53
Asia/Oceania	170,200	4987	86	6,412	287	17	14,425	465	45
Australia	73,300	2148	256						
CIS	114,500	3,355	369	15,975	669	28	57,123	1,841	70
Europe	25,200	738	99	2,648	111	10	5,866	189	18

2.5.2 Reserves and Resources of Fossil and Nuclear Fuels

Industrial democracies and some oil exporters have enjoyed a boost in the standard of living since the end of World War II. But how long will the party last? The answer depends on the evolution of the reserves and resources of primary energy.

Reserves are the occurrences of energy carriers that have been identified and measured and that are known to be technically and economically recoverable. Resources are all occurrences of energy carriers with less certain geological assurance and/or doubtful economic feasibility [29].

There is a static and a dynamic way of estimating the time after which the finiteness of reserves will become a problem.

The static method considers the reserves of a given year and divides them by the exploitation rates of that year. This gives the depletion times, after which the reserves will be totally gone, if the exploitation rate stays the same until the end. Table 2.5 presents the very unevenly distributed reserves of coal, oil, and gas, and the static depletion times. According to the latter, the global reserves of coal, oil, and gas would be exhausted in about 160, 40, and 60 years, respectively. The estimated global *resources*, which may be exploited – with advanced technologies – at higher cost than reserves are 4.1689×10^{12} t of coal equivalents, 82.056×10^9 t of oil equivalents, and 206.770×10^{12} m^3 gas. Thus, at the exploitation rate of 2005, the resources of *coal* would last for nearly *1,000* years.

The global reserves and resources of the nuclear fuel uranium are given in Table 2.6. The heating value of 1 t of uranium is between 0.4 and 0.7 petajoules (PJ), depending on the reactor type and the fuel cycle; here, we use 0.5 PJ (0.5×10^{-3} EJ). The reserves of natural uranium can be mined at cost up to US $40/kg. The resources of natural uranium can be mined at cost between US $40/kg and US $80/kg; 88% of them are in ten countries. Canada, Australia, Niger, and Russia are presently

Table 2.6 Global reserves, resources, and speculative resources of natural uranium, in million tons (Mt) [28] and exajoules (EJ), in 2005

Reserves		Resources		Speculative resources	
1.95 Mt	975 EJ	5.32 Mt	2,660 EJ	7.54 Mt	3,770 EJ

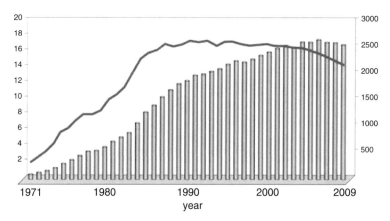

Fig. 2.8 Global nuclear electricity production (TWh/year) (*right ordinate, bars*) and percentage of total electricity production (*left ordinate, line*) from 1971 to 2009. (Source: World Nuclear Association)

the main suppliers of uranium. The speculative resources include uranium in stocks and especially in old nuclear weapons inventories; the latter contain more fissile uranium-235 (U^{235}) than natural uranium. The concentration of U^{235} in natural uranium is only 0.7%. Global annual uranium consumption was 65,000 t in 2007. About only half of that was mined. The rest was obtained from stocks and the conversion of nuclear warheads to reactor fuel rods.

The potential of thorium as a nuclear fuel is discussed in Sect. 2.6.

According to the World Nuclear Association, the share of nuclear power was 15% of global electricity generation in 2007, when the installed generating capacity of the 439 nuclear reactors was 371.7 $GW_{electric}$.[13] A decade before it was 17%. The growth of nuclear generating capacity and of the number of nuclear reactors between 1971 and 2009 is shown in Fig. 2.8.

The dynamic method of estimating the time after which the finiteness of Earth's energy reserves will cause trouble takes into account that industrial growth is coupled to the growth of energy consumption. An exception to this observation is periods during which energy conservation measures improve the overall energetic efficiency of economies. This happened especially after the first and the second oil price shocks of 1973–1975 and 1979–1981. But there are thermodynamic limits to energy conservation [30]. Once they are reached, increasing exploitation of energy reserves and resources is inevitable if industrial production grows. Given this

---

[13]http://www.world-nuclear.org/info/reactors.html.

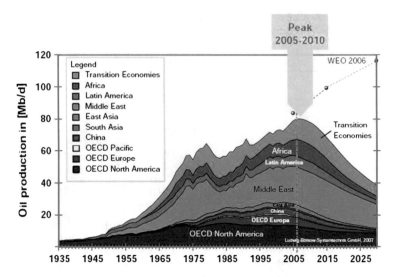

Fig. 2.9 Regional and global oil production in the past and as predicted by the "Peak Oil" theory and the World Energy Outlook 2006 [26]

situation, the Texan oil geologist Marion King Hubbert introduced the concept of "Peak Oil" in 1956 [32]. Since the late 1990s this concept has gained more and more attention. This means that oil production, as a function of time, will increase and then decline inevitably in the form of a roughly bell shaped curve. The maximum of oil production is "Peak Oil". In his book *The Last Oil Shock* David Strahan [33] tells the exciting story of how Hubbert predicted that the USA would reach Peak Oil around 1970, that experience proved him right, and how individuals and institutions with special interests in the oil business tried to suppress his method and findings. By now Hubbert's method is widely used by researchers and is also applied to energy sources other than oil. Figure 2.9 shows global and regional peak-oil scenarios and Fig. 2.10 plots the growth and decline of global fossil fuel and uranium reserves. These scenarios are based on a number of economic and geological assumptions that reflect past experiences. They are not predictions. If they worked as self-destroying prophecies, their authors would be more than happy. Chance is also that the great economic crisis, triggered by the burst of the American mortgage bubble in 2007, will shift the peaks from the near into a more distant future.

2.5.3 Renewable Energies

Biomass, water power, wind power, solar heat, and solar electricity are renewable sources of energy. They originate from solar radiation. Geothermal energy is usually counted as renewable energy too. It originates from radioactive decay of uranium and thorium in the interior of Earth.

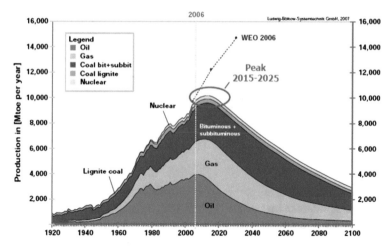

Fig. 2.10 Production of conventional energy carriers in the past and scenarios for the future [26]. *Mtoe* million tons of oil equivalents

With few exceptions, heat and electricity from renewables have been more expensive than heat and electricity from fossil fuels. But when investment costs for exploiting renewables will no longer exceed those for providing and burning fossil fuels, market penetration of renewables will rapidly increase beyond the scope reached so far.

By definition, economic potentials of renewables can be exploited with available technology at costs that are comparable to those of exploiting conventional energy sources. Technical potentials of renewables can be exploited with known technologies that have to be developed further for grand-scale operation. Thus, the economic and technical potentials of renewables correspond (more or less) to the reserves and resources of fossil fuels. The potentials of renewable energy sources are subject to environmental constraints, although probably to a lesser extent than the reserves and resources of fossil fuels.

Renewables generated by solar radiation have the greatest technical and economic potential. This is because of the sheer size of the solar input. Earth absorbs 1.2×10^{17} W of solar radiation power. This is 239 W/m^2 according to (2.12). The solar power influx exceeded global energy consumption of 427 EJ in 2004 by a factor of roughly 10,000, because 427×10^{18} W s/year $= 1.35 \times 10^{13}$ W. Thus, the fusion reactor the Sun is a huge resource that supplies energy free to Earth. It will last for several billion years. However, sunlight reaches the surface of Earth intermittently. One has the problem of storing its energy. Furthermore, one needs large areas. For instance, more than 3×10^6 m^2 is needed just to collect 800 MW of solar radiation. This area increases with the inverse of the efficiency of converting solar radiation into the required final energy, be it heat or electricity. At a first glance, this compares quite unfavorably with the roughly 600 m^2 occupied by a typical 800-MW steam turbine of a fossil-fuel-burning electric power plant. But the area required for

Table 2.7 Solar power farms for a hydrogen economy serving a world population of ten billion people. For two scenarios, the following are shown: the power provided per person, the photovoltaic collector area required, and the amount of steel required for scaffolding, reinforcement, etc. at $10\,kg/m^2$ [34]

Power per person (kW)	Collector area required (km^2)	Amount of steel required (Mt)
10	4×10^6	40,000
4	1.6×10^6	16,000

providing the fuel and for evaporating the waste heat may exceed the steam-turbine area by several orders of magnitude.[14] Thus, both the abundance of an energy source and the scarcity of areas required by the technology for its supply and utilization are crucial. Area scarcity differs with the type of energy source, geography, and climate.

2.5.3.1   A Global Scenario

An illustrative example of what might be possible under very favorable conditions is an estimate of the areas and materials required for satisfying future global energy demand by solar power. In this example, population growth on Earth is assumed to level-off at ten billion people. All energy demand is satisfied by photovoltaic sun farms in the subtropical desert belt of Earth. They convert sunlight to electricity. The electricity is used to produce hydrogen by electrolysis, thus storing solar energy in an energy carrier that, with some caution, can be handled and used like oil. The example was published in 1975 by Paul Erbrich [34] in an evaluation of the second report of the Club of Rome. Two scenarios were considered. The first one corresponds to a US-like per capita energy consumption of 10 kW. The second scenario considers a per capita consumption of 4 kW, the French consumption rate in the 1970s. The estimate is shown in Table 2.7.

It is still appropriate for giving an idea of the orders of magnitude involved in satisfying future energy demand by solar power. The basic assumptions are as follows: solar radiation density in the subtropical desert belt is $250\,W/m^2$; the efficiency of photovoltaic conversion of solar radiation to electricity is approximately 20%; the efficiency of electrolytic production of hydrogen is approximately 50%; the efficiency of conversion of hydrogen to energy services (including transportation) is approximately 80%. The total system efficiency is (optimistically) about 10%.

---

[14]A comparison of geothermal power plants with others by the US Department of Energy estimated the total area per megawatt of a coal-fired power plant to exceed 70,000 m^2 (probably assuming strip mining of coal); http://www1.eere.energy.gov/geothermal/geopower_landuse.html.

For other total system efficiencies one has to scale the numbers in the second and third columns of Table 2.7 correspondingly. Areas for comparison are those of the Sahara (9×10^6 km^2), Saudi Arabia (2×10^6 km^2), and Spain (0.5×10^6 km^2).

The International Iron and Steel Institute announced that world crude steel output reached 1.2395×10^9 t in 2006. This is less than one tenth of the steel demand of the 4 kW/capita scenario. Maybe one can replace steel by concrete. Cost estimates for this scenario were between US $20 trillion and US $50 trillion in 1975. Technological progress in the production of photovoltaic cells would decrease that cost, perhaps by an order of magnitude or more. At the 2004 conference on renewable energies in Bonn, the president of the International Solar Energy Society, Yogi Goswani, reported that in 1974 a solar cell cost US $30/W, whereas by 2004 the cost had dropped to US $3/W. In 2007 the firm First Solar announced a cost of US $1.12/W.

Annual photovoltaic electricity production would be 87.6×10^{12} kWh in the sun farm scenario of 4 kW/capita energy services from hydrogen. This would be more than five times the global electricity consumption in 2006, which was 16.379×10^{12} kWh. The globally installed generating capacity grew from 2,900 gigawatts (GW) in 1995 to 4,012 GW in 2006 [35].

2.5.3.2   Individual Potentials

Biomass, the product of photosynthesis, has the longest tradition as a renewable energy source. Its extracorporal energetic use dates back to the taming of fire some 400,000 years ago. Dry biomass with a carbon content of about 50% has a heating value of 17.6 MJ/kg. Annually, on the global average, 12 t of dry biomass is produced per hectare of green land area. Thus, annually the continents of Earth produce 120×10^9 t of dry biomass with a burning value of 2,000 EJ [36]. This is nearly five times the 427 EJ consumed globally in 2004. However, only 80 EJ of that may be combusted as biofuels, because the major part of biomass is needed as food and nonenergetic raw materials [36].

Wood, grain, oil seeds, oil palms, cane, other annual and biannual plants, and organic wastes, e.g., straw and manure, are biofuels. However, the production of biofuels such as palm oil in Colombia and Southeast Asia and ethanol from sugarcane in Brazil is causing considerable ecological, social, and human rights problems in these countries and regions. The combustion of cereals is problematical too, as long as people go hungry.

Water power has been used as a renewable energy since ancient times dating back to the ancient Greeks and Romans. Water mills, powered by running water in creeks and small rivers, have ground grain, lifted weights, cut stone, and driven mechanical, iron-forging hammers. The few water mills left are mostly used for electricity generation. Water power is provided by the Sun. About one fourth of the solar irradiation absorbed by Earth evaporates water, mostly in the warm oceans. The water vapor condenses in clouds. Clouds sail over the land and drop their water as rain, filling creeks, streams, and reservoirs of water power plants. When the water rushes down the tubes from a reservoir to the water turbines at the foot of the dam, the potential energy of the water in the reservoir is converted into the kinetic energy that drives the electricity-generating turbine wheels. The global capacity of hydroelectricity generation was $650\,GW_{electric}$ in 1995 and provided 2,560 terawatt-hours (TWh) in that year. It grew to about $710\,GW_{electric}$ in 2005, covering nearly 20% of global electricity demand, and accounting for more than 60% of electricity from renewable sources. It is estimated that by 2050 it should be possible to obtain 4,000–6,000 TWh of electric energy per year from water power globally [36]. However, expansion of water power means building huge dams in remote mountains, or flooding large areas of rain forests, or dislocating people from the banks of big rivers. The associated financial, social, and environmental costs are high.

Ocean thermal energy conversion and tidal power stations also generate electricity with the help of water, although on a much smaller scale than conventional hydroelectric power plants. Ocean thermal energy conversion uses solar energy stored in the oceans by operating a heat engine between the warm surface water (at about 15°C) and the cold water (at about 5°C) in greater depths of the oceans. The energy source of tidal power is not sunshine but rather gravitational and rotational energy of the Earth–Moon–Sun system.

Wind originates from rising warm air and from the condensation of water vapor to form clouds. Its kinetic energy, which is only a small percentage of the solar radiation received by Earth, has been used by windmills for 2000 years in China and the Middle East, and for 800 years in Europe [36]. In the early eighteenth century about 250,000 windmills operated in Europe, grinding grain, draining the land, and performing other mechanical work. However, by the end of the eighteenth century, windmills had been ousted by the coal-fired steam engine. Since the 1990s they have come back as low-emission, high-technology electricity generators. For instance, in Germany, the number of windmills and their generating capacity grew from about 1,000 and 110 MW in 1991 to about 20,000 and 22,250 MW in 2007 [37], thanks to the economic incentives of a feed-in tariff system established by legislation.[15] In 2006, the contribution of wind power to German electricity generation from an installed wind power capacity of 20,622 MW was 30.5 TWh, or 3.5% of the total. In the same year, global installed wind power capacity was 75 GW, of which Germany contributed 20.6 GW (a share of 27.5%), and Spain and the USA both contributed 12 GW (15.6% each) [38]. By 2009 global installed wind power capacity had grown to 158 GW, with 35.2 GW in the USA, 25.8 GW in Germany, and 25.1 GW in China [39]. Estimates of the global technical potential of wind power vary between 3,000 TWh/year [36] and more than ten times that [40].

Low-temperature solar heat is provided by thermal collectors that convert sunshine into the heat of water or another working fluid, often containing an antifreeze mix. In their simplest form they are uncovered plastic absorbers that warm swimming pools. Flat-plate collectors within insulating, transparent boxes are widely used to supply warm water and heat to buildings. Vacuum tube collectors

---

[15] *Erneuerbare Energien Gesetz* ("Renewable Energy Law").

achieve higher temperatures and efficiencies. They are especially well suited for colder climates. Normally, flat-plate and vacuum-tube collectors are installed on roofs and are combined with water tanks for daily, weekly, or seasonal storage of solar heat. Optimum integration into the total heating system is important for maximum efficiency. For instance, analysis of a Bavarian pilot project – a district heating system for 100 well-insulated housing units with an annual total heat demand of 616 MWh – yielded the following results for flat-plate collectors: collector areas between 1 and 2.5 m^2/(MWh annual heat demand) and water storage volumes between 1.2 and 4.2 m^3/(m^2 collector area) can cover 32–95% of the total heat demand [41]. Satisfying the total German heat demand for room heating and warm water, which was 700×10^9 kWh in 1995, would require a collector area of about 2,000 km^2 and seasonal heat stores with a total water-equivalent volume of about 14×10^9 m^3. It is possible to halve German room-heating demand by appropriate thermal insulation of buildings. Then, 1,200-km^2 collector area and 8×10^9-m^3 storage volume would be sufficient to provide all Germans with warm rooms and warm water via solar heat [36].

Photovoltaic electricity is generated by solar cells. Solar cells are made from semiconductors doped with tiny amounts of impurities that act as donors and acceptors. Doping works as in the field-effect transistor described in Sect. 2.3.2. Donors in the n-type region of the semiconductor produce electrons (negatively charged particles) in the conduction band, and acceptors in the p-type region produce positively charged holes in the valence band of the semiconductor. One has a p–n junction, in principle. Diffusion currents, driven by concentration differences, transport electrons from the n-type into the p-type region and transport holes from the p-type into the n-type region. This creates an internal electric field in and a potential drop across the interface layer between the two regions. When the potential drop has finally reached what is called the "diffusion voltage" $\mathscr{V}_d$, the diffusion currents vanish. Then the interface layer, also called the "space charge layer," is depleted of mobile carriers. Because of the internal electric field, the p–n junction is a diode, where current can only flow in one direction across the junction. If light quanta are absorbed by the solar cell, free, energy-rich electron–hole pairs generated in the space charge layer are separated immediately by the internal electric field. Electrons move into the n-type region and holes move into the p-type region. The p-type region is charged positively and the n-type region is charged negatively. The diffusion voltage is reduced, and a photovoltage $\mathscr{V}_p < \mathscr{V}_d$ develops between the ends of the diode. When the solar cell is connected to an external resistor, the photovoltage drives electric currents through the circuit, and electric work is performed.

Silicon in monocrystalline, multicrystalline, and amorphous forms is the most widely used base material of most solar cells. Gallium arsenide and cadmium sulfide are alternatives. The efficiency of converting sunlight into electricity is 12–15% in commercial solar cells. High-efficiency ultrathin solar cells have an efficiency of more than 20%. "If photovoltaic systems are to be able to compete with fossil fuels to generate electricity in the future, swift action need to be taken to make them not only steadily cheaper, but also more efficient. The mid-term aim is to go below

1 euros/W_p (W_p = generating capacity at peak power, in W) using silicon technology, and the long-term aim is to go below 0.5 euros/Wp with the aid of new solar technologies. Solar cells made from thin crystalline silicon on a substrate present great potential for reducing costs. This technology combines the high efficiency of thick film technologies with the cost advantages of thin films. The greatest potential for saving costs lies in printed photovoltaic (PV) technologies, in particular, organic solar cells, which currently have an efficiency of just under 8%" [42].

An array of many solar cells is a photovoltaic module. Each module typically has an area of 1 m^2. Commercial modules provide between 50 and 200 W in full sunshine under standard conditions (solar insolation about 1 kW/m^2). Photovoltaic modules can be assembled in arbitrarily large areas, fields, and installations. This flexibility is a characteristic and decisive advantage for the evolution of photovoltaics. Between 1993 and 2009 globally installed photovoltaic capacity grew by a factor of more than 400, namely, from 56 MW_p to 5,500 MW_p in 2005 [43] and 23,000 MW_p in 2009 [44]. The latter was about 0.57% of total electricity-generating capacity in 2006. Total installed photovoltaic power in Germany was about 2,500 MW_p in 2006 [45], and it had grown to 9,800 MW_p by the end of 2009 [44].[16] Germany is the major photovoltaic market worldwide, thanks to the "Renewable Energy Law," and China floods the German market with photovoltaic modules. The area required for photovoltaics was estimated in 1997 [36]: 800 km^2 of roof area is available for solar collectors in Germany; if one were to cover 100 km^2 by photovoltaic modules, thus installing a generating capacity of 10,000 MW_p, one could generate about 10 TWh of electric energy per year. This would be 1.6% of the 617.5 TWh of electric energy consumed in Germany in 2007.[17]

Solar thermal electricity is generated by mirrors that focus sunlight on an absorber. The absorber converts the sunlight into heat. The heat is transferred to a vapor or gas medium, which drives a generator. At a given generating capacity, the area of the expensive mirrors increases with the daily and seasonal variations of sunshine duration. Therefore, appropriate locations of solar thermal power plants are only in regions between 40°N and 40°S. The efficiency of solar thermal conversion of sunlight into electricity is between 20% and 35%. Therefore, one needs mirror areas of 3–5 m^2/(kW of electricity-generating capacity) during the times when solar energy flow is maximum and is about 1 kW/m^2.

Solar thermal power plants differ in the arrangement of the focusing mirrors. Parabolic troughs focus the light two-dimensionally on pipes containing the heat-transporting medium. In solar tower power plants a field of up to 1,000 mirrors, called heliostats, focuses the sunlight three-dimensionally on an absorber on the top of the tower. Paraboloid dishes with Stirling engines as generators serve as decentralized (mobile) power stations. Installed world capacity of solar thermal power plants was 354 MW in 2005 [43].

---

[16]International Energy Agency.

[17]Industriegewerkschaft Bergbau, Chemie, Energie, "Brancheninfo: Elektrizitätswirtschaft, Fakten und Daten zur deutschen Elektrizitätswirtschaft 2007 und Ausblick."

Geothermal energy supplied less than 1% of the world's 2008 energy demand,[18] but estimates of the technical potential go as far as 100% for hundreds of years.[19] On the other hand, a 1997 estimate of the global technical potential of geothermal energy that might be exploited within the next couple of decades was 1 EJ, at maximum [36]. The 2008 lead study of the German Ministry of Environmental Affairs (BMU) estimated that 1.8 TWh electricity and 8.2 TWh heat might be produced by German geothermal installations by the year 2020. The numerous studies of seismicity induced by hydraulic fracture in geothermal reservoirs should be continued in order to assess the risks associated with the extraction of large quantities of heat from the Earth.

2.5.3.3 Energy Payback Times, Harvest Factors, and Energy Return on Investment

An important aspect when estimating the technological potentials of renewables is the energy required for their production. This energy will be mostly taken from the energy sources presently available. The energetic payback time and the related harvest factor are relative measures of the quantity of energy that has to be used to produce the installations that exploit a renewable energy source. The energetic payback time says how long a system needs until it has supplied as much energy as was needed for its construction. The harvest factor is the total amount of energy supplied by the system during its average lifetime divided by the amount of energy required for the system's production. Typical numbers [46] are:

- Wind power plants

  Lifetime: 20 years
  Average wind speed 4 m/s: payback time 7–22 months; harvest factor 11–36
  Average wind speed 5.5 m/s: payback time 4–11 months; harvest factor 21–63
  Average wind speed 7 m/s: payback time 2–7 months; harvest factor 31–93

- Solar thermal collector for warm water supply

  Lifetime: 20 years
  Substitution of an average heating system and 56% solar supply rate: payback time 5 months; harvest factor: 48
  Substitution of a gas-fired condensing boiler and 78% solar supply rate: payback time 30 months; harvest factor 8

- Photovoltaic electricity generation

  Lifetime of silicon solar cells: 30 years

---

[18] 2008 IEA Key World Energy Statistics.
[19] "The Future of Geothermal Energy", Massachusetts Institute of Technology 2006; http://geothermal.inel.gov/publications/future_of_geothermal_energy.pdf.

Monocrystalline silicon, efficiency 14.5–15.5%: payback time 48–75 months; harvest factor 4.8–7.4

Polycrystalline silicon, efficiency 12–14%: payback time 25–57 months; harvest factor 6.2–14

Amorphous silicon, efficiency not available: payback time 17–41 months; harvest factor 8.6–21

Energetic payback times and harvest factors of the various forms of biomass depend on how and where biomass is being produced. Estimates are controversial and go down to small numbers, especially with respect to palm oil production.

A concept that should be included in all estimates of how things change with alterations of the energy supply system is the energy return on investment (EROI) [47]. ("Investment" means "invested energy.") EROI is the ratio of the energy that is provided by a process to the energy that is used directly and indirectly in that process. "If the EROI of a fuel is high, then only a small fraction of the energy produced is required to maintain production, and the majority of that energy produced can be used to run the general economy. On the other hand, if the EROI is very low, … very little net energy is available to do useful economic work. High EROI fuels are vital to economic growth and productivity" [48].

When process chains are analyzed it is important to use the same order of energy analysis at each step and define system boundaries precisely.[20] If this is done properly, the EROI concept allows comparison of nonrenewable and renewable energy sources and conversion technologies. Table 2.8 shows EROIs for a number of processes and different steps of the process chain. The differences between the minimum EROI and the maximum EROI are partly due to different system boundaries, and partly they are due to the spread in process efficiencies at each level.

The numbers of energy units gained from one energy unit invested in extracting fossil fuels from their natural deposits decline as the easily accessible oil and gas fields are being depleted. In [48] it is estimated that, on the global average, the EROI of oil at the wellhead was roughly 26:1 in 1992, increased to 35:1 in 1999, and then decreased to 18:1 in 2006. Furthermore, according to Table 2.8 and [49], the energy returns from energy invested in collecting renewables from their natural flows are substantially below those of exploiting oil and gas wells in the twentieth century.

---

[20]Where to draw the system boundaries properly is sometimes controversial. For instance, there are people who argue that photovoltaic cells will never reproduce the energy invested in their production, because in this energy one should include the fuel used by workers in factories producing photovoltaic cells, and during boat-trip vacations in the Caribbean, for example. Arguments of this quality are perhaps responsible for the persisting rumors that the harvest factors of photovoltaic cells and even wind power installations are less than 1.

Table 2.8 Energy return on investment (*EROI*) for energy carriers and steps of process chains. *CCS* carbon (dioxide) capture and storage, *POU* point of use. (Source: Hannes Kunz, Institute of Integrated Economic Research, http://www.iier.ch)

	EROI (min)	EROI (avg)	EROI (max)
Coal			
Mine mouth	50	70	90
Coal to plant	45	65	86
Coal to electricity	14	28	43
Electricity from coal after CCS	4	14	24
Oil			
Well	12	56	100
Transportation of crude	10	52	95
Refinement	8	49	90
Gasoline/diesel at pump	7	44	81
Industrial fuels at POU	7	47	86
Direct heating from oil at POU	5	43	80
Natural gas			
Well	30	40	50
Pipeline/truck to power plant	24	36	48
Gas to electricity	7	17	26
Direct heating from gas at POU	18	29	40
Nuclear			
Mine	30	115	200
Enrichment	12	91	170
Power plant	4	36	68
Solar			
Electricity from photovoltaics	4	8	12
Electricity from solar concentration	4	7	10
Autonomous solar use (with battery)	1	3	5
Direct solar thermal energy (warm water)	30	40	50
Direct solar thermal energy (heating)	5	8	10
Wind			
Electricity from wind	10	18	25
Autonomous wind use (with battery)	3	8	13
Wind to grid	7	15	23

2.5.4   Energy Conservation

Grand-scale use of renewable energies requires large investments and big areas. Integration of renewables into the existing energy systems at minimum cost calls for careful system analysis. The study *Long-Term Integration of Renewable Energy Sources into the European Energy System* has looked into scenarios of phasing out fossil and nuclear energies by 2050 in Europe, replacing them by renewables in combination with energy conservation measures. The latter are supposed to reduce

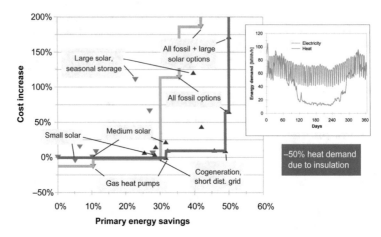

Fig. 2.11 Energy savings and cost increases for satisfying the fluctuating electricity and heat demand of "Würzburg." *Upright triangles* are are for 50% reduction of room heat demand by thermal insulation of buildings; *inverted triangles* are for 0% reduction of room heat demand. The *inset* shows the weekly fluctuations of the average daily electricity demand (MW) (*upper curve*) and the seasonal change of heat demand (MW) (*lower curve*). *CHP* combined heat and power (Source: Thomas Bruckner, University of Leipzig)

energy demand by more than 50%. "However ... the future cost figures do not include efficiency measures which involve substantial restructuring of the economy and could, thus, not be calculated for this book" [50], so the cost estimates in this study only refer to renewables.

Smaller, less complex systems such as cities allow computation of all costs for optimized technology combinations. Since the potentials of energy conservation and renewables do not add up, but are interdependent in not-at-all obvious ways, advanced system-analytical tools are required to find out how deeply, and at what costs, they can be exhausted under given climatic conditions and energy demand profiles. Thermoeconomics provides the basic analytical and conceptual framework [51]. One such tool is the *d*ynamic *e*nergy *e*mission and *c*ost *o*ptimization model *deeco*. To illustrate competition and synergy between energy technologies when restructuring an energy system, and to compute the associated costs as well, it has been applied to the model city of "Würzburg" [52]. This model city has the fluctuating room-heating and electricity demand shown in Fig. 2.11. These demands, and sunshine and temperature variations as well, were measured for the real city of Würzburg during a representative year. This city of about 135,000 inhabitants is located in central Germany on 50°N latitude. Reductions of energy consumption and emissions, and the associated cost reductions/increases, were computed for a number of scenarios that combine energy-saving and solar-heating technologies.

The specific costs of all technologies comprise annuities of investments as well as operation and maintenance costs. The total cost of an energy conversion technology is calculated by multiplying the specific cost with the maximum capacity

of the technology needed according to the optimization results. The total cost of
the optimized energy system is calculated as the sum of the costs of all energy
conversion and transportation technologies, plus the cost of fuel and the cost of the
electricity imported from the national grid.

In the reference scenario all heat is supplied by oil boilers, and all electricity
comes from the national grid. Energy prices are those of the mid-1990s. The energy
supply technologies are grouped in scenarios. Each scenario contains a subset
from the following technologies, in addition to those of the reference scenario:
gas-fired (conventional and condensing value) boilers, small and medium-sized
gas-fired cogeneration (combined heat and power) units, district heat from a large
coal-fired cogeneration plant with back-pressure turbine, gas-fired ambient air heat
pumps, solar heating systems of small, medium, and large size, combined with daily,
weekly, and seasonal heat storage, covering 7%, 25%, and 79% of total heat demand,
respectively. Appropriate conversion efficiencies are used for each technology.

Vector optimization with respect to primary energy consumption, emissions,[21]
and cost was done for a great number of scenarios. An overall picture of trade-
offs between primary energy savings and cost increase for various technology
combinations is shown in Fig. 2.11. More detailed results for some characteristic
scenarios (identified by the names given in [52]) are as follows[22]:

1. In the scenario *MCogen*, medium-sized cogeneration units, feeding short-range
   district grids, cover 57% of the heat demand and 86% of the electricity demand.
   The remaining 43% of the heat demand is contributed by gas-fired boilers. Thirty
   percent of the primary energy used in the reference scenario is saved and costs
   are reduced by 2%.
2. In the scenario *AllNonSolar*, it is possible to choose among all non-solar-
   energy technologies during optimization. In this case, small and medium-sized
   cogeneration units supply 73% of heat and 86% of electricity, whereas gas-
   fired condensing value boilers meet nearly all of the remaining heat demand.
   Nevertheless, the resulting energy savings of 36% are only 6% points higher than
   in *MCogen* and, thus, much less than the sum of the individual savings potentials
   of the technologies considered. Furthermore, the small additional savings are
   bought with a cost increase of 114% with respect to the reference scenario. This
   example shows how aggregated savings potentials cannot be fully realized owing
   to the competition between different technologies. The investment costs are spent
   for all technologies, but many contribute only little to savings.
3. In the scenario *All + Solar*, large solar-thermal installations, including seasonal
   heat storage, are available in addition to all energy conservation technologies.
   Here 48% of the heat is contributed by solar installations and 39% by
   cogeneration. The latter also meets 47% of electricity demand, the rest of which
   is covered by imports from the grid. Primary energy savings increase to 42%,

---

[21]CO_2, NO_x, SO_2, and dust.

[22]Emission reductions roughly follow energy savings.

but costs rise by 186% compared with the reference case. Again, competition between technologies shrinks the total primary energy conservation potential and drives up costs.

The maximum energy savings at a given cost increase without additional heat insulation are reached for those scenarios (not all identified here) that are placed near the left trade-off plot in Fig. 2.11. Additional thermal insulation of buildings increases potential energy savings substantially in almost all scenarios, whereas cost increases remain moderate, as shown by the right trade-off plot. A further finding, not shown in Fig. 2.11, is that a combined energy and carbon dioxide tax of US $30 per barrel of oil equivalent, as applied in the 1993 IEA CO_2 tax scenario [53], would reduce relative cost increases in the scenarios *AllNonSolar* and *All + Solar* by about 50% and lower the cost of the *MCogen* scenario by more than 10% below the cost of the reference scenario [52].

The special-case analyses for "Würzburg" show that care has to be taken when estimating potentials and costs of integrating energy conservation with renewable energy utilization. One needs detailed information on energy demand and climate in the system considered, appropriate modeling of technology interactions, and dynamic optimization techniques.

The thermodynamic limits of energy conservation are easier to compute, because one is entitled to use all physically thinkable, although technically extremely idealized, simplifying assumptions for this exercise. The limits have been estimated for the industrial energy demand of The Netherlands, Germany, Japan, and the USA by thermoeconomic optimization [30,31]. The Simplex algorithm yields, at best, the possibility of saving 50% of primary energy, with unchanged energy services, for system-integrated heat exchanger networks, heat pumps, and cogeneration plants. Vector optimization with respect to primary energy and cost indicates that energy and cost minimization become parallel objectives only at energy prices that are 3–4 times higher than those in the 1980s. It turns out that the given exergy demand structures of the production processes in the countries considered determine the maximum potentials of energy conservation. Improvements in energy efficiency are limited by basic thermodynamics to factors that do not exceed 2.

2.6 Technological Perspectives

Exploitation of the great potentials of wind power, photovoltaics, and solar heat in northern industrial countries is handicapped by fluctuations of wind and sunshine. Occasionally, so much wind power is fed into the German grid that overload can only be avoided by paying neighboring countries to take the surplus electricity.[23]

---

[23]The law (*Erneuerbare Energien Gesetz*) that stimulates the enormous growth of wind power and photovoltaics entitles windmill owners to feed their electricity into the grid at guaranteed feed-in tariffs, no matter whether the grid can take it or not.

Therefore, progress in storing electricity (and heat) is decisive for expanding the use of fluctuating renewable sources. Fission, fusion, and solar farms in deserts and in space are not handicapped by stochastic fluctuations. Of course, they have other problems.

> Experience shows that problems with and public opposition to an energy technology grow with the quantity of energy this technology provides.

2.6.1   Fission

The controlled fission of uranium and thorium in nuclear reactors is so far the only peaceful method for converting mass m into energy E according to Einstein's equation $E = mc^2$. Here m is the difference between the mass of the uranium-235 (^{235}U) nucleus and the smaller sum of the masses of the fission fragments into which the ^{235}U nucleus decays after having absorbed a thermal neutron. Fission can only be produced by thermal neutrons, which move so slowly that their kinetic energy is of the order of $k_B T$. However, the neutrons that are part of the fission fragments, and whose number must exceed 1 in order to maintain a chain reaction, are very fast. They must be slowed down by a moderator. Water and graphite are the moderators used in existing nuclear power plants.

In water-moderated reactors the water also serves as the coolant, from which the steam is produced that drives the steam turbine. Pumps circulate the cooling water so that it can transport the heat away from the fuel rods in the reactor core. In the case of an accidental breakdown of the cooling system, the core heats up, and vapor bubbles form in the boiling water. Since the density of water in the vapor bubbles is very low, neutron moderation weakens. The fast neutrons are no longer slowed down in sufficient quantities, and the chain reaction comes to an end. One says that the reactor has a negative temperature (void) coefficient. Nevertheless, in the worst-case scenario the heat that is generated in the fuel rods by β decay, for hours or even days after reactor shutdown, may cause a core meltdown. A partial core meltdown occurred during the Three Mile Island reactor accident near Harrisburg, USA, in 1979. On March 11, 2011, after an earthquake of strength $M\_w = 9$ in the Tohoku region of Japan and the follow-up tsunami that destroyed the cooling systems in several of the Fukushima-Daiichi nuclear power plants, core meltdowns occurred in 3 reactors, and large quantities of radioactivity were released. More than 100,000 people had to be evacuated, probably permanently. After this accident opposition to nuclear power increased. The German government made a U-turn in its energy policy. Only 8 months before the Fukushima accident it had extended the legal operation time for German nuclear power plants substantially. But on

May 30, 2011, it was decided to shut down the last German nuclear reactor in the year 2022. Plans are now to replace nuclear and coal power plants with large windparks in the North Sea in combination with high-power transmission lines and gas-fired power stations, with photovoltaics, and with storage facilities for the fluctuating energy from wind and sunshine. In the preceding discussions one cabinet member was asked how to finance the necessary huge investments without too much cost for electricity consumers and the tax payer. He replied: "More competition in the energy market will take care of that."

Graphite is the moderator in RBMK reactors, which were developed and installed in the Soviet Union and some of its satellite states. The fuel rods are imbedded in the moderating graphite. Cooling is by water. In contrast to water-moderated reactors, RBMK reactors have a positive temperature (or void) coefficient. This is because water also absorbs neutrons, and the reactor has to be designed correspondingly. If the reactor's cooling system fails, the graphite and the cooling water heat up, and water evaporates. In this case the very low water density in vapor *reduces* neutron absorption drastically, whereas moderation by graphite continues. This dramatically alters the balance of thermal neutron production, causing a runaway condition, in which more and more thermal neutrons accelerate the chain reaction. If this is not stopped by neutron-absorbing control rods, the core melts down. This happened on April 26, 1986, at reactor 4 of the Chernobyl power plant close to Kiev, when, in a badly organized safety experiment, the positive temperature coefficient, in combination with deliberately broken safety rules by insufficiently trained operators, produced the worst nuclear power plant disaster before the Fukushima disaster.

Since then nuclear power has been seen by many people as an energy option with intolerable risks. This is at least the situation in many wealthy industrial nations, which have not suffered from energy scarcity for several decades. Things are different elsewhere. For instance, there were still 12 RBMK reactors operational in 2009, despite their positive temperature coefficient, although with improved safety features: one in Lithuania and 11 in Russia. The Lithunian reactor was shut down at the beginning of the year 2010.

The other argument raised against nuclear energy is that nuclear-power-using societies have not yet agreed on appropriate sites for safe, long-time disposal of burned-out nuclear fuel rods. The problem is that spent fuel from nuclear reactors still contains considerable amounts of ^{235}U. In addition, significant amounts of plutonium-239 (^{239}Pu) are generated during fission.

Disposal of long-lived radioactive waste in solid rock is an option that comes close to what nature did when it accumulated natural uranium in sites such as the granite rock formations of the Black Forest in southwest Germany [25]. One would have to enclose the radioactive waste in appropriate containers, insert these containers in holes, deeply drilled into the rock, and completely fill all remaining voids at the disposal site with water-impermeable material. This option has not yet been realized by any of the nuclear-energy-using countries. One reason may be the associated cost. It is estimated to be between 2 and 3 euro cents per kilowatt-hour of electricity from nuclear power stations [25]. Presently, German utilities, generating electricity from nuclear power, are legally obliged to hold 0.5 euro cents per kilowatt-hour electricity in reserve for the disposal of radioactive waste and the decommissioning of nuclear power plants.

The energy requirements for removing burned-out nuclear fuel rods from the biosphere by accelerating them to escape velocity via electromagnetic mass drivers or rockets [54] and the resulting waste heat burden on the environment, are calculated in Sect. 3.6.3.

Reprocessing plants can separate ^{235}U and ^{239}Pu from the other components of spent fuel. This would significantly address two major concerns, in principle. It would greatly reduce the long-lived radioactivity of the residue, and it provides purified ^{235}U and ^{239}Pu as reactor fuel. Using uranium in fast breeder reactors, cooled by liquid metals such as sodium, would have similar benefits. But both technologies have raised considerable safety and environmental concerns and are used in a few countries only. Especially in France, where nuclear power covers more than 70% of electricity demand, people value nuclear risks much lower than people in most other west European countries.

Pebble bed high-temperature reactors are an alternative to water-moderated and RBMK reactors. Since their safety and waste-disposal risks are much lower than those of the other reactor types, they represent an option for what some people call "green" fission. These reactors use thorium besides ^{235}U as a nuclear fuel. Although not fissile itself, thorium-232 (^{232}Th) will absorb slow neutrons to produce uranium-233 (^{233}U), which is fissile (and long-lived). Hence, it is fertile, like uranium-238 (^{238}U). In one significant aspect ^{233}U is better than ^{235}U and ^{239}Pu, because of its higher neutron yield per neutron absorbed. Utilizing thorium as a nuclear fuel has the advantage that it is more abundant in Earth's crust than uranium. Also, all of the mined thorium is potentially usable in a reactor, as compared with 0.7% of natural uranium. Thus, about 40 times the amount of energy per unit mass might theoretically be available, without recourse to fast breeder reactors. Global total thorium resources are estimated to be about 2.5×10^6 t [55].

Thorium was used as a fuel in the first thorium high-temperature reactor (THTR-300), which was developed in Germany. It was a pebble bed reactor, where the nuclear fuel is enclosed in spherical fuel elements called "pebbles." The reactor core contained approximately 670,000 pebbles. These tennis-ball-sized pebbles are made of pyrolytic graphite (which acts as the moderator), and they contain thousands of microscopic fuel particles. These microscopic fuel particles consist of a fissile material (such as thorium and uranium) surrounded by a coated ceramic layer of silicon carbide (SiC) for structural integrity. The durability of the ceramic enclosure of the radioactive material after removal from the reactor is estimated to be at least one billion years. Thus, storing radioactive waste from pebble bed reactors is less problematical than storing nuclear fuel rods from conventional water-cooled reactors. Furthermore, in the case of a terrorist attack, e.g., by a rocket, the fuel elements would fly apart and suffer (and do) much less damage than scattered fuel rods, because the microscopic fuel particles are extremely hardened by the SiC coating. Pebble bed reactors are cooled by convection (i.e., natural circulation) of inert or semi-inert gases such as helium, nitrogen, or carbon dioxide. Because of their passive safety and their negative temperature coefficient – if the temperature in the pebble bed reactor goes up, the reaction rate goes down – a core meltdown, due to the breakdown of an active cooling system, is impossible.

The official history of the THTR-300, as told by the Web-based encyclopedia Wikipedia and other sources, goes like this. The THTR-300 was a thorium high-temperature nuclear reactor rated at 300 $MW_{electric}$ (THTR-300). The German state of North Rhine–Westphalia and Hochtemperatur-Kernkraftwerk financed construction of the THTR-300. Operations started on the plant in Hamm-Uentrop, Germany, in 1983, and it was shut down on September 1, 1989. The reactor was synchronized to the grid for the first time in 1985 and started full-power operation in February 1987. The THTR-300 cost 2.05 billion euros and is expected in 2011 to cost an additional 1 billion euros in decommissioning and other associated costs. On September 1, 1989, the THTR-300 was deactivated because of its cost and increased public scrutiny following both the Chernobyl accident and the THTR-300 fuel pellet event of May 4, 1985, in which a fuel pellet became lodged in a fuel feed pipe to the core. On October 10, 1991, the 180-m-high dry cooling tower, which at one time was the highest cooling tower in the world, was explosively dismantled, and from October 22, 1993 to April 1995 the remaining plant was decommissioned. The THTR-300 technology was sold to China in 1991. Safe enclosure, including the prestressed-concrete reactor vessel, was finished in September 1996 [56].

The high-temperature reactor technology is being developed further by MIT, the South African company PBMR, General Atomics (USA), the Dutch company Romawa, Adams Atomic Engines, Idaho National Laboratory, and the Chinese company Huaneng. In June 2004, it was announced that a new pebble bed reactor would be built at Koeberg, South Africa, by Eskom, the government-owned electric utility. But on September 17, 2010 the South African Minister of Public Enterprises announced the closure of the PBMR.

The unofficial story of why Germany gave up the most advanced and safest nuclear technology, developed at high cost to the taxpayer, was given in a private communication from Hermann Josef Werhan. Werhan belongs to a prestigious German industrial dynasty and is a son-in-law of Konrad Adenauer, the first chancellor of the Federal Republic of Germany. In the late 1990s we met accidentally at breakfast in a Berlin hotel, discovered that we had been at different energy meetings the evening before, and started talking energy business. Werhan opened his wallet and showed me two photographs, side by side: "This is my wife, and that is Rudolf Schulten of Forschungszentrum Jülich, who developed the pebble bed reactor." When I responded, "What a pity that Germany gave up this technology," he asked me, "And do you know why?" I told what I knew from the press: "After the Hamm-Uentrop prototype had started operations, the government asked the utilities to finance further development of the reactor to full market maturity, because enough taxpayer's money had gone into the prototype development. But the big power companies refused to do so, because they had already water-moderated reactors developed and running and were afraid of the financial risks of further THTR development." Werhahn laughed. Then he told me his insider story. He had been present at the board meetings of one of the big German utilities when the decisions against the THTR-300 were taken. According to Werhahn, the real reason for killing this innovative technology was that safe, small-scale high-temperature reactors could be operated even by small municipal utilities right within the cities, producing electricity and heat in a very decentralized way. This was against the commercial interests of the big utilities, who wanted to dominate the market with large, centralized power stations, including water-moderated nuclear reactors in the 1,000-MW$_{electric}$ range.[24]

Werhahn keeps advocating pebble bed reactors. Because of their safety features and the billion-year enclosure time of the radioactive particles by the SiC coatings, he calls them with some justness "green" fission reactors. In fact, Edward Teller, the "father of the fusion bomb," had a reactor with the properties of the high-temperature reactor in mind when he commented at the 1955 United Nations conference in Geneva on the peaceful use of nuclear energy in the following sense. "The aim of a peaceful use of nuclear energy for world-wide electricity generation requires the development of completely new nuclear power stations – free from the risk of an accident that releases large quantities of radioactivity. Only when such risk-free power stations will be available the people of our countries will accept the peaceful use of nuclear energy in the long run" [25].

2.6.2   Fusion

Solar fusion of hydrogen to helium described in Sect. 2.2 has an extremely small overall reaction rate. Nevertheless, the gravity of the huge solar mass of 2×10^{27} t holds the solar protons and electrons together long enough for sufficient reactions to occur and compresses them to a plasma whose density exceeds that of water by a factor 150 in the solar core. Fusion occurs in this spherical core of radius 140,000 km at temperatures of 10×10^6–15×10^6 K and pressures of about 10×10^9 atm. The resulting power is 300 W/m^3 of plasma.

Fusion has occurred on Earth in hydrogen bomb detonations, so far. In thermonuclear weapons the energy of a fission bomb is used to compress and heat the fusion fuel, which consists of the hydrogen isotopes deuterium and tritium, or lithium deuteride. No thermonuclear weapon has been used in warfare. The last time scientists set off a hydrogen bomb was in 1991 under the Nevada desert. It is hoped that thermonuclear detonations will never again occur under any circumstances.

Research into controlled fusion for electricity generation has been going on since the early 1950s. Lacking gravitational assistance in producing high plasma densities, one has to find ways that differ from that of the Sun and stars.

One method tries to confine an extremely hot plasma of very low density in a magnetic bottle for sufficiently long times so that fusion reactions can occur. Promising parameters are as follows: plasma temperature 150×10^6–200×10^6 K,

---

[24]In fact, 20 years later, Germany is the country with the highest electricity prices in Europe.

density 4×10^{-9} g/cm^3, pressure 6 atm, confinement time 1–1,000 s. The expected fusion power is 2 MW/m^3. Different from the Sun, terrestrial fusion fuel is not simply hydrogen (H), but is a mix of the heavier hydrogen isotopes deuterium (D) and tritium (T). The reaction

$$D + T \rightarrow {}^4 He + n + 2.8 \times 10^{-12} \, W \, s$$

generates neutrons (n), which transport useful heat. They also produce fresh fuel supply, radiation damage, and radioactivity.[25] The ^{4}He nuclei (α particles) heat up the plasma to the temperatures needed for self-supporting fusion.

Another way to achieve fusion is inertial confinement by energy-rich laser or particle beams. These beams are directed on a pellet that usually contains a mixture of deuterium and tritium. Shock waves from the exploding outer layer of the pellet compress the remainder of the target within extremely short times of a few nanoseconds into the extremely high plasma density of about 200 g/cm^3. The plasma heats up to about 100×10^6 K. At a frequency of 20 implosions per second, one expects a thermal fusion power of about 2 MW/(m^3 of reaction volume) [36].

Stellerators and tokamaks are machines that can confine fusion plasmas magnetically. Plasma is confined in a stellerator by ring-shaped, twisted magnetic field lines, produced by a current-carrying coil of quite complicated geometry. A stellerator could operate continuously. "Wendelstein 7-X," the largest stellerator under development at Greifswald, Germany, is supposed to test whether the stellerator principle is suitable for power stations. Initial plasma heating to fusion temperatures has to be done by an outside source. A tokamak produces a toroidal magnetic field for confining the plasma. Initial heating to fusion temperatures is done by inducing electric currents in the plasma. This is a pulsed process. The international ITER project, the largest fusion experiment worldwide, is building a tokamak in Cadarache, France. The projected technical data are as follows: total radius 10.7 m, overall height 30 m, plasma radius 6.2 m, plasma volume 837 m^3, plasma mass 0.5 g, magnetic field 5.3 T, maximum plasma current 15×10^6 A, heating power and current operation 73 MW, fusion power 500 MW, energy enhancement factor 10, average plasma temperature 100×10^6 K, duration of burning more than 400 s. Construction costs were estimated at €5 billion in 2001, and the operation cost was estimated at €265 million annually during the projected 20 years of testing.[26]

The biggest problem with controlled fusion is material stability under neutron irradiation. One does not expect to have the first fusion reactor that supplies

---

[25]Häfele et al. [57] estimated that fusion reactors will produce about as much radioactive waste as fast breeder reactors. However, the radioactivity of the confinement material, which must be replaced periodically because of damage by neutron bombardment, dies off much more rapidly than the radioactivity of spent fuel rods from fission reactors.

[26]Source: Max-Planck-Institut für Plasmaphysik, http://www.ipp.mpg.de/.

electricity to the public grid before 2050. Despite that, much money is spent on fusion research, because there is a practically inexhaustible supply of terrestrial fusion fuel.

The deuterium content of water is 0.1%. Thus, the oceans represent a reservoir of 5×10^{13} t deuterium. Radioactive tritium has a half-life of 12 years. It can be bred from lithium, embedded in the reactor walls, by bombardment with neutrons, which are produced in the fusion plasma. Two kilograms of lithium results in 1 kg of tritium. The 2-km-thick layer below the surface of Earth's land masses contains about 2×10^{13} t of lithium [36]. To satisfy the global electricity demand of 16,379 TWh in 2006 by fusion power stations, one would need 655 t of lithium and 197 t of deuterium annually.

2.6.3   Solar Power from Deserts and Space

2.6.3.1   Solar Thermal Power Plants in Deserts

The Deutsche Physikalische Gesellschaft (DPG; German Physical Society)[27] has analyzed the potential for growth of solar thermal power plants in Earth's sun belt between 40°N and 40°S. In its 2005 study *Climate Protection and Energy Supply in Germany 1990–2020*, the German Physical Society stated [58]: "Seen from a physical and technical point of view, there can be no doubt that solar thermal power plants in southern latitudes represent one of the best options for supplying ... large quantities of CO2-free electricity. The relevant research and development activities have been in progress for about 25 years, and have reached a stage where it is time to energetically pursue their commercialisation. The Deutsche Physikalische Gesellschaft appeals to all the parties involved – industry, energy providers and the appropriate government bodies – to do everything in their power to promote the launch of the outlined program to create a market for solar thermal power plants in Earth's equatorial sunbelt. We can justifiably hope for new inventions. Providing incentives that will encourage successful research and creating the appropriate research infrastructure must also be seen as a valuable investment towards finding solutions to global warming. ...

A promising first step would be to make solar energy available to the populations actually living in the sunbelt of Earth near the equator (North Africa, the Middle East and Central America). The process heat generated concurrently with the solar power could then be used for desalination of sea water to meet the growing demand for drinking water. Given that the people living in these regions themselves account for approximately 15% of the world's energy requirements, this in itself would go a long way towards protecting the global climate. The calculated potential electricity yield of solar thermal power plants in the regions close to the equator is tremendous,

---

[27]The German Physical Society has more than 55,000 members and is the biggest physical society in the world.

far exceeding local demand. In North Africa, for example, it amounts to about 200–300 GWh$_{el}$ per square kilometer per annum. In other words, Germany's entire electricity demand could be satisfied with a built-over surface area of $45 \times 45\,km^2$ (equivalent to 0.03% of all suitable areas in North Africa). The next step in this direction must therefore be to set up an efficient electricity network between North Africa and Europe. This can be achieved with high-voltage DC (HVDC) transmission lines of the type already in operation for transmitting electricity over distances up to several thousand kilometers. At today's prices, it would cost approximately 2.5 billion euros to build a high-voltage DC transmission line of the kind that such an electricity network would require, with a capacity of 2,000 MW and covering a distance of 3,000 km. This means that it would cost 1.5–2 cents/kWh to transfer solar power from North Africa to Central Europe. Assuming a dynamic market introduction of solar thermal power plants, from about 2015 onwards, it should be possible to attain a price level of about 10 cents/kWh for imported solar electricity in Germany. This level should ultimately drop to about 5.5 cents/kWh. ...

These are major projects involving high investment costs and a correspondingly high financial risk which would have to be cushioned by State guarantees in view of the urgency of the climate problem. Moreover, the facilities would have to be built at the southernmost tip of Europe or in North Africa, which calls for the appropriate international contacts on the part of the electricity distributors – and possibly even at government level – providing the necessary transit arrangements. The hesitant attitude of the energy industry and the appropriate government bodies may perhaps be explained by the initial optimism of the 1990s in Germany, when we were confident of being able to cope with the CO_2 problem at home. However, the extremely slow rate at which CO_2 emissions are being reduced, as demonstrated once again in this study, should be proof enough that we urgently need to import solar power. Although a number of solar thermal projects exist at the planning stage in various parts of the world, their implementation has been postponed year after year despite promised financial support from the World Bank (Global Environmental Facility)."

Four years after this declaration of the DPG the public was surprised to learn that big international companies are considering building large solar thermal power plants in the deserts of Earth. The cost of building power plants in North Africa and HDVC transmission lines there and to Europe is estimated to be about €400 billion. The plan has the name DESERTEC. The following is a communication that announces the first formal steps toward producing solar thermal electricity for the EUMENA region, consisting of Europe, the Middle East, and North Africa.

"On the 30th of October 2009, the articles of association for the DESERTEC industrial initiative 'DII GmbH' (DII) have been signed in Munich by twelve companies and the DESERTEC Foundation. The long-term goal is to satisfy a substantial part of the energy needs of the MENA countries and meet as much as 15% of Europe's electricity demand by 2050. On 13th of July the founders already signed a Memorandum of Understanding with the aim to realise the DESERTEC Concept in the EUMENA Region. More information and the official press release can be read and downloaded . . . (from): http://www.DESERTEC.org.

DESERTEC Facts, History and Future:

(1) The DESERTEC Concept was developed since 2003 by the German Club of Rome and TREC, a network of experts around the Mediterranean (incl. His Royal Highness Prince Hassan bin Talal of Jordan, former president of the international Club of Rome).

(2) Three studies (2004–2007) under the lead of the German Aerospace Center (DLR) confirmed the feasibility of DESERTEC.

(3) In 2008 the DESERTEC Foundation has been formed to coordinate the activities of the DESERTEC networks and to create alliances to realize DESERTEC worldwide. . . .

(6) At October 1st 2008 founders of the DESERTEC Foundation convinced MunichRe to initiate jointly the DII to realise DESERTEC in the EU-MENA region. The DESERTEC industrial initiative has been by DESERTEC Foundation in cooperation with MunichRe.

(7) The group of DII founding members consists of 12 companies and the independent NGO DESERTEC Foundation. Shareholders of the DII are ABB, ABENGOA Solar, Cevital, DESERTEC Foundation, Deutsche Bank, E.ON, HSH Nordbank, MAN Solar Millennium, Munich Re, M+W Zander, RWE, SCHOTT Solar and SIEMENS. It is envisaged that further companies from different countries, preferentially from North Africa and the Middle East, will join the DII. . . .

(10) The DESERTEC Foundation plans to initiate further DIIs in other regions (e.g. USA, India, China and Australia) to achieve its mission of realising the DESERTEC Concept Clean Power from Deserts for a world with 10 billion people. . . .

(14) On the 30th of October 2009 the articles of association for the 'DII GmbH' (DII) were signed in Munich, where the headquarters of the DII will be based. Until 2012 the DII will focus on: Acceleration of implementation of the DESERTEC Concept, by creating of a favourable regulatory and legislative environment, analysis of concrete reference projects, developing of a roll-out plan for such projects, and additional or more detailed studies, where needed.

(15) Paul van Son will be the CEO of the DII . . . Mr. van Son is also Chairman of the European Federation of Energy Traders (EFET) and Chairman of the Energy4All Foundation which is active in Africa."

Nearly 40 years after the first vision of solar farms for 10 billion people, sketched in Table 2.7, plans for large-scale solar power from Earth's deserts are beginning to materialize.

2.6.3.2 Solar Power Satellites and Space Industrialization

Solar power satellites (SPS) are an alternative to terrestrial sun farms. They were first proposed and patented by Peter E. Glaser [59–62]. They are to be stationed in geosynchronous Earth orbit (GEO), always above the same point on the equator at a maximal distance of 35,785 km. Like terrestrial sun farms and solar thermal power plants, they convert sunlight into electric energy, either directly by photovoltaic cells or by solar thermal dynamic systems such as Brayton, Rankine, or Stirling generators. Klystrons convert the electric energy into microwaves of about 3-GHz frequency, which are beamed from a phased-array transmitting antenna of the satellite to a large receiving antenna on Earth. There, the microwave energy is reconverted into electricity, which is fed into the public grid. Typical generating capacities of SPS are 5000–10,000 MW at a bus bar on Earth.

A 10,000-MW solar cell SPS, as proposed by Glaser, consists of two solar "paddles," each having an area of 60 km^2 covered with silicon or gallium arsenide photovoltaic cells. The total weight is between 34,000 and 86,000 t, depending on the construction materials.

Boeing proposed SPS using thermal electric conversion [63]: mirrors focus sunlight on cavities where a circulating gas such as helium is heated and drives heat engines that power electricity generators. A 10,000-MW SPS based on this principle has the following dimensions: length 15 km, area of the reflectors that concentrate sunlight on the gas-heating cavities 50 km^2, waste-heat radiator 1 km^2.

In both SPS versions, the diameters of the transmitting antenna of the satellite and of the receiving antenna on Earth are about 1 and 10 km, respectively.

SPS receive 4–11 times more sunlight per area than installations in the sunshine-richest regions of Earth. And this energy is almost always available. Earth over-shadows a satellite in GEO during such short intervals such that satellite output is reduced by only 1%, on an annual average, compared with a satellite in permanent sunshine. The umbra of the satellite does not reach Earth. Other advantages over solar Earth-based systems are that structures can be up to 1,000-fold lighter and have a longer life, and there is need neither for collector cleaning nor for energy storage.

Since a satellite in GEO does not move relative to Earth's surface, the sharply focused microwave beam can be directed to receiving antennas that are close to large energy consumers. Thus, energy losses in long transmission lines can be avoided. Microwaves in the 3-GHz frequency range penetrate the atmosphere and clouds with small losses and are converted into electric energy by the terrestrial rectifying antenna (rectenna) with an efficiency of 90%. The maximum intensity in the center of the microwave beam is 230 W/m^2. This is comparable to the average solar energy flow on Earth, whereas the maximum solar intensity is about 1,000 W/m^2. Outside the rectenna the microwave intensity is 0.3 W/m^2. In the case of satellite perturbations, dephasing of the transmitting antenna and defocusing of the microwave beam disperses the microwaves to intensities much below that. The rectennas pose some land-use problems and are about 10% of the overall cost. There are plans to make them from a wire mesh, which absorbs most of the microwave energy but allows transmission of sunlight and water to the ground under it for sustenance of the existing flora and fauna.

Of course, the big obstacle to SPS is the unsolved problem of how to get them into GEO, and at what cost. The US Department of Energy and NASA conducted the first comprehensive studies on SPS in the 1970s. The summary of the first study,

"Satellite Power System – Concept and Evaluation Program" [64] states: "The Reference System description emphasizes technical and operational information required in support of environmental, socioeconomic, and comparative assessment studies. Supporting information has been developed according to a guideline of implementing two 5 GW SPS systems per year for 30 years beginning with an initial operational date of 2000 and with SPS's being added at the rate of two per year (10 GW/year) until 2030. . . . the Reference system concept, which features gallium–aluminum–arsenide (GaAlAs) and silicon solar cells options . . . utilizes a planar solar array (about 55 km^2) built on a graphite fiber reinforced thermoplastic structure. The silicon array uses a concentration ratio of one (no concentration), whereas the GaAlAs array uses a concentration ratio of two. A 1-km diameter phased array microwave antenna is mounted on one end. The antenna uses klystrons as power amplifiers with slotted waveguides as radiating elements. The satellite is constructed in geosynchronous orbit in a six-month period. The ground receiving stations (rectenna) are completed during the same time period. The other two mayor components of an SPS program are (1) the construction bases in space and launch and mission control bases on earth and (2) fleets of various transportation vehicles that support the construction and maintenance operations of the satellites." The final report on all studies [65] gives details on transportation and bases: 425-t single launch capacity to low Earth orbit (LEO; 200 km from Earth's surface); eight 425-t launches per week, 400/year; a 6,400-t construction base at LEO; electric propulsion (by ion or plasma engines) from LEO to GEO; Earth–GEO personnel transfers 32 times per year, 75–80 passengers per transfer; 3-month individual stay time. On the basis of this system and 1978 prices, the estimated capital investment needed to just get the system started is of the order of US $250 billion, with a cost of US $1,400 to US $7,000 per kW (electric) generating capacity. Twenty-five percent to 30% of this cost is for space transportation, 80% of which is for Earth–LEO transfer [66].

The US National Research Council and the Congressional Office of Technology Assessment examined this SPS concept and concluded that it might be technologically feasible but programmatically and economically unachievable.

An alternative transportation and production scheme was not considered in these studies, although its exploration had been supported by NASA in the mid-1970s. "The Low (Profile) Road to Space Manufacturing" [67] by Princeton physics professor *Gerard K. O'Neill* describes the construction of SPS with a minimum of materials and energy from Earth, relying mostly on energy from the Sun and materials from the Moon. It was seen as the first step in a grandiose plan for the colonization of space [68], which was further elaborated in the vision of *The High Frontier* [69].

The low profile road rests on two key technological elements: (1) the Space Shuttle, or a follow-up Earth-to-LEO transportation system, and (2) the electromagnetic mass driver. The first Space Shuttle made its first orbital test flight on April 12, 1981, and the Space Shuttle program is scheduled for retirement in 2011. The next-generation American spacecraft, named Orion, is targeted for the first manned launch in 2014 at the earliest. In between, all manned spaceflight will depend on Russian vehicles. Space transportation systems in various stages of development are reviewed in [66].

The mass driver, prototypes of which were developed and built at MIT and Princeton University, works according to the principle of the electric linear motor. Such motors also power the magnetically levitated trains built in Japan and Germany. Roughly speaking, a mass driver consists of a tube surrounded by coils that carry pulsed currents. Within the tube, "buckets" are accelerated by electromagnetic forces. These forces pull on "handles" that consist of superconducting coils that surround the buckets and carry permanent currents. The buckets are filled with material, and within short intervals they are accelerated one after another within the tube at high velocities. At the end of the acceleration distance they are slowed to a halt, and their material flies off by inertia – just like water is released from a swaying bucket. The buckets are reloaded with material and recirculated to the initial position for another acceleration run.

If the mass driver is used as a rocket engine, the ejected masses provide thrust according to the recoil principle.

Stationed on the Moon, the mass driver can serve as a catapult for lunar material to space manufacturing facilities at stable Lagrange (libration) points of the Earth–Moon system. There, according to Lagrange's solution of the three-body problem, a small body would remain forever at equal distance from the Earth and Moon.[28] However, the presence of the distant but gravitationally powerful Sun changes Lagrange's three-body problem into a four-body problem. Its solution shows that the stable libration points L4 and L5 change to stable regions that move in orbits of very large dimensions about L4 and L5. In each of these regions a great number of habitats could be located, circulating about their Lagrange points on a slow, 89-day cycle [69]. Each habitat could accommodate many times the workforce required for building SPS [68].

The Apollo missions to the Moon brought samples of lunar material to Earth, which show the following typical soil composition (in weight percentages): 20% silicon, 12% aluminum, 4% iron, 3% magnesium, 40% oxygen. These elements are well suited for the construction of SPS and habitats. There has been speculation that hydrogen may have accumulated around the poles of the Moon. If this is not the case, hydrogen will be the only essential element of life-supporting systems that has to be supplied from Earth.

O'Neill's "low (profile) road" scenario for SPS construction was based on the following building blocks and estimates [67]:

1. The Space Shuttle transports a payload of about 29 t per flight. Six flights can deliver the components of a 170-t mass driver to LEO. There they are assembled

---

[28]The mathematician Joseph Lagrange discovered five special points in the vicinity of two orbiting masses where a third, smaller mass can orbit at a fixed distance from the larger masses. The Lagrange points mark the positions where the gravitational pull of the two large masses precisely provides the centripetal force required to rotate with them. Of the five Lagrange points, three are unstable and two are stable. The unstable Lagrange points – labeled L1, L2, and L3 – lie along the line connecting the two large masses. The stable Lagrange points – labeled L4 and L5 – form the apex of two equilateral triangles that have the large masses at their vertices.

into a heavy-duty space transporter powered by electricity from solar cell arrays. The only throwaway component of the Space Shuttle is the 35-t external tank, which is usually jettisoned shortly before LEO is reached. If one forgoes a small percentage of payload, however, the external tank can be brought into LEO and ground there into a fine powder that serves as reaction mass for the mass driver. Sixty Space Shuttle flights per year accumulate 1,700 t of payload and 2,100 t of reaction mass. The mass driver of the heavy-duty space transporter accelerates the pulverized reaction mass to 10,000 m/s and carries a payload of 1,300 t during a 200-day trip from LEO to an orbit around the Moon. Moon landing of the payload is by chemical rockets.

2. The lunar base to be built consists essentially of a mass driver, length 4,320 m, which accelerates lunar material to lunar escape velocity of 2.4 km/s. After traveling 63,000 km through space, the material is caught by a "catcher" in the Lagrange libration point L2 and transported from there by simple freighters to the first space manufacturing facilities for SPS and habitats. The annual material throughput is initially 30,000 t and later 600,000 t. The Moon Base also comprises solar power stations, service stations, waste heat radiators and housing units for initially 50 persons, and after testing and assembly of the mass driver ten persons may be sufficient to maintain routine operations. The total mass of the base is estimated to be 1,085 t, including 180 t of a 1-year food supply. To soft-land this mass on the Moon one needs the same mass of fuel (liquid oxygen and hydrogen). Furthermore, one needs installations such as the "catcher" and quarters for space workers close to the Moon. All in, about 100 Space Shuttle flights must bring 2,800 t of payload into space to initiate material transport from the surface of the Moon to some point in space at a rate of 30,000 t/year.

3. The first space manufacturing facility for chemical processing of lunar material has a mass of 150 t. A 46-MW solar cell power station, having a mass of 245 t, supplies the necessary energy. The required workforce is 150 people. Annually, 9,000 t of metal and silicon and about 7,000 t of oxygen are produced from 30,000 t of lunar raw material. The oxygen is used as breathing air for people and as liquid reaction mass for the heavy-duty space transporters, so that the Space Shuttle external tanks must no longer be pulverized but can serve as the first orbital living quarters.

   Fresh supply of 1,800 t/year should be delivered to LEO by new heavy-lift vehicles with a carrying capacity of 87 t, to be developed from the Space Shuttle. The number of personnel in space should increase by 150 persons per year, transported on 13 Space Shuttle flights. Then, the production capacity of the system would grow by 150,000 t annually. Seven to nine years after the program start, 630,000 t of raw materials and 190,000 t of finished products would be produced annually. This corresponds to the mass of two 10,000-MW SPS. After that about two SPS could be produced per year.

The cost estimates were based on numbers available in the 1970s for very complex and not-so-complex space systems: research and development $28,000–$62,000/kg, construction cost $1,100/kg, transportation cost $230–$690/kg, interest

10%. Assuming an SPS value of $500 per kW power delivery at a bus bar on Earth, the financial break-even point could be reached within 10 years of the start after having invested $_{1978}50 billion to $_{1978}60 billion. The Apollo Moon Program would have cost that much in the mid-1970s. Although the cost estimate was too optimistic in view of the cost accumulated by the Space Shuttle program, O'Neill's "Low (Profile) Road to Space Manufacturing" indicates a way of building SPS mostly from extraterrestrial resources that should not be forgotten.

O'Neill died from leukemia on April 27, 1992, at the age of 65. During the Seventh Princeton University Conference on Space Manufacturing Facilities in 1985, the year his leukemia was diagnosed, he told me: "I have mortgaged all my property up to the last sock in order to found and operate Geostar Corporation." On the basis of a patent for a satellite position determination system, granted to O'Neill in 1982, Geostar was supposed to generate the funds for continuing research into space manufacturing. After the first oil price shock in 1973–1975 public funding for such research was good. But when oil prices plummeted to nearly pre-1973 levels between 1981 and 1985, public funds dried up. One year before O'Neill's death, Geostar went bankrupt. The Space Studies Institute (http://ssi.org/), formerly at Princeton (NJ, USA), now at Mojave (CA, USA), preserves O'Neill's legacy.

When energy prices increase and concerns about emissions from fossil fuel combustion mount, new legislative initiatives may remember the "House Concurrent Resolution 451," presented to the 95th Congress of the United States of America by Representative Olin Teague on December 15, 1977, and referred to the Committee on Science and Technology. It states [67]:

"*Whereas* historically it is an inherent genius of the American people that we vigorously reach out to explore, to fulfill and enhance the resources of new and challenging frontiers, for the benefit of all humanity; and
Whereas the magnificent achievements of our explorations into space in the past twenty years have proved decisively that this tiny Earth is not humanity's prison, is not a closed and dwindling resource, but is in fact only part of a vast expanding system rich in extraterrestrial opportunities as yet far beyond our comprehension, a "high frontier" which irresistibly beckons and challenges the American genius; and
Whereas our ventures into space, though daring, have not been rash, but have in fact succeeded only because of rigorous, disciplined, careful analysis, planning, training and skilled performance, thus establishing standards and precedents which must continue to guide all further national policy decisions and efforts in space; and
Whereas ... many Americans seem for the moment beset and confused by complex problems, discouraged by alleged "limits to growth" and by careless waste of the Earth's resources ... ; and
Whereas the "High Frontier" of Space does provide valid opportunities whereby we can conserve and enhance humanity's existence on Earth, including but not limited to such social and economic benefits as greater employment, a cleaner environment, new energy sources, new knowledge and understanding ... : Now, therefore be it

Resolved by the House of Representatives (the Senate concurring), That the Congress hereby finds and declares the following national policy:

(1) It is vital to the well-being of the American people, and all the people on this Earth, that every feasible means now shall be mobilized to explore and assess the resources of the "high frontier" of outer space, to better understand and to make practical, beneficial uses of these resources.
(2) As immediate priorities, research efforts shall be intensified to reveal and to better understand the solar system and the universe beyond it, to develop a practical, efficient transportation system in space ...
(3) As long range, high priority national goals, it is anticipated that by the year 2000 these explorations will have opened the resources and environment of extraterrestrial space to an as yet incalculable range of other positive uses, including but not limited to, international cooperation for the maintenance of peace, the discovery and development of new sources of energy and materials, industrial processing and manufacturing ... and, conceivably, the establishment of self-sustaining communities in space.

The Congress hereby encourages and instructs all pertinent legislative committees and executive agencies to determine how they may most effectively act ... to achieve these urgent national goals.

To assist in these efforts the Office for Technology Assessment specifically is requested to organize and manage a thorough study and analysis to determine the feasibility, potential consequences, advantages and disadvantages of developing as a national goal for the year 2000 the first manned structures in space for the conversion of solar energy and other extraterrestrial resources to the peaceable and practical use of human beings everywhere."

The Advanced Concepts Team of the European Space Agency[29] has been considering solar power from satellites and terrestrial sun farms. In 2004 it compared space concepts with terrestrial solutions based on equally advanced technology and equal economic conditions for the time frame 2020–2030 in terms of energy payback times, final euros per kilowatt-hour generation cost, adaptability to different energy scenarios, reliability, and risk [70]. The conclusions were: "While terrestrial solar power plants will already play an increasing part in European electricity production in the next 20 years, solar power satellites will technically and economically reach their maturation phase only at the end of the considered time frame. The competitiveness of the space option increases with increasing total plant sizes. Under the given assumptions, space options are not competitive with terrestrial plants for relatively small solar power plants (depending on the type from 0.5 to 50 GW_{el}). Earth-to-orbit transportation is the single most important factor requiring a decrease of more than one order of magnitude compared to current launch costs. Depending on the plant size, launch costs between 155 and 1615 euro per kg_{LEO} (kilogram into Low Earth Orbit) for peak load and around 600–700 euro per kg_{LEO}

---

[29]http://www.esa.int/gsp/ACT/publications/index.htm .

for base-load supply scenarios are necessary to be competitive with terrestrial solar power plants. ... Both, space and large terrestrial solar power plants have very attractive, low energy payback times. Both laser and microwave power transmission concepts show energy payback times of only a few months. Almost all space and terrestrial concepts produce within less than one year more energy than was needed to produce and operate them, based on detailed material flow analysis."

In 2007, the National Space Security Office authored the study *Space-Based Solar Power As an Opportunity for Strategic Security* [71]. It states in its Executive Summary: "The magnitude of the looming energy and environmental problems is significant enough to warrant considerations of all options, including revisiting a concept called Space Based Solar Power (SBSP) first invented in the USA almost 40 years ago." Its recommendations conclude with: "The study group recommends that the US Government should become an early demonstrator/adopter/customer of SBSP and incentivice its development."

Appendix 1: Basic Forms of Energy

The many manifestations of energy occur in a great variety of systems. Physical systems consist of components, constraints, and boundaries. Components are "particles" (with and without mass) in any number, and the forces that act on them. Constraints restrict the motions and interactions of the components. Boundaries – material or immaterial ones – separate systems from one another.

System Energies

System energies are *properties* of systems. They differ from energy forms that are *exchanged* between systems. It is important to discriminate between these two forms of energy in energy conversion processes:

1. *Kinetic energy* is the energy of motion. Its simplest form shows in a system that consists of configuration space spanned by the three Cartesian coordinates x, y, z and a body of mass m, whose center moves with velocity $\mathbf{v}$ relative to an observer at rest. If, as we will always assume from here on, the velocity of the mass is much less than the speed of light, i.e. ,$|\mathbf{v}| \ll c$, the kinetic energy of that body measured by the observer is

$$E_{kin} = \frac{1}{2}m\mathbf{v}^2 = \frac{\mathbf{p}^2}{2m},\qquad(2.18)$$

where $\mathbf{p} = m\mathbf{v}$ is the momentum of the body.[30]

---

[30]More generally $E_{kin} = (m - m_0)c^2 \approx (1/2)m_0\mathbf{v}^2 + (3\mathbf{v}^2/8c^2)m_0\mathbf{v}^2 + \dots$. This becomes important when $|\mathbf{v}|$ approaches c, so $m = m_0/(1 - \mathbf{v}^2/c^2)^{1/2}$ differs substantially from m_0.

Kinetic energy creates wealth in hammers, sickles, windmills, steam turbines, gas turbines, and other tools and machines. In general, each tiny mass element Δm of these bodies, in its position $\mathbf{r} \equiv (x, y, z)$, has its own velocity $\mathbf{v}(\mathbf{r}, t)$, which changes with time t if forces act on the bodies, and its kinetic energy is given by (2.18), with Δm in place of m. The total kinetic energy of each body is the sum of the kinetic energies of all its elements. Kinetic energy also works destructively in swords, arrows, bullets, shell fragments, tsunamis, and hurricanes. In water and air the mass elements are molecules.

2. *Potential energy in a gravitational field.* Consider a body of mass m, whose center of mass has been elevated against the pull of Earth's gravity from the ground to a level at height h above the ground. It has acquired the potential energy

$$E_{pot} = mgh \tag{2.19}$$

relative to the ground, where $g = 9.81$ m/s^2 is the acceleration in the gravitational field of Earth.

Water dams accumulate the potential energy of water provided by the Sun. Conversion of this potential energy into kinetic energy, when the water rushes through the turbines of water power stations, provided nearly 20% of the 16,379 TWh of electricity generated globally in 2006.

A more general case of gravitational potential energy is realized in a system that consists of two masses m_1 and m_2 (e.g., a sun and its only planet) whose centers are located at positions $\mathbf{r}_1$ and $\mathbf{r}_2$. The two masses attract each other gravitationally, and the potential energy of mass m_1 in the gravitational field of mass m_2, and vice versa, is given by Newton's gravitational law

$$E_{grav} = -\gamma \frac{m_1 m_2}{|\mathbf{r}_1 - \mathbf{r}_2|}; \tag{2.20}$$

$\gamma = 6.672 \times 10^{-11}$ N m^2/kg^{-2} is the gravitational constant. The negative sign takes into account that apples fall from trees, like the famous one of physics legend that fell upon Isaac Newton's head and put the gravitational law into it: $|\mathbf{r}_1 - \mathbf{r}_2|$ is the distance between the center of mass of the apple on the tree and the center of Earth. When the apple broke off from the tree and this distance decreased by a few meters, E_{grav} became more negative. The lost potential energy became kinetic energy transferred from the apple to Newton's head, stimulating the idea that led to Newtonian classical mechanics.

3. *Coulomb energy.* Let a system consist of two pointlike charges q_1 and q_2 that occupy positions $\mathbf{r}_1$ and $\mathbf{r}_2$. The potential energy of charge q_1 in the electric field of charge q_2 (and vice versa) bears the name of Coulomb and has the magnitude

$$E_{coul} = \frac{1}{4\pi\varepsilon_0} \frac{q_1 q_2}{|\mathbf{r}_1 - \mathbf{r}_2|}; \tag{2.21}$$

$\varepsilon_0 = 8.8542 \times 10^{-12}$ A s/V m is the dielectric constant of vacuum.

Coulomb energy (2.21) and gravitational energy (2.20) look alike: charges correspond to masses and the *absolute* magnitude of energy increases as the distance $|\mathbf{r}_1 - \mathbf{r}_2|$ between the interacting partners decreases. But although there is always attraction between masses, which is taken care of by the minus sign in (2.20), only charges of opposite sign attract each other. Charges of the same sign repel each other; their potential energy is smaller the farther they are apart. Furthermore, the smallest mass elements in all systems, where nuclear reactions do not matter, are atoms, which differ in size and mass. The smallest charge elements, however, the positively charged protons and the negatively charged electrons, have exactly the *same* absolute magnitude of charge. This is the elementary charge $e = 1.602 \times 10^{-19}$ A s. Although the mass of the proton, $m_{\mathrm{p}} = 1.672 \times 10^{-27}$ kg, exceeds the mass of an electron, $m_{\mathrm{e}} = 9.108 \times 10^{-31}$ kg, by more than a factor of 1,000, it is justified in most cases to treat both as pointlike charges.

The similarity of (2.21) and (2.20) is one justification of Rutherford's planetary model of the atom: nearly all the mass of the atom is concentrated in the positively charged core of protons and neutrons, whereas the light electrons encircle this core in orbits similar to the planetary orbits around the Sun. However, this model contradicts the laws of classical physics, according to which an orbiting electron would emit electromagnetic radiation; this energy loss would more and more reduce the electron's distance to the nucleus (the right-hand side of (2.21) is negative for q_1 positive and q_2 negative) so the orbit would not be stable. Rather, the electron would spiral down into the nuclear core. But quantum mechanics has replaced classical physics in the description of atoms. Quantum mechanically there are (only) probabilities of finding an electron with stable, quantized energies at certain distances from the nucleus without any radiative energy losses being involved.

In thunderstorms, clouds and earth accumulate opposite charges, building up huge amounts of Coulomb energy. In a lightning flash, about 100 kWh is liberated by a principal discharge. This energy is the same as the amount of energy required to lift a mass of 4.1 t from sea level to the peak of Mount Everest at 8,848 m. Coulomb energy is the interaction energy in all many-particle systems consisting of electrons and stable nuclei. These systems are discussed below in the context of chemical energy.

4. *Energy of electric and magnetic fields.* The energy density of electric and magnetic fields $\mathbf{E}$ and $\mathbf{H}$ in a vacuum is

$$\eta = \frac{1}{2}\varepsilon_0 \mathbf{E}^2 + \frac{1}{2}\mu_0 \mathbf{H}^2, \tag{2.22}$$

where $\mu_0 = 4\pi \times 10^{-7}$ V s/A m is the permeability of a vacuum. Thus, a system that contains these fields within a certain volume V of (practically empty) space has the electromagnetic energy that is the integral of the energy density η over the volume:

$$E_{\mathrm{elmag}} = \int_V \eta \mathrm{d}V. \tag{2.23}$$

Information in telecommunications is imprinted on propagating electromagnetic waves with oscillating $\mathbf{E}$ and $\mathbf{H}$ vectors. These waves are emitted by oscillating charges in the antennas of mobile phones, radio stations, and TV stations. Lasers that emit rays of high energy density η are used as cutting and welding tools in surgery and laser-assisted machining of materials.

5. *Photon energy.* The energy quanta of an electromagnetic wave, whose electric and magnetic field vectors $\mathbf{E}$ and $\mathbf{H}$ oscillate with angular velocity ω, are photons of energy

$$E_{\text{photon}} = \hbar\omega. \tag{2.24}$$

Here $\hbar = 1.0546 \times 10^{-34}$ W s^2 is Planck's constant divided by 2π.

Photons carry the Sun's radiation to Earth.

6. *Radiation energy inside a cavity with wall temperature T.* A cavity of volume V, whose walls are maintained at the absolute temperature T, is a system we encounter in blast furnaces, combustion chambers of power stations, and rooms. The electromagnetic radiation that exists in thermal equilibrium as an assembly of photons inside such a cavity has an energy density $u_0(T)$ that increases with the fourth power of T:

$$u_0(T) = \frac{\pi^2}{15} \frac{(k_B T)^4}{(c\hbar)^3}, \tag{2.25}$$

where $k_B = 1.3807 \times 10^{-23}$ W s/K is Boltzmann's constant. Thus, the radiation energy contained in the volume V is the integral of the energy density $u_0(T)$ over this volume,

$$E_{\text{rad}} = \int_V u_0(T)\mathrm{d}V, \tag{2.26}$$

and is proportional to T^4.

7. *Internal energy (chemical energy) of an interacting many-particle system.* Solids, fluids, and gases are formed by atomic nuclei and electrons that interact with each other via the Coulomb energy of (2.21). Let us consider a system of many particles that are enclosed in a box with rigid, insulating walls, which form the system boundaries. The particles are L nuclei and N electrons. The nuclei at positions $\mathbf{R}_i$ have masses M_i, momenta $\mathbf{P}_i$, and positive charges $Z_i e$. The electrons at positions $\mathbf{r}_k$ have masses m, momenta $\mathbf{p}_k$, and charges $-e$. The total energy of the system, called a Hamiltonian, is the sum of all kinetic and Coulomb energies:

$$\mathscr{H} = \sum_{i=1}^{L} \frac{\mathbf{P}_i^2}{2M_i} + \frac{1}{2}\sum_{i\neq j}^{L} \frac{1}{4\pi\varepsilon_0} \frac{e^2 Z_i Z_j}{|\mathbf{R}_i - \mathbf{R}_j|} + \sum_{k=1}^{N} \frac{\mathbf{p}_k^2}{2m} + \frac{1}{2}\sum_{k\neq l}^{N} \frac{1}{4\pi\varepsilon_0} \frac{e^2}{|\mathbf{r}_k - \mathbf{r}_l|}$$
$$- \sum_{i,k}^{L,N} \frac{1}{4\pi\varepsilon_0} \frac{e^2 Z_i}{|\mathbf{R}_i - \mathbf{r}_k|}. \tag{2.27}$$

An experimenter has prepared the system in such a way that its total energy is in a narrow energy interval between E and $E + \delta E$. The system is completely isolated and in equilibrium. For a macroscopic system, even a very small energy interval δE contains a huge number of many-particle states. These states are characterized either classically by all particle positions and momenta compatible with the boundaries, or quantum mechanically by as many quantum numbers as independent coordinates are needed for the description of the system. (The number of these coordinates is called the "number of degrees of freedom.") In order to count the states they are labeled by, say, r. For $1\,\mathrm{cm}^3$ of any metal, or $1\,\mathrm{m}^3$ of air, the label r runs from 1 to numbers so large that infinity is a good approximation. The corresponding many-body states are symbolized by $|\Phi_r >$. Let $\Omega(E)$ denote the number of many-particle states $|\Phi_r >$ whose energies[31] E_r lie between E and $E + \delta E$. The fundamental postulate of statistical physics says that the system occupies with equal probability any one of these $\Omega(E)$ states; see also Chap. 3. Accordingly, it can be found with equal probability in any one of the $\Omega(E)$ states $|\Phi_r >$. Thus, the *internal energy* U of the system, defined as the average of all energies E_r between $E_r = E$ and $E_r = E + \delta E$, is the sum of all energies in this range divided by the number of states in this range:

$$U = \frac{1}{\Omega(E)} \sum_{|\Phi_r >; E}^{E+\delta E} E_r. \tag{2.28}$$

The minimum internal energy is that of the ground state. A system would occupy its ground state at the zero point of absolute temperature, i.e., at $T = 0$ K. Internal energy is also called *chemical energy*. Computation of the internal energy of an interacting many-particle system by field-theoretical methods [3, 4] is a frontier of modern physics.

Industrial production has been powered for 200 years by the chemical energy of fossil fuels. Solar photons may be converted into and stored by the chemical energy of novel fuels such as hydrogen.

8. *Energy stored in capacitors and current rings.* A capacitor with capacitance C, carrying charges $+q$ and $-q$ on its two conductors, stores the electrostatic energy

$$E_{\mathrm{elst}} = \frac{q^2}{2C}. \tag{2.29}$$

A current-carrying circuit stores the magnetic energy

$$E_{\mathrm{mag}} = \frac{1}{2} L I^2, \tag{2.30}$$

---

[31]Quantum mechanics computes E_r as the expectation value of the Hamiltonian (2.27) with $|\Phi_r >$: $E_r = < \Phi_r | \mathscr{H} | \Phi_r >$.

where I is the current in and L is the self-inductance of the circuit. If the circuit is superconducting, the current circulates permanently without any energy dissipation. Electrostatic and magnetic storage of electric energy may complement the presently dominating chemical storage in batteries. Low-cost, efficient energy storage is decisive for utilizing the energy of fluctuating wind and sunshine fully.

Work and Heat

Work and heat are the energy forms people are most familiar with, because they are directly accessible to our senses. They originate in energy conversion processes from changes of system energies and are conceptually different from them: They are energy forms that *transgress* system boundaries.

Work

If a system interacts with its environment, mechanical work W is performed by a force $\mathbf{F}$ on a moving system boundary. Such a boundary can be the surface of a plow or of a piston, or the handle of a weight to be lifted, or any other contact between a system and the force that acts upon the system. An infinitesimal displacement $d\mathbf{r}$ of the boundary requires an infinitesimal amount δW of mechanical work performed by the force *on* the system. This work is equal to the scalar product of the force vector $\mathbf{F}$ with the path element $d\mathbf{r}$:

$$\delta W = \mathbf{F}\, d\mathbf{r}. \tag{2.31}$$

The total work performed *on* the system *by* the environment during a finite displacement of the system boundary from space point 1 to space point 2 is given by the integral

$$W_{12} = \int_{1}^{2} \mathbf{F}\, d\mathbf{r}. \tag{2.32}$$

If W_{12} is positive (negative), the system receives (transfers) energy from (to) the environment.

Two special cases of *mechanical* work are external mechanical work and work of volume change. *External* mechanical work W^{MEx} is done *on* a system, if the velocities $\mathbf{v}$ of the elements dm of total system mass m are changed, and if the position z of the center of mass is displaced in a gravitational field that produces the acceleration $\mathbf{g}$ in negative z direction:

$$W_{12}^{\mathrm{MEx}} = \frac{1}{2}\left(\int \mathbf{v}^2\, dm\right)_2 - \frac{1}{2}\left(\int \mathbf{v}^2\, dm\right)_1 + mg\,(z_2 - z_1). \tag{2.33}$$

The integrals represent the difference of kinetic energy in the system states 2 and 1 after and before W_{12}^{MEx} has been performed on the system, and the remainder is the change of potential energy of the system. If rotatory motions with different velocities of the mass elements are involved, states 2 and 1 are characterized not only by changed positions of the system boundaries but also by angular changes. Mechanical work of *volume* change W^{MV} is done *on* a system that rests as a whole if forces from the environment act perpendicularly to the system boundaries and change the system volume V. Consider a fluid or a gas enclosed in a cylinder with a movable piston. The area A of the piston is the system boundary. At rest the piston is kept in equilibrium by an external force $\mathbf{F}$ that balances the force from the pressure p of the fluid or gas:

$$\mathbf{F} = -p\mathbf{A}. \tag{2.34}$$

The direction of $\mathbf{F}$ is antiparallel to the area vector $\mathbf{A}$, which is directed outwardly. A displacement of the piston by the distance $\mathrm{d}\mathbf{r}$ changes the volume of the fluid or gas by $\mathrm{d}V = \mathbf{A}\,\mathrm{d}\mathbf{r}$. If the piston movement displaces the external force $\mathbf{F}$ so slowly that it remains in equilibrium with the force $-p\mathbf{A}$, the work performed in this *quasistatic* process *on* the system is according to (2.31) and (2.34)

$$\delta W^{\mathrm{MV}} = \mathbf{F}\,\mathrm{d}\mathbf{r} = -p\,\mathrm{d}V. \tag{2.35}$$

During compression $\mathrm{d}V < 0$ and $\delta W^{\mathrm{MV}} > 0$; the system receives energy *from* the environment. During expansion $\mathrm{d}V > 0$ and $\delta W^{\mathrm{MV}} < 0$; the system supplies energy *to* the environment.

During piston motion, pressure p and volume V change with time t. The computation of $p(t)$ and $V(t)$ is a difficult problem in fluid dynamics, in general. However, as long as the piston velocity is small compared with the velocity of sound in the gas or fluid, the quasistatic approximation of assuming a unique dependence $p = p(T, V)$ of pressure p on volume V and temperature T gives good results. Then (2.35) can be integrated between the initial position 1 and the final position 2 of the piston and one obtains the total work of volume change performed *on* the system by the environment as

$$W_{12}^{\mathrm{MV}} = -\int_1^2 p\,\mathrm{d}V. \tag{2.36}$$

Electric work W^{El} is performed, if a system is connected to a voltage source, and an electric potential difference $\tilde{V}$ drives a current I through the system. The electric work done *on* the system by the voltage source between the times t_1 and t_2 is

$$W_{12}^{\mathrm{El}} = \int_{t_1}^{t_2} I \times \tilde{V}\,\mathrm{d}t. \tag{2.37}$$

In summary, if external parameters of a many-particle system do not remain fixed, but change, while the system is thermally isolated from its environment, work W_{12} is performed *on* the system. This work changes the system's internal energy by

$$\Delta_{\mathrm{W}} U \equiv U_2 - U_1 = W_{12}. \tag{2.38}$$

If $W_{12} > 0$, the system receives energy from the environment. If $W_{12} < 0$, the system transfers energy *to* the environment.

Whereas (2.28) says how to calculate internal energy theoretically, (2.38) says how to measure internal energy differences by the measurement of W_{12}.

Heat

One observes that there are processes in which (2.38) does not correctly describe the change of internal energy of a given system. Such processes occur if the system is in thermal contact with its environment. Then, another form of energy (also) crosses the system boundaries. This energy form is called *heat*. Quantitatively, in an arbitrary process that changes external parameters while there is *no* thermal isolation, the total change of the system's internal energy is

$$\Delta U = U_2 - U_1 \equiv W_{12} - Q_{12}. \tag{2.39}$$

The quantity Q_{12} thus introduced is a measure of the internal energy change *not* due to the change of external parameters. Equation (2.39) is the *definition* of the heat Q_{12} given off *by* the system *to* the environment in an arbitrary process that carries the system from state 1 to state 2. Heat, the disorderly motion of particles, is transferred by heat conduction in solids and liquids, convection in gases, and randomly oscillating electromagnetic fields. Active cooling is often required to avoid system destruction by overheating.

Heating and cooling is important for computers. Transistors in computers process digital information by blocking (0) or letting pass (1) electric currents. They are sketched in Sect. 2.3.2. The voltage that drives the current performs electric work, given by (2.37). This work eventually ends up in heat, $W_{12} = Q_{12}$ in (2.39), because the internal energy of a transistor is the same before switching on and after switching off the computer, supposing that all electrically produced Joule heat can be given off to the environment by thermal conduction.[32]

---

[32]Heat production is inconvenient for further evolution of computers, which has been characterized during the last four decades by a doubling of the density of transistors on a microchip every 18 months. If this trend and current trends of power consumption continue, the computer industry could possibly face the so-called Problem 2020, when the temperature of a miniaturized computer would be equal to the Sun's temperature, because the Joule heat could no longer escape sufficiently rapidly out of the densely packed compound of transistors.

Work and heat are energy forms that *cross* system boundaries. Infinitesimal quantities of them are *not* total differentials and are written as δW and δQ. Internal energy U, on the other hand, is a system *property*. It is described by a *state* function, which depends only on the system variables in the actual state and not on the path by which the system has arrived at this state. Its infinitesimal change is the total differential dU.

Enthalpy and Exergy

Enthalpy

Consider two systems of different atoms and molecules that are initially isolated from each other. Their Hamiltonians are $\mathscr{H}_1'$ and $\mathscr{H}_1''$. Their corresponding internal energies, which are measured at constant volumes, are U_1' and U_1'', and the total initial internal energy is $U_1 = U_1' + U_1''$. Then the systems are brought in contact and interact with each other. During chemical reactions the electrons and nuclei are rearranged. When the total system has reached a new equilibrium, its internal energy is U_2.

Most chemical reactions occur at constant pressure p, not constant volume. If the total volume of the two isolated systems is initially $V_1 = V_1' + V_1''$ and the total final volume after the reaction is V_2, work $p(V_2 - V_1)$ is exchanged between the total system and the environment. This environment is the atmosphere in many cases.

The difference $U_2 - U_1$ between the final and the initial internal energies is then

$$U_2 - U_1 = -p(V_2 - V_1) - Q_{12}. \tag{2.40}$$

The chemical reaction is said to be exothermal (endothermal) if $Q_{12} > 0$ ($Q_{12} < 0$).

If one is only interested in the heat balance of chemical reactions, it is convenient to introduce a new state function, called *enthalpy*. Enthalpy H is defined as

$$H \equiv U + pV. \tag{2.41}$$

With that (2.40) can be written as

$$H_1 - H_2 = Q_{12}. \tag{2.42}$$

If both the initial and the final states of a chemical reaction are solids or fluids, the difference between enthalpy H and internal energy U is small. If, however, gaseous states are involved, as happens during the combustion of fossil fuels, one must calculate the produced heat from (2.42).

Exergy

Energy quantity, measured in enthalpy units such as joules (or tons of oil/coal equivalents, or kilowatt-hours, or Board of Trade Units) is not sufficient to characterize the usefulness of an energy carrier, in general. Energy *quality* is important too. According to Karlsson [73] and van Gool [74], energy quality is defined as

$$\text{Quality} = \text{exergy/enthalpy.}$$

Exergy is the share of an energy quantity that can be completely converted into physical work. It is complemented by useless anergy in the energy conservation equation (2.2).

In all real-life energy conversion processes, useless anergy grows at the expense of useful exergy.

Examples of exergy are:

1. The kinetic energy of a mass m with velocity $\mathbf{v}$ is 100% exergy. The same is true for the potential energy of this mass in a gravitational field.
2. Electric energy is 100% exergy.
3. The chemical energy stored in coal, oil, and gas is practically 100% exergy. The same is true for the energy obtained from mass conversion according to $E = mc^2$. This is because very high temperatures can be obtained, in principle, from fossil-fuel combustion and mass-to-energy conversion. Hence, the Carnot efficiency (2.17) and the quality of heat, as given in example 5, can be very close to 1.
4. Solar radiation is practically 100% exergy. For electromagnetic radiation Karlsson [73] has shown that the quality of a quasi-monochromatic beam of incoherent radiation in an angular frequency range between ω and $\omega+d\omega$, falling perpendicularly onto the surface of a black body at temperature T_0, is

$$\text{Exergy/enthalpy} = 1 - T_0/T + [\exp(\hbar\omega/k_{\mathrm{B}}T) - 1]$$
$$\times(k_{\mathrm{B}}T_0/\hbar\omega)\ln\frac{[1 - \exp(-\hbar\omega/k_{\mathrm{B}}T)]}{[1 - \exp(-\hbar\omega/k_{\mathrm{B}}T_0)]}. \qquad (2.43)$$

Here,

$$T = \hbar\omega/\{k_{\mathrm{B}}\ln[(2\hbar\omega^3/c^2 P_{\mathrm{E}}) + 1]\} \qquad (2.44)$$

is the equivalent temperature of a black body that emits power $P_{\mathrm{E}}(\omega)d\omega$ per unit area in the frequency range between ω and $\omega + d\omega$. If T is equal to the effective solar surface temperature of 5,777 K, and T_0 is a temperature of the order of 288 K, the average surface temperature of Earth, the quality of the corresponding black-body radiation is close to 1 [73].

5. Heat of quantity Q at temperature T, in an environment of temperature T_0, contains the exergy $E_X = Q(1 - T_0/T)$. Thus, its quality is given by the Carnot efficiency $1 - T_0/T$, defined in (2.17).

6. A many-particle system of internal energy U, entropy S, and volume V, which is out of equilibrium with its environment of temperature T_0 and pressure p_0, and which can exchange heat and work – but not matter – with the environment, contains the exergy [5,72]

$$E_X = (U - U_0) + p_0(V - V_0) - T_0(S - S_0). \tag{2.45}$$

Here, U_0, V_0, and S_0 are internal energy, volume, and entropy when the system has come to equilibrium with its environment. The internal energy U is given by (2.28). Entropy S, the physical measure of disorder in the system, is defined statistically as

$$S = k_B \ln \Omega. \tag{2.46}$$

Section 3.4.1 goes into this in more detail. Ω, the number of many-particle states accessible to the system within a given energy range, has already been used in (2.28), which defines internal energy U.

7. A thermodynamic system of volume V at pressure p, with a stationary current of mass m and kinetic energy $m\mathbf{v}^2/2$ entering it at one place and leaving it at another place, having the potential energy mgz at height z above a reference point in the gravitational field of acceleration g, contains the exergy [5,72]

$$E_X = (U + pV - U_0 - p_0 V_0) - T_0(S - S_0) + m\mathbf{v}^2/2 + mgz.$$

8. Consider a combustion process that produces a many-particle system of internal energy U, entropy S, and volume V. The system is out of equilibrium with its environment of temperature T_0 and pressure p_0. It can exchange heat, work, and matter with the environment. The concentration of the combustion products in the combustion chamber is higher than that in the environment. In principle, work can be obtained from their diffusion into the environment. To compute this work, which contributes to exergy, one pretends that immediately after combustion the system components are mixed and already in thermal and mechanical equilibrium with the environment, but that they have not yet left the combustion chamber. Then they leave the combustion chamber and diffuse. Thus, if the system consists of N different sorts i of particles, with n_i being the number of particles of component i, and if μ_{i0} and μ_{id} are the chemical potentials of component i in thermal and mechanical equilibrium before and after diffusion, respectively, then the exergy content of the system is [72]

$$E_X = (U - U_0) + p_0(V - V_0) - T_0(S - S_0) + \sum_{i=1}^{N} n_i (\mu_{i0} - \mu_{id}); \tag{2.47}$$

here the kinetic and potential energies of the components have been disregarded.

Thermodynamic Potentials

What is the difference between exergy and the "free energies" represented by the thermodynamic potentials Helmholtz free energy and Gibbs free energy?

Thermodynamic potentials are used in equilibrium thermodynamics when *quasistatic* processes are described. Quasistatic processes occur so slowly that the relaxation mechanisms in the many-particle system maintain equilibrium practically always. Nevertheless, equilibrium thermodynamics can also describe results of nonequilibrium processes if one is only interested in changes of thermodynamic potentials. These are state functions of equilibrium systems, and state functions (like internal energy or enthalpy) are independent of the process that has brought the system into its actual state.[33]

There are four thermodynamic potentials that describe many-particle systems according to their interaction with the environment or reservoirs: internal energy U, enthalpy H, Helmholtz free energy F, and Gibbs free energy G. They are related to each other by *Legendre transformations*.

Isolated System: $U(S, V)$

To work out the basic relations, we first consider a homogeneous system that is isolated from its environment; its particle number does not change. Its internal energy has been defined in (2.28). With (2.46), which defines entropy S, internal energy U becomes

$$U = \exp(-S/k_{\mathrm{B}}) \sum_{|\Phi_r>;E}^{E+\delta E} E_r. \tag{2.48}$$

Entropy is determined by preparing the system in the energy range between E and $E+\delta E$ with the corresponding number of accessible states $\Omega(E)$. Internal energy U also depends on external parameters. These parameters fix the boundary conditions, which determine the energy eigenvalues E_r of the many-particle states $|\Phi_r >$ of a given Hamiltonian $\mathscr{H}$. Enclosing the system within a volume V fixes the boundary conditions. Thus, entropy S and volume V can be chosen freely by an experimenter who prepares the system in a state of his liking. They are the independent variables of the state function

$$U = U(S, V). \tag{2.49}$$

---

[33]For instance, if one puts a pot of cold water on a hot plate, the heating process is not quasistatic, but the heat given off by the hot plate to the water can be simply calculated as the difference between the enthalpies of the water in the hot and in the cold state.

We recall that state functions are uniquely determined by their independent variables. Thus, their infinitesimal changes are total differentials, and their second-order mixed derivatives are equal.

The statistical definition (2.46) of entropy implies the statistical definition of absolute temperature T:

$$\frac{1}{T} = \left(\frac{\partial S}{\partial U}\right)_V. \tag{2.50}$$

This is shown by the derivation of (3.14) in Chap. 3. With that the total differential

$$dU = \left(\frac{\partial U}{\partial S}\right)_V dS + \left(\frac{\partial U}{\partial V}\right)_S dV \tag{2.51}$$

becomes

$$dU = TdS + \left(\frac{\partial U}{\partial V}\right)_S dV. \tag{2.52}$$

We write (2.40) for quasistatic transitions between two infinitesimally close states 1 and 2. Then $U_2 - U_1 \equiv dU$, $V_2 - V_1 \equiv dV$, and $-Q_{12} \equiv \delta Q$. By convention and definition δQ is the infinitesimal amount of heat the system *absorbs* in a quasistatic process. With that (2.40) changes into

$$dU = \delta Q - pdV. \tag{2.53}$$

In equilibrium thermodynamics, this equation is called the first law of thermodynamics.

Comparison of (2.52) and (2.53) yields

$$dS = \frac{\delta Q}{T} \tag{2.54}$$

and

$$p = -\left(\frac{\partial U}{\partial V}\right)_S. \tag{2.55}$$

In equilibrium thermodynamics, (2.54) is the phenomenological definition of entropy, whereby absolute temperature T plays the role of an integrating factor, which produces the total differential dS from the infinitesimal δQ. The standard form of the first law of thermodynamics for quasistatic mechanical work of volume change and heat exchange with the environment is obtained by combining (2.54) and (2.53):

$$dU = TdS - pdV. \tag{2.56}$$

According to its derivation, it is only appropriate for changes that can be approximated by infinitely slow, reversible processes.

Since the second-order mixed derivatives of $U(S, V)$ must be equal, i.e.,

$$\frac{\partial^2 U}{\partial S \partial V} = \frac{\partial^2 U}{\partial V \partial S},\tag{2.57}$$

and since $\left(\frac{\partial U}{\partial V}\right)_S = -p$, and $\left(\frac{\partial U}{\partial S}\right)_V = T$, we obtain the first of the so-called Maxwell relations:

$$-\left(\frac{\partial p}{\partial S}\right)_V = \left(\frac{\partial T}{\partial V}\right)_S.\tag{2.58}$$

Subsequently, we consider systems in contact with a reservoir. By definition, a reservoir is so large that its temperature, pressure, and chemical composition remain practically unchanged during interactions with the systems under consideration. This is the case if the number of degrees of freedom in the reservoir is very much larger than that in the systems. Of course, the natural environment satisfies the reservoir conditions.

System in Contact with a Reservoir at Constant Pressure: $H(S, p)$

We have seen above that enthalpy H is the appropriate state function of a system in contact with a reservoir that maintains a constant pressure p on the system. Thus, pressure p replaces volume V as the independent variable to be chosen by the experimenter. Formally, H is obtained from U by the Legendre transformation

$$H = U + pV.\tag{2.59}$$

Then the total differential

$$dH = dU + p\,dV + V\,dp = T\,dS + V\,dp\tag{2.60}$$

has S and p as the independent variables of H, where (2.56) has been used. Reasoning like that which leads to the first Maxwell relation (2.58) yields the second Maxwell relation:

$$\left(\frac{\partial T}{\partial p}\right)_S = \left(\frac{\partial V}{\partial S}\right)_p.\tag{2.61}$$

System in Contact with a Reservoir at Constant Temperature: $F(T, V)$

If a system with fixed volume V is brought into thermal contact and equilibrium with a reservoir at constant temperature T, the independent variables that determine the system properties are T and V. The Legendre transformation to the state function Helmholtz free energy F is

$$F = U - TS.\tag{2.62}$$

The total differential is

$$dF = dU - T\,dS - S\,dT = -S\,dT - p\,dV, \tag{2.63}$$

where, again, (2.56) has been used. Thus, $F = F(T, V)$, and the third Maxwell relation results from the equality of the second-order mixed derivatives of $F(T, V)$ as

$$\left(\frac{\partial S}{\partial V}\right)_T = \left(\frac{\partial p}{\partial T}\right)_V. \tag{2.64}$$

In equilibrium, F is minimum.

System in Contact with a Reservoir at Constant Temperature and Pressure: $G(T, p)$

If a system is brought into thermal contact and equilibrium with a reservoir at temperature T, and if its volume may change so that the system assumes the same pressure p as the reservoir, the independent variables that determine the system properties are T and p. The Legendre transformation to the state function Gibbs free energy G is

$$G \equiv U - TS + pV. \tag{2.65}$$

The total differential of G becomes, in combination with (2.56),

$$dG = dU - T\,dS - S\,dT + p\,dV + V\,dp = -S\,dT + V\,dp. \tag{2.66}$$

The fourth Maxwell relation results from the equality of the second-order mixed derivatives of $G(T, p)$ as

$$-\left(\frac{\partial S}{\partial p}\right)_T = \left(\frac{\partial V}{\partial T}\right)_p. \tag{2.67}$$

In equilibrium, G is minimum.

If the system absorbs heat from the reservoir and performs work against the constant pressure of the reservoir quasistatically, the maximum work (*other* than the work done on the pressure reservoir) is given by the change of $G(T, p)$. This is the reason why G is called a "free energy."

Summary of Formal Aspects

Internal energy U, enthalpy H, entropy S, Helmholtz free energy F, and Gibbs free energy G characterize system *properties*. They are described by *state functions*.

Infinitesimal changes of state functions are total differentials. The integral of a total differential between an initial state 1 and a final state 2 is *independent* from the path of integration. Exergy, on the other hand, is *not* a state function of the system, because it depends not only on internal properties of the system but also on properties of the environment. The exergy of a system with internal energy U, entropy S, and volume V that is not in equilibrium with its environment of constant temperature T_0 and pressure p_0 is, according to (2.45), $E_X = (U - U_0) + p_0(V - V_0) - T_0(S - S_0)$. This can be written as the difference of two terms that formally resemble the Gibbs free energy. With the definition $G_0(U, S, V) \equiv U - T_0 S + p_0 V$, the exergy of the system relative to its environment becomes

$$E_X = G_0(U, S, V) - G_0(U_0, S_0, V_0). \tag{2.68}$$

Appendix 2: Energy Units

Magnitudes

Symbol	Abbreviation	Number	Word
μ	Micro	10^{-6}	
m	Milli	10^{-3}	
k	Kilo	10^3	Thousand
M	Mega	10^6	Million
G	Giga	10^9	Billion
T	Tera	10^{12}	Trillion
P	Peta	10^{15}	Quadrillion
E	Exa	10^{18}	Quintillian

SI Energy Units

SI is an abbreviation of *Système International* (French for "International System").

One joule (J) $= 1$ watt second (W s)

1 megajoule	$= 10^6$ J	$= 1$ MJ
1 gigajoule	$= 10^9$ J	$= 1$ GJ
1 terajoule	$= 10^{12}$ J	$= 1$ TJ
1 petajoule	$= 10^{15}$ J	$= 1$ PJ
1 exajoule	$= 10^{18}$ J	$= 1$ EJ

One million (metric) tons of coal equivalents (tCE) $=1$ MtCE $= 29.3$ PJ
One million tons of oil equivalents (tOE) $=1$ MtOE $= 41.9$ PJ
One ton of oil equivalents $= 7.3$ barrels of oil equivalents (1 barrel $= 159$ liters)
Historical unit: calorie (cal); 1 cal $= 4.19$ J

Energy Conversion Factors

Units	MJ	kWh	tCE	tOE
1 MJ	1	0.278	0.000034	0.000024
1 kWh	3.6	1	0.000123	0.000086
1 tCE	29,304	8,140	1	0.700
1 tOE	41,868	11,630	1.429	1

Power Units

One watt (W) $= 1$ J/s
One horsepower (hp) $= 0.7355$ kW

References

1. Ostwald, W.: Die Energie, Verlag von Johann Ambrosius Barth, Leipzig (1908)
2. Lindner, A.: Grundkurs Theoretische Physik, p. 235. Teubner, Stuttgart (1994)
3. Fetter, A.L., Walecka, J.D.: Quantum Theory of Many-Particle Systems. McGraw-Hill, New York (1971)
4. Dreizler, R. M., Gross, E. K. U.: Density Functional Theory. Springer, Berlin (1990)
5. Baehr, H. D.: Thermodynamik, 5. Ed. Springer, Berlin, Heidelberg (1984)
6. Fricke, J., Schüssler, U., Kümmel, R.: CO_2-Entsorgung. Phys. Unserer Zeit, **20**, No. 2, 56–61 (1989)
7. Ullmanns Encyclopädie der Technischen Chemie, 14. Verlag Chemie, Weinheim (1977)
8. Giovanelli, R.G.: Secrets of the Sun. Cambridge University Press, Cambridge (1984)
9. Berthomieu, G., Cribier, M. (Eds.): Inside the Sun, Kluwer, Dordrecht (1990)
10. Dearborn, D. S. P.: Standard Solar Models. In: [11], pp. 159–174
11. Sonett, C.P., Giampapa, M.S., Mathews, M.S.: The Sun in Time. The University of Arizona Press, Tucson (1991)
12. Stix, M.: The Sun. Springer, Heidelberg (1989)
13. German Bundestag (ed.): Protecting the Earth 's Atmosphere, Bonn (1989); Fig. 8, p.359
14. Eddy, J. A.: Variability of the present and ancient Sun: A test of solar uniformitarianism. In: [15]
15. Stephenson, F.R., Wolfendale, A.W. (Eds.): Secular Solar and Geomagnetic Variations in the Last 10 000 Years, Kluwer, Dordrecht (1988)
16. Labitzke, K.: On the interannual variability of the middle stratosphere during northern winter. J. Meteor. Soc. Japan **60**, 124–139 (1990)

17. Wigley, T.M.L.: The climate of the past 10 000 years and the role of the Sun. In: [15], pp. 209–223
18. Schönwiese, C.-D., Walter, A., Brinckmann, S.: Statistical assessments of anthropogenic and natural global climate forcing. An update. Meteorol. Z. **19 (1)**, 003–010 (2010)
19. Sybesma, C.: Biophysics. Kluwer, Dordrecht (1989)
20. Sieferle, R. P.: Das vorindustrielle Solarenergiesystem. In: Brauch, H. G. (ed.) Energiepolitik, pp. 27–46. Springer, Berlin (1997)
21. Wikipedia, the free encyclopedia
22. Heinloth, K.: Energie und Umwelt. B.G. Teubner, Stuttgart (1993)
23. Institut der deutschen Wirtschaft Köln: Deutschland in Zahlen 2006: Wirtschaftszahlen, Internationale Vergleiche, Primärenergieverbrauch, 12.22, online service.
24. Institut der deutschen Wirtschaft Köln: Deutschland in Zahlen 2006: Wirtschaftszahlen, Internationale Vergleiche, Bevölkerung, 12.1, online service.
25. Heinloth, K.: Klimaverträglichkeit von Arten der Energiebereitstellung für Nahrung, Wärme, Strom, Treibstoffe. In: Nordmeier, V., Grötzebauch, H. (eds.) Beiträge zur MNU-Tagung, Regensburg 2009, MNU/M_09_02/M_09_02.pdf. Lehmanns Media, Berlin (2009)
26. Kroy, W., Ludwig Bölkow Stiftung: Können Erneuerbare Energieformen unseren Energiebedarf in der Zukunft sichern? Talk presented on October 10, 2008, at the founding Symposium of the "Denkwerk Zukunft" in the Margarethenhof/Tegernsee.
27. Bundesministerium für Wirtschaft und Technologie, Energiedaten 2005: Tables 40, 41, 42, online service.
28. Bundesanstalt für Geowissenschaften und Rohstoffe, 2006, quoted by: "Welt der Physik, Uranreserven", edited by Deutsche Physikalische Gesellschaft and Bundesministerium für Bildung und Forschung, http://www.weltderphysik.de
29. Blok, K.: Introduction to Energy Analysis. Techne Press, Amsterdam (2006).
30. Groscurth, H.-M., Kümmel, R., van Gool, W.: Thermodynamic Limits to Energy Optimization. Energy—Intntl. J. **14**, 241-258 (1989).
31. Groscurth, H.-M., Kümmel, R.: The Cost of Energy Conservation: A Thermoeconomic Analysis of National Energy Systems. Energy—Intntl. J. **14**, 685–696 (1989). Groscurth, H.-M.: Rationelle Energieverwendung durch Wärmerückgewinnung. Physica-Verlag, Heidelberg (1991)
32. King Hubbert, M.: Nuclear Energy and the Fossil Fuels. American Petroleum Institute, 1956. One can read the entire paper at http://www.hubbertpeak.com/hubbert/1956/1956.pdf
33. Strahan, D.: The Last Oil Shock, John Murray, London (2007)
34. Erbrich, P.: Ernährung und Energiegewinnung—Ergebnisse aus dem zweiten Bericht des Club of Rome. Orientierung **39**, 79 (1975)
35. Energy Information Administration: International Energy Annual 2006, posted on December 8, 2008.
36. Heinloth, K.: Die Energiefrage. Vieweg, Braunschweig (1997)
37. Bundesverband Windenergie, quoted by "Welt der Physik", edited by Deutsche Physikalische Gesellschaft and Bundesministerium für Bildung und Forschung, http://www.weltderphysik.de
38. "Welt der Physik", see [37]
39. http://www.gwec.net/fileadmin/documents/PressReleases/PR_2010/Annex%20stats%20PR%202009.pdf
40. Wiese, A., Kaltschmitt, M.: Stand und Perspektiven der Windkraftnutzung in Deutschland. In: Brauch, H.G. (ed.) Energiepolitik, pp. 87–100. Springer, Berlin (1997)
41. Lindenberger, D., Bruckner, T., Groscurth, H.-M, Kümmel, R.: Optimization of solar district heating systems: seasonal storage, heat pumps, and cogeneration. Energy—Intntl. J. **25**, 591–608 (2000).
42. ZAE Bayern (Bavarian Center for Applied Energy Research): Annual Report 2009, p. 34. ZAE, Würzburg, (2010)

43. Luther, J.: Solar Energy Conversion—Solar Electricity Generation, Photovoltaic Energy Conversion. Fraunhofer Institut für Solare Energiesysteme, Freiburg; http://www.ise-solar. info.
44. Forschungsverbund Erneuerbare Energien (FVEE) (Renewable Energy Research Association): Beitrag des FVEE zum 6. Energieforschungsprogramm der Bundesregierung. October 2010 (http://www.fvee.de/fileadmin/politik/fvee-input_6.efp_2010.pdf)
45. German Solar Industry Association, as quoted by L. Wissing in the "National Survey Report of PV Power Applications in Germany 2006", Forschungszentrum Jülich
46. Institut für Elektrische Energietechnik, Fachgebiet Erneuerbare Energien, Technische Universität Berlin: Energetische Amortisation und Erntefaktoren regenerativer Energien, and references therein; http://www.herzo-agenda21.de/_PDF/emsolar.ee.pdf
47. Hall, C., Powers, R., Schoenberg, W.: Peak oil, EROI, investments and the economy in an uncertain future. In: Pimentel, D. (ed.) Biofuels, Solar and Wind as Renewable Energy Systems, pp. 113-136. Elsevier, London (2008)
48. Gagnon, N., Hall, C., Brinker, L: A Preliminary Investigation of Energy Return on Energy Investment for Global Oil and Gas Production. Energies 2, 490–503 (2009); doi:10.3390/en20300490
49. Murphy, D., Hall, C.: Year in review—EROI or energy return on (energy) invested. Ann. N.Y. Acad. Sci. 1185 102–118 (2010)
50. The LTI-Research Group (Ed.): Long-Term Integration of Renewable Energy Sources into the European Energy System. Research Department Environmental and Resource Economics, Logistics, ZEW.—Physica-Verlag, Heidelberg (1998)
51. Kenney, W.F.: Energy Conservation in the Process Industries. Academic Press, Orlando, (1984)
52. Bruckner, T., Groscurth, H.-M., Kümmel, R.: Competition and synergy between energy technologies in municipal energy systems. Energy—Intntl. J. 22, 1005–10014 (1997).
53. International Energy Agency (IEA): World Energy Outlook. Paris (1993)
54. Kümmel, R., Schüssler, U.: Heat equivalents of noxious substances: a pollution indicator for environmental accounting, Ecol. Econ. 3, 139–156 (1991)
55. World Nuclear Association, July 2008; http://www.world-nuclear.org/
56. Dietrich, G., Neumann, W., Roehl, N.: Decommissioning of the thorium high temperature reactor (THTR 300). In: Technical committee meeting on technologies for gas cooled reactor decommissioning, fuel storage, and waste disposal. Juelich (Germany) 8-10 Sep 1997, pp. 9–15. International Atomic Energy Agency, Vienna. IAE-TECDOC-1043
57. Häfele, W., Holdren, J.P., Kessler, G., Kulcinski, G.L.: Fusion and Fast Breeder Reactors. International Institute of Applied System Analysis (IIASA), Laxenburg (1977)
58. Deutsche Physikalische Gesellschaft (German Physical Society): Climate Protection and Energy Supply in Germany 1990–2020. Bad Honnef (2005) (http://www.dpg-physik.de/gliederung/ak/ake/studien/energiestudie_engl.pdf)
59. Glaser, P.E.: The Future of Power from the Sun. In: IECEC 1968 Record, IEEE Publication 68C21-Energy, pp. 98–103, (1968); Power from the Sun; its future. Science 162, 857–861 (1968)
60. Glaser, P.E.: Method and Apparatus for Converting Solar Radiation to Electrical Power, US Patent 3,781,647 December 23, 1973.
61. Glaser, P.E.: Perspectives of Satellite Solar Power. Journal of Energy, March/April 1977.
62. Glaser, P.E.: Solar Power from Satellites. Phys. Today, February 1977, pp. 30–38
63. Boeing Aerospace Co.: System's Definition—Space Based Power Conversion Systems. NASA, MSFC, Contract NAS8-31628, Fourth Performance Briefing, August 11, 1976
64. US Department of Energy and the National Aeronautics and Space Administration: Satellite Power System. Reference System Report, October 1978, DOE/ER-0023. National Technical Information Service, US Department of Commerce, Springfield (1979)
65. Koomanoff, F.A.: Satellite power system concept development and evaluation program. Space Solar Power Review 2, 163–168 (1980)
66. Lior, N.: Power from Space. Energy Convers. Manage. 42, 1769–1805 (2001)

67. O'Neill, G.K.: The Low (Profile) Road to Space Manufacturing. Astronautics and Aeronautics **16**, Special Section, pp. 18–32 (1978)
68. O'Neill, G.K.: The Colonization of Space. Phys. Today, September 1974, pp. 32-40
69. O'Neill, G.K.: The High Frontier—Human Colonies in Space. William Morrow & Co., New York (1977)
70. Summerer, L., Ongaro, F.: Solar Power from Space—Validations of Options for Europe. http://www.esa.int/gsp/ACT/doc/POW/ACT-RPR-NRG-2004-ESA-SPS_Validation_of_options_for_Europe.pdf
71. National Space Security Office: Space-Based Solar Power As an Opportunity for Strategic Security. Phase 0 Architecture Feasibility Study, 10 October 2007
72. Fricke, J., Borst, W.L.: Energie, 2nd Edn. Oldenbourg, Munich (1984)
73. Karlsson, S.: The Exergy of Incoherent Electromagnetic Radiation. Phys. Scr. **26**, 329 (1982).
74. van Gool, W.: The Value of Energy Carriers. Energy—Intntl. J. **12**, 509 (1987)

Chapter 3
Entropy

Entropy was encountered in Chap. 2 as a system's property that reduces exergy, thus diminishing the system's capacity of performing useful work. But this is just one aspect of entropy. There are more and troublesome facets:

> Entropy is the physical measure of disorder. All energy conversion processes produce entropy. Entropy production is coupled to emissions of heat and particles.

The concept of entropy is even more elusive than that of energy. Looking at entropy from different angles may help to understand it.

3.1 No Free Lunch

Fool's paradise has been sought for a long time by learned people. Between the thirteenth century and the eighteenth century Oriental and Occidental authors described machines that would produce mechanical work without any input of energy. But they never managed to build such a machine. Others, even in our days, give talks, write papers, and offer shares in companies that claim that they could build machines that perform mechanical or electrical work by extracting energy from the environment without changing anything else. Unfortunately, again, no such device is actually working – and never will be.

Mother Nature's refusal to give her children a free lunch is formulated by the first law and the second law of thermodynamics. These laws sum up humanity's experiences with change. They reveal the ties between energy and entropy and are so fundamental that they are sometimes called the "constitution of the universe."

R. Kümmel, *The Second Law of Economics: Energy, Entropy, and the Origins of Wealth*, 113
The Frontiers Collection, DOI 10.1007/978-1-4419-9365-6_3,
© Springer Science+Business Media, LLC 2011

As stated in Chap. 2, energy is conserved quantitatively in energy conversion processes: it consists of valuable *exergy* and useless *anergy*, the sum of which is constant. Exergy can be converted into any form of work, whereas no work whatsoever can be obtained from anergy. The assertion of energy conservation is the most general formulation of the first law of thermodynamics. In conversion processes energy loses some or all of its ability to perform work: anergy increases at the expense of exergy. This is what is meant by "energy consumption." Energy depreciation is part of an irresistible drive in all natural and technical systems to produce entropy "when something happens." Systems that undergo change produce entropy by spreading out their components as evenly as possible in space and over the states of motion. This is the inevitable consequence of the second law of thermodynamics. The price of ordering things locally is enhanced disorder in the environment and loss of energetic quality. People have felt that since ancient times. In the Bible, *Kohelet* (200 BC) bemoans: "All the things run to exhaustion."

The science of thermodynamics was developed in the nineteenth century in order to understand heat engines and prevent them from exploding. Its basic laws, the first law and the second law of thermodynamics, summarize the experiences of all patent offices and their frustrated clientele of would-be perpetual motion machine inventors.

In a technical formulation, the first law of thermodynamics says:

> It is impossible to construct a machine that performs work without any input of energy.

Such a machine would be a perpetual motion machine of the first kind.

The technical formulation of the second law of thermodynamics, in the words of *William Thomson* (Lord Kelvin), is: It is impossible, by means of inanimate material, to derive mechanical effect from any portion of matter by cooling it below the temperature of the coldest of the surrounding objects.

Shorter, in the (modified) words of *Max Planck*, the second law of thermodynamics states:

> It is impossible to construct a cyclically operating machine that does nothing other than performing physical work and cooling down a heat reservoir.

Such a machine would be a perpetual motion machine of the second kind. It is sketched in Fig. 3.1.

Fig. 3.1 Impossible heat
engine: perpetual motion
machine of the second kind.
Heat Q from a reservoir at
temperature T would be
received by a cyclically
operating machine M and
converted completely into
work W

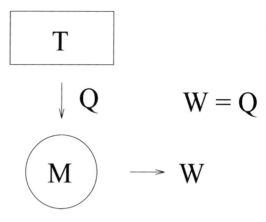

By definition, a reservoir is a many-body system whose number of degrees
of freedom tremendously exceeds that of all other systems interacting with it.
Therefore, any exchange of heat, work, and matter with the other systems leaves it
in an undisturbed equilibrium at constant finite temperature, pressure, and chemical
composition.

The experience summarized by the second law of thermodynamics teaches us
that a heat engine that receives an amount of heat Q from a donating reservoir
at temperature T must always reject a part Q_0 of that heat to a second, receiving
reservoir at a lower temperature $T_0 < T$. Thus, it can only perform the work $W =
Q - Q_0$. In the case of the ideal (Carnot) heat engine, discussed in Sect. 2.3, the
efficiency $\eta = W/Q$ of heat-to-work conversion is the Carnot efficiency

$$\eta_C = 1 - T_0/T. \tag{3.1}$$

The scheme of a real, possible heat engine is indicated in Fig. 3.2. Thus, the natural
environment of temperature T_0, including the huge mass of the oceans,[1] cannot

---

[1] The total mass of the oceans is 1.4×10^{21} kg [1]. The energy content of the ocean layer between
0- and 700-m depth is estimated to be about 10^{23} J [2]. This is more than 230 times world energy
consumption in 2004. But only a small fraction of that could be used permanently by ocean thermal
energy conversion. In ocean thermal energy conversion the Sun is the source that puts the energy
into the ocean and maintains the temperature difference between the surface and deeper water
layers.

Fig. 3.2 A real heat engine. The machine M receives heat Q from a reservoir at temperature T, rejects a part Q_0 of that heat to a second reservoir at lower temperature $T_0 < T$, and performs the work $W = Q - Q_0$ on an external device

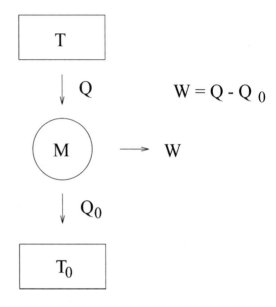

be used as the receiving *and* the donating reservoir for work-performing heat engines: $\eta_C = 0$ for $T = T_0$.[2,3] Furthermore, any heat Q_0 that is rejected to the natural environment becomes useless too. This is the unavoidable process of

---

[2]Nevertheless, even university professors outside the natural sciences, with the best of intentions, ask for public funding of proposals to convert the quantum-mechanical zero-point oscillations of the vacuum into electricity. The energy quanta of these oscillations would constitute a "heat reservoir" at the zero of absolute temperature, and the machine that would convert their energy into work would be the ultimate perpetual motion machine of the second kind. The promoters of such a machine are part of the "vacuum field energy" movement. They believe that the Casimir force from the field fluctuations of the vacuum can be converted into work performed by a cyclically operating machine without any energy input. As "proof" they show videos of machines, of course always connected to the public grid, where one watt meter shows a small electricity input and another watt meter shows a huge electricity output. The suggestion to charge batteries with the net electricity output, sell them, become rich and famous, and build the new vacuum field power system all over the world is met with indignation.

[3]Someone might have a somewhat more sophisticated proposal: With the help of a freezing machine let us create a many-body system in its ground state of minimum internal energy at the absolute zero of temperature and let it serve as the receiving reservoir with $T_0 = 0\,$K for a heat engine; then this engine will convert all the heat from a finite temperature reservoir completely into work W, because the Carnot efficiency of (3.1) is unity in this case. This proposal is of no help either. One would have to expend at least the same amount of exergy to cool down the reservoir to $T \approx 0$ K as one would gain by letting a machine work between it and, say, the ocean. This is because even the tiniest amount of absorbed heat δQ_0 would lift the many-body system out of its ground state into any one of its many exited states of equal energy which become accessible to it by the absorption of δQ_0; this raises the temperature of the system to some finite value, and all further operations of the heat engine would further raise the temperature until it is back to that of the ocean, and the efficiency in (3.1) becomes zero. If one could gain more exergy in the process of heating up the low-temperature reservoir than one had invested in cooling it down, one could reinvest part

energy dissipation and depreciation in any heat-to-work conversion, for which
respiration was the first example encountered in Sect. 2.2.4. There are many other
formulations of the first law and the second law of thermodynamics, and the variety
of formulations indicates the wide range of experiences on which the two laws are
based. This may also explain why it seems to be so difficult to understand them
properly.

3.2   Equipartition, the Toddler, and Entropy

The best way to understand entropy and the second law of thermodynamics is paved
by statistical physics and its fundamental postulate of equal a priori probabilities.[4]
It says:

> An isolated system in equilibrium is equally likely to be in any of its
> accessible states.

This postulate is equivalent to the statement that there is no perpetual motion
machine of the second kind: The impossibility of such a perpetual motion machine
means that there is no way for an isolated system to change periodically into a
situation where it is not equally likely to be in any of its accessible states and from
where it can perform work. For instance, imagine a cylinder with air molecules,
sealed-off by a movable piston, in thermal and mechanical contact with a reservoir
of temperature T and pressure p. The system "reservoir plus cylinder" is thermally
isolated and in equilibrium. Then, it is fantastically unlikely that periodically, let us
say, every other second, all the air molecules in the cylinder give off their kinetic
energy to the reservoir and accumulate at the bottom of the cylinder so that the
external pressure pushes the piston down, and then, with the air molecules absorbing
heat from the reservoir, the air expands again to its former volume, pushing the
piston up, and, via a gear, the moving piston drives an external electricity generator.

If a system is equally likely to be in any of its accessible states, then there is
equipartition of its components over all positions in configuration and momentum
space. ("Configuration space" is the conventional space we live in, and "momentum

---

of it in cooling down the reservoir anew and continue the cycles of net exergy generation. This
would be another impossible perpetual motion machine of the second kind.

[4] Although Liouville's theorem in classical mechanics and Boltzmann's H theorem in its quantum-
mechanical generalization make the postulate of equal a priori probabilities mathematically very
plausible, its basis is experience. And no experience whatsoever has ever contradicted this postulate
and the conclusions derived from it.

space" is spanned by momenta parallel to the x, y, and z axes of conventional space. The position of material particles in momentum space can also be indicated by their velocity vectors in configuration space, as shown by Fig. 3.4.)

The general belief in the validity of the postulate of equal a priori probabilities is demonstrated every Saturday night on German TV, when the winning numbers of the public lottery are drawn: 49 white balls, identical except for the numbers painted on them, are mixed mechanically in a transparent globe. After a sufficiently long time the mixing stops, an exit opens, and one ball with a winning number drops out. This happens seven times. Then the lucky winners have been determined.[5] If the general public, and especially the law makers (although, maybe, not all of the systematic lottery players), did not believe that each configuration of ball positions relative to the exit is equally likely after sufficient mixing, they would not tolerate the lottery because of the possibility of fraud.

This simple mechanical example with the thoroughly mixed and randomly positioned lottery balls illustrates an important equilibrium property of many-particle systems subject to statistical forces: the equilibrium state, where each particle configuration is equally likely, is a state of maximum randomness and disorder.

There is change in a system as long as it is out of equilibrium. Closed systems left alone strive for equilibrium. In equilibrium there is no change of macroscopic parameters such as volume, pressure, and temperature, and there is equipartition and maximum disorder of the individual system components. Thus, one expects that a closed universe would suffer "heat death" in some distant future, when it will have finally reached equilibrium. Then all radiation energy would be at constant temperature everywhere, and matter would be spread as evenly as possible in space and over all states of motion. All the energy of the universe would have become degraded to thermal energy. Nothing could happen anymore. But who knows whether our universe is closed or open.

An everyday experience of the trend toward equipartion and disorder in nonequilibrium systems is made by parents of young children. Watch a toddler. After a good night's sleep and a healthy breakfast, he marches into his room and starts to play with his toys. The first thing he usually does is pull them out of the racks one after another. He fumbles them, tastes them, senses their shapes, sees their colors, and tries out what he can do with them beyond the purposes for which they are designed. Throwing them around is a favorite experiment. Occasionally they break and he learns something about the consistency of things. He also enjoys producing art work in books and on walls with crayons, and then drops them. Usually, after not too long a time, most of the movable things are more or less evenly scattered throughout the room. It is then mother's and father's business to restore order at the end of a successful day. The success is in the brain of the child: playing has formed new connections between the brain's neurons, storing in it new information

---

[5]Stopping the mixing and letting one ball drop out is like taking a snapshot of the ball's position during the mixing process. The mechanical ball mixing is statistically equivalent to the thermal mixing of gas particles in equilibrium within an isolated box.

A monster 's there that always shows,
the more you work the more it grows.

Fig. 3.3 Many natural scientists never become really familiar with entropy. (Modified cartoon from [3])

about the world, and improving comprehension. Thus, order has increased in the brain at the expense of disorder in the domestic environment. Thanks to mummy's and daddy's work of putting the toys back into the rack and repairing somehow the broken ones, the youngster can continue the next day with playing and learning. Later in life, quite a number of teenagers love to mess up their room while studying for school.

The state of the toddler's room before his parents' intervention resembles the state on the desk shown in Fig. 3.3. Furthermore, there is a huge number of similar states that can be produced by *changing* the positions of the toys in the room or of the objects on the desk. And the overwhelming majority of all these states convey the impression of disorder. Now, let Ω be the number of different states that can be produced from a given quantity of toys in the room or objects on the desk. Obviously, the larger this quantity is, the larger will be Ω and the number of states that look disorderly. On the other hand, if Ω is considerably larger than 1, a further increase of it by some amount $\delta\Omega$ increases messiness by less than that amount: adding another pencil to the objects on the desk in Fig. 3.3 hardly changes the picture. A function that reflects this, because it increases more slowly than Ω, is the natural logarithm (ln) of Ω. These examples from everyday experience make it perhaps intuitively clear that

the entity that is proportional to ln Ω, and which is called entropy S, is the physical measure of disorder.

The reasoning of statistical physics [4] that leads to the precise definition of entropy S follows. *The reader who is content with what has been said about entropy so far, can go from here directly to (3.16).*

3.3   States of Systems

A system consists of particles, energy, and system boundaries. The boundaries may be external ones which, e.g., isolate the system completely from its environment so that it cannot exchange energy and matter with it. They may also be internal constraints that subdivide the system into subsystems, which may or may not interact with each other via the exchange of heat, work, and matter. The particles of the system may be elementary charges or all sorts of atoms and molecules in any shape or state of matter: think of electrons in a transistor, gas in a box, wine in a bottle, or a power station with its fossil fuel inputs and exhaust gases.

The study of systems consisting of many particles is one of the most active areas of modern science. The forces between the systems' atoms essentially involve well-understood electromagnetic interactions. (Nuclear forces, present in atom-changing nuclear reactions, are different, but are not considered here.) Therefore, one can, in principle, write down the (classical or quantum-mechanical) equations of motion for all system particles. But the *complexity* of many-particle systems is often so great that it makes the task of deducing any useful consequences or predictions extremely difficult. "The difficulties involved are not just questions of quantitative detail which can be solved by brute force application of bigger and better computers. Indeed, even if the interactions between individual particles are rather simple, the sheer complexity arising from the interaction of a large number of them can often give rise to quite unexpected *qualitative* features in the behavior of a system" [4].

A famous example is superconductivity. Below a certain critical temperature T_C metals such as aluminum, lead, mercury, niobium, and zinc, or the modern cuprates such as $YBa_2Cu_3O_{6+x}$, lose all their electrical resistance *and* become perfect diamagnets, which screen not too strong magnetic fields out of their interior. At T_C, which is below some 20 K for conventional metals and goes up to over 90 K for the cuprates, they make a (phase) transition into a completely new, more highly ordered state of matter with many unexpected properties. In this macroscopic quantum state, in which pairs of electrons perform exactly the same motion no matter where they are in the superconductor,[6] the laws of quantum mechanics, known from the microscopic world of atoms, manifest themselves directly on a macroscopic scale, e.g., in superconducting magnets that lift high-speed trains. The phenomenon of superconductivity was discovered in 1911 by *Kammerlingh Onnes*,

---

[6]This correlated motion cannot be broken up by the statistical oscillations of the metal atoms. Since there is no scattering of the paired electrons, the electrical resistance vanishes.

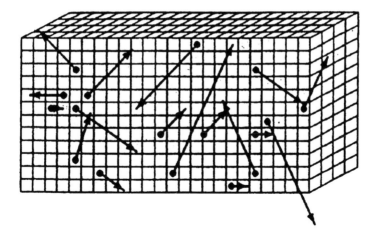

Fig. 3.4 One many-body state (macrostate) of an ideal gas in a box of volume V, subdivided into elementary cells. The *points* indicate the particle positions in three-dimensional configuration space, and the *arrows* represent the momenta (velocities×mass) of the gas particles. ("Configuration space" means the conventional, three-dimensional space we notice with our senses)

but it took nearly 50 years before it was explained by the Nobel-prize-winning theory of *Bardeen, Cooper, and Schrieffer* [5] in 1957. An adequate theory of high-temperature superconductivity in the cuprates, discovered in 1986 by *Bednorz and Müller*, is still lacking.

Common to all many-body systems is the huge number of states available to them within a tiny energy interval δE.

This can be most easily illustrated by looking into the simplest many-body system, the ideal gas. By definition, an ideal gas consists of pointlike particles of mass m_i that do not interact with each other. (It can be approximately realized by a very dilute real gas at high temperatures.) Figure 3.4 shows *one* single macrostate of an ideal gas of N particles in a box of volume V. Each elementary cell within this volume is a tiny cube of edge length Δl. The center of a cube has the spatial coordinates (x, y, z). One can also define a momentum space, spanned by the three Cartesian momentum components (p_x, p_y, p_z) of a single particle. The elementary cells within this momentum space are tiny cubes of edge length Δp, and the center of a cube is characterized by (p_x, p_y, p_z). (Heisenberg's uncertainty principle in quantum mechanics requires that the size of the product $\Delta l \Delta p \equiv h_0$ not be smaller than $\hbar$, which is Planck's constant divided by 2π, in order to be physically meaningful.) If one of the N particles, say, particle i, is at point $\mathbf{r}_i = (x_i, y_i, z_i)$ of configuration space and has momentum $\mathbf{p}_i = (p_{ix}, p_{iy}, p_{iz})$, one characterizes its *single* particle state by indicating the two respective elementary cells in configuration and momentum space that contain $\mathbf{r}_i$ and $\mathbf{p}_i$. One *many-body* state of the N-particle system is then characterized by indicating these elementary

cells for *all* N particles. The total (internal) energy E_N of the ideal gas is the sum
of the kinetic energies $\mathbf{p}_i^2/2m_i$ of all the individual particles:

$$E_N = \sum_{i=1}^{N} \frac{\mathbf{p}_i^2}{2m_i}. \tag{3.2}$$

Let us assume that an experimenter has prepared the gas in such a way that
this energy E_N lies in the range between E and $E + \delta E$. (It is impossible to
prepare the system with vanishing δE because the very act of preparation involves
energy transfers between the experimenter and the system before the system is
isolated. These energy transfers, due to unmeasurable residual interactions during
preparation, introduce the energy uncertainty δE.) Then there are fantastically many
different combinations of single particle states, each combination forming one
many-body state, whose total kinetic energy E_N falls into the range between E and
$E + \delta E$: The displacement of a *single* particle from one elementary cell in Fig. 3.4
into another one already leads to another *many-body* state. Detailed calculation
for N ideal gas particles in volume V shows [4] that the number $\Omega(E)$ of many-
body states in the energy range between E and $E + \delta E$ grows extremely rapidly
with E and gas volumes V of a few cubic centimeters, where N is huge under
environmental conditions[7]:

$$\Omega(E) = cV^N E^{3N/2}. \tag{3.3}$$

Here c is a constant independent of volume and energy.

The description of real, interacting many-body systems, such as superconductors,
and the calculation of their energy states require the apparatus of modern many-body
quantum mechanics. However, a rough estimate of $\Omega(E)$ can do without that [4].
This estimate suffices to give an idea of how extremely unlikely it is to find a system
in just *one* of its many accessible macrostates. The corresponding tiny probability,
which is due to the fantastically large $\Omega(E)$, is derived in the following passage. If
one skips the small print, one will not miss too much important information.

Consider a system the description of which requires f independent coordinates; f is also
called the "number of degrees of freedom" of the system. To specify one many-body state of the
system quantum mechanically one needs f quantum numbers. (Classically one many-body state
is characterized by the f coordinates and the f momenta of its particles.) Let E be the energy
of the system measured from its lowest possible energy, which is the energy of the quantum-
mechanical ground state. $\Phi(E)$ is the total number of possible many-body states with energies less

---

[7]Quantum mechanically the many-body state of an ideal gas in a box of volume V is only
characterized by the quantized values of the momenta $\mathbf{p}_i$ in (3.2). But for a single particle the
quantum-mechanical density of states, that is, the number of states in a tiny energy range, is also
proportional to V. Thus, the classical and the quantum-mechanical $\Omega(E)$ both grow with V^N.
They only differ in the counting of states occupied by identical particles.

than E. Obviously, $\Phi(E)$ increases as the energy E increases. To see how rapid this increase is, we consider first just one typical degree of freedom of the system. $\Phi_1(\varepsilon)$ denotes the total number of possible values that can be assumed by the quantum number associated with this particular degree of freedom when it contributes an amount of energy ε or less to the system. Let $\Delta\varepsilon$ be the mean spacing between the possible quantized energies associated with the typical degree of freedom. If ε is not too small, a crude approximation for Φ_1 is

$$\Phi_1(\varepsilon) \approx \left(\frac{\varepsilon}{\Delta\varepsilon}\right)^\alpha, \tag{3.4}$$

where α is a number of the order of 1 in the widest sense. In another crude approximation, the energy per degree of freedom, ε, is estimated from the total energy E of the many-body system with f degrees of freedom by

$$\varepsilon \approx \frac{E}{f}. \tag{3.5}$$

The total number $\Phi(E)$ of many-body states with energies below E is equal to the product of the numbers of $\Delta\varepsilon$ spacings pertaining to all f degrees of freedom. Combining (3.4) with (3.5), and considering the result as valid for all degrees of freedom, one arrives at the estimate

$$\Phi(E) \approx [\Phi_1(\varepsilon)]^f \approx \left[\left(\frac{E}{f\Delta\varepsilon}\right)^\alpha\right]^f. \tag{3.6}$$

This is the total number of states when the system has energy E or less. The number of states $\Omega(E)$ in the range between E and $E + \delta E$ is then

$$\Omega(E) = \Phi(E + \delta E) - \Phi(E). \tag{3.7}$$

This expression can be evaluated approximately, observing that f is a huge number in ordinary many-body systems. For an ideal gas of N particles, this number is $f = 3N$. A mole of gas contains 6.022×10^{23} particles (the Avogadro number), and the number of electrons contained within $1\,\text{cm}^3$ of a metal is similar. Then one can disregard the logarithm of f against f ($\ln 10^{23}$ is less than 53). With approximations of that kind, one can show [4] that $\Omega(E)$ and $\Phi(E)$ are of the same order of magnitude. (This has a simple geometrical explanation: nearly all of the volume of an f-dimensional sphere of radius $E + \delta E$ is in its surface shell between E and $E + \delta E$ if f is extremely large.) Therefore, the number $\Omega(E)$ of states accessible to a many-body system whose energy is in the range between E and $E + \delta E$ results in

$$\Omega(E) \approx [\Phi_1(\varepsilon)]^f \approx \left[\left(\frac{E}{f\Delta\varepsilon}\right)\right]^{\alpha f}, \tag{3.8}$$

where $\alpha \approx 1$. Since $\frac{E}{f\Delta\varepsilon}$ is a number larger than unity according to the definitions of ε and $\Delta\varepsilon$, and f is of the order of 10^{23}, the number $\Omega(E)$ of accessible states is fantastically large.

If a many-particle system is in equilibrium, it can be found in any one of its $\Omega(E)$ many-body states – let us label these states by r – with the same probability

$$P_r = \frac{1}{\Omega(E)} \propto \frac{1}{E^{\alpha f}}, \quad \alpha \approx 1, \tag{3.9}$$

according to the fundamental postulate of equal a priori probabilities. In the macroscopic systems encountered in daily life, such as air in a room, water in a cup, or ions and electrons in a piece of metal, the number f of degrees of freedom is of the order of 10^{23}. For such f, the probability P_r is extremely small. Therefore, one has never observed the situation where, say, half of the air molecules in a large room accumulate below a chair and lift it up. This many-body state *is* accessible to the system. But it is very unlikely that it is occupied once during the lifetime of the universe.

One may ask whether it might not in principle be possible to prepare at least a simple many-body system in one specific many-body state, say, r_0, where it remains forever – although one cannot precisely fix its energy during the preparation. Then P_{r_0} would be unity, and all other P_r would be zero. The answer is that no system will remain forever in any one particular state accessible to it, because it will always be disturbed by residual internal interactions, heat leakages through never perfect thermal insulations, sound waves coupling to its boundaries from the outside world, or the gravitational pull of a remote star at the edge of the Milky Way. However tiny these perturbations are, they will be sufficient to throw the system out of one state of energy E into any other one of the $\Omega(E)$ accessible states, because energetically these states are all so extremely close to each other [6].

3.4   The Way Things Change

3.4.1   Driving Toward Disorder

The number of states accessible to a system, $\Omega(E)$, characterizes the equilibrium state of a many-body system that is prepared in the energy range between E and $E + \delta E$. It must be related to a thermodynamic state function of the system that complements the state function "internal energy U" defined by (2.28). This additional state function was introduced ad hoc in Appendix 1 of Chap. 2 as *entropy S*. It must be an extensive quantity, like U. This means that if a system A consists of two subsystems I and II, the entropies add: $S_A = S_I + S_{II}$.

The relation between entropy S and $\Omega(E)$ results as the answer to the question of how two initially isolated systems (I and II) come into equilibrium after they have been brought into contact and interact with each other. The formal answer is given first. Then it will be illustrated by a heat-engine example.

The number Ω_A of states accessible to system A consisting of two subsystems (I and II) is the product of the number of states accessible to each of the two subsystems [4]:

$$\Omega_A = \Omega_I \times \Omega_{II}. \tag{3.10}$$

Let E_A be the energy of the total system A, and E_I and E_{II} be the energies of subsystems I and II. Then $E_A = E_I + E_{II}$, or

$$E_{II} = E_A - E_I. \tag{3.11}$$

The fundamental postulate of equal a priori probabilities asserts that in equilibrium system A is equally likely to be found in any one of its accessible states. Therefore, the probability $P(E)$ of finding the combined system A in a configuration where subsystem I has its energy E_I in the range between E and $E + \delta E$ is proportional to the number of states accessible to system A in this situation:

$$P(E) = C \times \Omega_A(E) = C \times \Omega_I(E) \times \Omega_{II}(E_A - E), \qquad (3.12)$$

where C is a constant of proportionality independent of E. (This constant is determined by the requirement that the sum of the probabilities $P(E)$ over all energy states E must be 1.) Since subsystems I and II are both systems of very many degrees of freedom f, $\Omega_I(E)$ increases extremely rapidly with E, whereas $\Omega_{II}(E_A - E)$ decreases extremely rapidly with E. This can be inferred from (3.8) and (3.9). Consequently, the product of these two factors, and thus $P(E)$, exhibits an extremely sharp maximum for some particular value $\tilde{E}_I$ of E. In equilibrium, the overwhelming number of states accessible to the total system A have E_I extremely close to $\tilde{E}_I$, and E_{II} extremely close to $E_A - \tilde{E}_I \equiv \tilde{E}_{II}$.

The position of the maximum of $P(E)$ is the same as the position of the maximum of

$$\ln P(E) = \ln C + \ln \Omega_I(E) + \ln \Omega_{II}(E_A - E). \qquad (3.13)$$

Hence, to find the condition that subsystems I and II are in equilibrium, we have to determine the value $E = \tilde{E}_I$ for which $\ln P(E)$ is maximum. In the maximum, $\ln P(E)$ does not change when E varies by infinitesimally small amounts. Thus, the variation of $\ln P(E)$ with E, which is $\partial \ln P(E)/\partial E$, must vanish for $E = \tilde{E}_I$. The detailed calculation follows in small print. The result is given by (3.14) and (3.15).

> The condition for thermodynamic equilibrium of the two subsystems is
>
> $$\frac{\partial \ln P(E)}{\partial E}\Big|_{\tilde{E}_I} = \frac{\partial \ln \Omega_I(E)}{\partial E}\Big|_{\tilde{E}_I} + \frac{\partial \ln \Omega_{II}(E_A - E)}{\partial E}\Big|_{\tilde{E}_I} = 0.$$
>
> Since $E_{II} = E_A - E$ (with $E \le E_I \le E + \delta E$), we have
>
> $$\frac{\partial \ln \Omega_{II}(E_A - E)}{\partial E}\Big|_{\tilde{E}_I} = -\frac{\partial \ln \Omega_{II}(E_{II})}{\partial E_{II}}\Big|_{\tilde{E}_{II}}.$$
>
> Here $\tilde{E}_{II} = E_A - \tilde{E}_I$ is the energy of subsystem II in the maximum of $P(E)$. With the definitions
>
> $$\beta(E) \equiv \frac{\partial \ln \Omega}{\partial E} \equiv \frac{1}{k_B T},$$
>
> the combination of the last two equations results in the equilibrium condition
>
> $$\beta_I(\tilde{E}_I) = \beta_{II}(\tilde{E}_{II}).$$

With the statistical definition of absolute temperature T,

$$\frac{1}{T} \equiv k_B \frac{\partial \ln \Omega(E)}{\partial E}, \tag{3.14}$$

where $k_B = 1.38 \times 10^{-23}$ J/K is the Boltzmann constant, one finds that the total system has reached its equilibrium if the temperatures T_I and T_{II} of subsystems I and II have become equal:

$$T_I = T_{II}. \tag{3.15}$$

According to (3.14), the infinitesimal change of $\ln \Omega(E)$ determines the state property "absolute temperature" of a system in equilibrium, where $\Omega(E)$ is the number of accessible many-body states in the energy range between E and $E + \delta E$. The function that is proportional to $\ln \Omega(E)$, and which enters the defining equation (3.14) of absolute temperature, is a state function. It is given the name *entropy* and the symbol S. From (3.14) we see that

$$S = k_B \ln \Omega(E). \tag{3.16}$$

This statistically derived definition of entropy is written on the tombstone of *Ludwig Boltzmann* (1844–1906). It says that the entropy of a system whose total energy is in the range between E and $E + \delta E$ increases with the natural logarithm of the number $\Omega(E)$ of macrostates that are accessible to the system. In equilibrium, where this number and entropy are maximum, the system can be found with equal probability in any one of these macrostates. Then, with overwhelming probability, one will encounter the system in macrostates where its particles, i.e., atoms and energy quanta, are as randomly distributed in space and over the states of motion as the constraints on the system allow. In this sense, entropy is the physical measure of disorder.

In equilibrium a system does not undergo any macroscopic changes whatsoever; there are only microscopic fluctuations. For a macroscopic process to occur, be it respiration in a living cell or the performance of work by a heat engine, the total system must be out of equilibrium. Let us consider a heat engine that couples a heat source, e.g., a pressurized tank with superheated steam of initial temperature T_2, to a heat sink, e.g., a second tank of cold water with initial temperature $T_1 < T_2$. As long as the heat flow through the heat engine has not equalized the temperatures of the sink and the source, the total system "source, heat engine, and sink" can only be found in subsets of all the states Ω_{tot} that have become accessible by the coupling. This is not an equilibrium situation according to the fundamental postulate of statistical physics. The drive toward equilibrium pushes heat through the engine and lets it perform work, e.g., lift a weight. If – before equal tank temperatures have been reached so that (3.15) is satisfied – the coupling to the sink is interrupted, the number of states accessible to the total system is, say, Ω_a. If coupling is reestablished for some time and then interrupted again, the number of states then accessible will have grown to Ω_b, and it will continue to grow, if the process continues in the same way, until it has reached Ω_{tot} in the equilibrium state of equal tank temperatures. Then, no more work can be performed, the total energy of the system stays constant, and entropy is maximum. (In real-life heat engine operations the second "tank" is usually the environment; equilibrium would mean "no combustion.")

In summary, if a thermally isolated many-body system is not in equilibrium, its entropy grows. Entropy reaches a maximum in equilibrium when the system is equally likely to be in any of its accessible many-body states. At finite temperatures, the overwhelming majority of these macrostates resemble the state that is shown in Fig. 3.4, where the individual particles which make up the many-body system are randomly distributed in space and over their states of motion.[8]

Growing entropy means growing overall disorder, or randomness. The fact that entropy measures the quantity of randomness within a system was emphasized by *Ruelle* [7] and nicely illustrated by *Stahl*.[9]

The physical manifestation of randomness within a system of given entropy S may differ from system to system [8]:

- In a gas there is random motion of molecules with random instantaneous positions.
- In heat radiation there are randomly oscillating electromagnetic fields.
- The flow of heat, exchanged between two bodies at different temperatures, is driven by the random forces which mediate the interaction between the two bodies across their interface.

The natural unit of entropy is the *bit*. One bit of randomness corresponds, e.g., to tossing a coin with equal probability for "number" or "picture." In computers, the bit is the elementary unit of information, which is stored in a cell with two possible states. These states are the codes for the dual numbers "zero" and "one." The relation between the entropic and the information *bit* is as follows: One bit of entropy is given if in the storage cell 0 or 1 can be found with equal probability. The bit is a very small unit; it fits the description of entropic changes on the molecular level. In the photosynthesis reaction described in Sect. 2.2.4 the entropy removed from the system is 40 bits per molecule of glucose. These 40 bits correspond to the loss of freedom of motion of the atoms which are bound to their fixed places in the sugar molecule; in this situation they can perform less random motions than in the original substances. One may also interpret the 40 bits as the information required

---

[8]From a macroscopic point of view a "random distribution" appears as a "uniform distribution."

[9]The following passages until the end of this subsection are translated excerpts from Arne Stahl's article "Entropiebilanzen und Rohstoffverbrauch"[8].

Table 3.1 Ecologically relevant entropy production data [8]

Process	Macrobit/s ($\doteq$ W/K)
Influx of solar radiation entropy per square meter of Earth's surface	< 0.1
Infrared entropy export into cold space per square meter of Earth's surface	1.2
Transport entropy in Earth's greenhouse per square meter of Earth's surface	0.2
Entropy reduction by photosynthesis per square meter of Earth's surface (daily average in moderate climate)	-1.3×10^{-3}
Physiological entropy production of an adult human being	0.5
Entropy production per capita by economic activities	
World average	10
USA	35
Germany	20
India	2
Entropy production by water pollution per capita in Germany	1–2

to assign the fixed places in the glucose molecule to the atoms. In macroscopic systems, typical entropic changes are of the order of 10^{23} bits; 10^{23} bits may be appropriately called a "macrobit." One macrobit is just the amount of randomness transported across a real or imaginary boundary during 1 s by a heat current of 1 J/s at the temperature of 1 K. The usual thermodynamic entropy unit of 1 J/K therefore corresponds to 1 macrobit. More precisely, inserting $\Omega = 2^{10^{23}}$ into (3.16), one finds that 1 macrobit corresponds to the entropy $S_{\text{Mb}} = 1.38 \times 10^{-23}$ J/K $\times \ln 2^{10^{23}} = 0.956$ J/K. Table 3.1 shows the order of magnitude of entropy changes per unit time in some ecologically relevant processes.

3.4.2 Haste Makes Waste: Irreversible Processes

Changes occur in a system when system constraints are removed. Then macroscopic system parameters readjust to the reduced number of constraints. If the system is so isolated that it cannot exchange heat and work with any other system, the parameters tend to readjust in such a way that the number of accessible states approaches a maximum. If the final equilibrium situation is actually reached, the system occupies each of its Ω_f accessible final states with the same probability. If Ω_f is larger than the initial number Ω_i of accessible states before the constraint removal, entropy has increased to its maximum possible value. Then, simply restoring the constraints without doing any work on the system does not restore the initial situation. There is nothing to cause the system to restrict itself spontaneously to occupying only a subset of its accessible states. The changes caused by the constraint removal are said to constitute an *irreversible* process. On the other hand, if in an isolated system a change of constraints does *not* change the number of states accessible to the system so that the initial and the final situation are identical from a macroscopic point of view, one says that the process is *reversible*.

A simple example is a gas in the left half of a box, separated by a partition from the vacuum in the right half of the box. This partition is a constraint on the volume of the gas. If one pulls out the partition, the volume of the gas increases from its initial value $V_i = V/2$ to the total volume V of the box. If the gas can be approximated by an ideal gas of N particles, the number of final accessible states is $\Omega_f \propto V^N$, whereas the number of initial states before removal of the partition was $\Omega_i \propto (V/2)^N$, according to (3.3). Once the molecules are in equilibrium and uniformly distributed throughout the box, the simple act of replacing the partition does not change the macroscopic situation. The molecules still remain uniformly distributed throughout the box. Therefore, the original removal of the partition has triggered an irreversible process. On the other hand, if one inserts a partition into a box filled by a gas in equilibrium, the change is reversible, because the gas particles are distributed as uniformly in the box after the insertion of the partition as before.

Another example is the two subsystems (I and II) at different temperatures T_I and T_{II} considered in Sect. 3.4.1. If they are brought into thermal contact, heat flows between them until the temperatures are equal and the total system (system A $\equiv$ subsystem I + subsystem II) has come to equilibrium. Then the particles in both subsystem I and subsystem II are as uniformly distributed in space and over their states of motion as the constraints permit, and entropy is maximum. Reestablishing a thermal insulation between the two subsystems does not return them to the initial state of unequal temperatures. The process of heat transfer is irreversible. On the other hand, if two systems at equal temperatures are initially isolated thermally from each other and the constraint "thermal insulation" is removed, no macroscopic system parameter changes, and the number of states accessible to the total system stays constant. The process of removing the thermal insulation between two systems of equal temperatures is reversible.

This leads to processes important to textbook equilibrium thermodynamics. If processes occur so slowly that the system is practically always in equilibrium during experimental observation times t_{exp}, they are also considered as reversible. Consider two systems at only infinitesimally different temperatures which are brought into thermal contact. Then the heat flow between them, which is driven by the temperature difference, is infinitely slow, so during finite observation times both systems always appear to be in equilibrium. In general, one can *approximate* a real process by a reversible process if the relaxation time τ during which the system comes back to equilibrium after a perturbation is very much smaller than t_{exp}.[10] For instance, the relaxation time of a hot gas in the cylinder of a heat engine is very much less than the time t_{rot} during which the piston completes one cycle; τ may be of the order of 10^{-10} s, whereas $t_{rot} = t_{exp}$ may be longer than 10^{-2} s in, e.g., a diesel engine. Therefore, for rough calculations, the gas can be treated as if it were always in internal equilibrium during the cycle.

Of course, in the strict sense, all natural and technical processes we observe are irreversible processes, because they are not infinitely slow. Thus, entropy increases in them. This, however, does not mean that after an irreversible process a system can never be restored to the initial situation of lower entropy. It can, *provided* that the system is *not* kept isolated but is allowed to interact with other systems. But the price to be paid for that is entropy increase in the other systems.

Take the above example of the particles that have spread into the empty half of the box after the removal of the partition in the middle of the box. One can replace the right wall of the box by a thin piston. This piston is moved by an external device, for instance, a falling weight, to the middle of the box, thus doing work on the gas, pushing it back into the left half of the box against the pressure exerted by the gas. After compression, the volume of the gas has been restored to its original value $V_i = V/2$, and the right half of the box is again empty. But the energy of the gas is greater than it originally was because of the work done on the gas during compression. To reduce it to its original value, one has to bring the gas into thermal contact with another suitable system to which it can give off just the right amount of heat. After that the gas in the box has been restored to its initial state and entropy. But the systems which have interacted with it have changed, and the total entropy of all three systems – gas in the box, falling weight, and heat-receiving system – has increased.

In summary, all irreversible processes produce entropy. Although there are irreversible processes without energy conversion, such as pulling out a partition that separates a gas from the empty half of a box, all energy conversion processes in nature and society remove constraints, occur during finite times, are irreversible, and produce entropy. In this sense, entropy production is the twin of energy conversion.

3.5 Arrow of Time

We notice the passing of time by events that occur within and around us. Events are irreversible processes in nonequilibrium situations. In equilibrium nothing would happen, except for usually tiny microscopic fluctuations, in which all memory of the past is being lost within relaxation times that are tiny fractions of a second.

---

[10]In the opposite limit $\tau \gg t_{exp}$, where equilibrium is achieved very slowly compared with experimental times, one can also treat the system as if it were in equilibrium while one observes its behavior.

(Such is, by the way, the vision of the above-mentioned "heat death" of the universe held by those who believe in a closed universe which one day would come into perfect equilibrium: when the stars will have burned their nuclear fuel, scattered their matter in gravitational collapses, and a dull, uniform radiation field will fill all space. But we need not worry. If the universe expands, as most people believe, it may never have a chance to come into equilibrium. If it is stationary, it will still have a long time to go. As we saw in Chap. 2. our Sun will shine for at least another five billion years. For the foreseeable future we will have to worry about other heat problems.)

There is something very special about time. Joel Lebowitz, in his prize speech on the occasion of receiving the 2007 Max Planck Medal of the German Physical Society, described it like this: "In the world about us the past is distinctly different from the future. Milk spills but doesn't unspill, eggs splatter but do not unsplatter, waves break but do not unbreak, we always grow older, never younger. These processes all move in one direction in time – they are called 'time-irreversible' and define the arrow of time. It is therefore very surprising that the relevant fundamental laws of nature make no such distinction between the past and the future. These laws permit all processes to run backward in time. This leads to a great puzzle – if the laws of nature permit it why don't we observe the above-mentioned processes run backwards? Why does a video of an egg splattering run backwards look ridiculous? Put another way: how can time-reversible motion of atoms and molecules, the microscopic components of material systems, give rise to observed time-irreversible behavior of our everyday world?" [9].

The flow of time has a direction for us because all events evolve in such a way that the result is more randomness and equipartition as long as no forces from the outside prevent that.[11] There are always efforts necessary in the form of exergy inputs in order that "something new happens." For instance, if an experimenter wants to study a nonequilibrium process, he first must prepare the system under investigation in a nonequilibrium state; e.g., using exergy, he must suck all the molecules of a gas within one half-chamber of a box through a valve in a dividing wall into the other half-chamber by means of a vacuum pump. If he then opens the valve, the gas rushes back into the empty half-chamber, and in equilibrium the molecules spread evenly over all volume elements; the number of states accessible to the system has increased dramatically (compared this with the situation of the empty half-chamber), all states are occupied with equal probability, and entropy has increased in the course of time.

> The increase of entropy in nonequilibrium processes determines the thermo-dynamic arrow of time.

---

[11]Recently, one has also begun to discuss an arrow of time associated with self-organization. A broad analysis of the physical basis of the direction of time is given in [10].

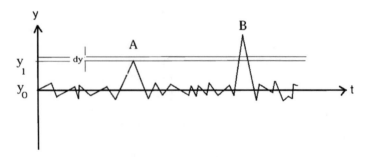

Fig. 3.5 The time dependence of a parameter $y(t)$ fluctuating about its mean value y_0 in thermodynamic equilibrium

This is perfectly compatible with time-reversal symmetry of the fundamental laws of nature: When the experimenter prepares a nonequilibrium state, he picks a situation which is extremely unlikely to occur via statistical fluctuations in equilibrium [11]; the constraints relaxed, e.g., the valve opened (or the egg splattered), the system goes into the extremely more likely configuration of maximum randomness. Let us illustrate that in some more detail by a simple example. Consider a variable y whose fluctuations in equilibrium are shown schematically in Fig. 3.5. There is no magnetic field present and y is some quantity, for instance, a displacement, which is invariant under time reversal. If, starting at some time t_0, one looks into the future at times $t_0 + \theta$, where $\theta > 0$, or into the past at times $t_0 - \theta$, the situation is quite indistinguishable. This is the microscopic reversibility of the process under consideration. However, suppose that one *knows* that the parameter lies in the small range between y_1 and $y_1 + dy$, where y_1 is appreciably different from the mean value y_0 about which y fluctuates, because one has *prepared* the system in this range. Then y must lie near one of the very few peaks like peak A, which correspond to such improbably large fluctuations that y attains a value near y_1 (see Fig. 3.5). If one then removes the external constraint, with whose help one has prepared the system in y_1, the situation will be just the same as if y had attained the value y_1 as a result of prior spontaneous fluctuations, and y will practically *always* decrease toward y_0. The situation where y has a value near y_1 and yet *increases* would correspond to the occurrence of a peak like peak B, which is even larger than peak A; on the rising side of this peak, y would indeed *increase*. But the occurrence of peaks as large as peak B is enormously less probable than the already very improbable occurrence of peaks as large as peak A. Hence, if y had been prepared in y_1, it will practically always *decrease*, and approach closer to y_0, after the removal of the constraints. Such is the flight of the arrow of time.

More complex systems such as our living bodies are also prepared, and maintained for some time, in extremely unlikely states from an equilibrium point of view: first by our parents in the act of procreation and then by the input of high-quality energy in the form of food, which is transformed into the low-quality bodily wastes and ejected.

Dissipation and increasing disorder are inevitably associated with our very existence. The best we can hope for is that this disturbs only the environment beyond the biosphere.

This, in fact, was more or less the case before the Industrial Revolution, when all creatures on Earth lived on the annual budget of high-quality energy provided by the radiation of the Sun and photosynthesis. Essentially all entropy produced on the Earth and in the Sun went as thermal radiation into space. Consequently, the technological situation was quasistationary. It only changed at the rate at which people learned to use the available solar energy to a larger extent, e.g., by building more and more sophisticated sailing ships which between the end of the Middle Ages and the Industrial Revolution cruised the oceans in increasing numbers and initiated the grip of European civilization on the world. Until the end of the Middle Ages hardly any civilizational change was felt on the material level. Technologically, the faithful who built the Romanic and Gothic cathedrals of Speyer, Reims, and Coventry lived in nearly the same world as the worshippers in the temples of ancient Rome, Athens, and Jerusalem.

In the days of the Old Testament, Kohelet (the preacher) expressed the general feeling of living in stagnation, where all activities are just meaningless fluctuations, by his famous lament: "Vanity, vanity, and nothing but vanity. What is left to man after all his labor under the Sun? Generations come and go, but Earth remains the same forever. The Sun rises and the Sun sinks....the wind blows here and there..... All the things run to exhaustion.... There is nothing new under the Sun... . All memory will be lost of those who have been and of those who will come..." [12].

Progress in the sense of a directed change was not noticed in those days. It could hardly be felt that there was any change, with a direction given by the increase of population and of per capita energy consumption and the associated irreversible processes of human activity: The increase was too slow. As we saw in Sect 2.5.1, per capita energy use grew from 14 kWh/day 7,000 years before the present to just 30 kWh/day in AD 1400, i.e., by a factor of 0.0019/year. In contrast, since the beginning of the Industrial Revolution the USA's daily per capita energy use grew to 76 kWh in 1850 and 270 kWh in 1995 (Germany: 89 kWh in 1900 and 133 kWh in 1995), i.e., by a factor of 1.34/year, that is, 700 times faster than during the time span before the Middle Ages. In addition, the US population increased from 23 million to 260 million. The change of the world due to increasing, energy-driven, human-activated irreversible processes is being felt more and more intensely, and the concept of progress has become commonplace. But in our days more and more people are having second thoughts about whether this sort of change still deserves the positive attribute of "progress." And their doubts are justified, because now the other effect of the proliferating irreversible processes is becoming more and more visible in the form of the growing environmental burden from emissions coupled to entropy production.

3.6 Entropy Production and Emissions

The previous sections dealt with entropy from an atomistic and statistical point of view, where the number of states accessible to a many-body system is the key quantity. Since relatively few many-body systems and their states have been analyzed completely by appropriate quantum-mechanical calculations, the reasoning was mostly qualitative. Nevertheless it should have shown that entropy is the measure of disorder. To better understand the economic relevance of this disorder, we must look into the *production* of entropy. Phenomenological nonequilibrium thermodynamics reveals the relation between energy-converting activities, entropy production, and emissions. This relation can be framed like this:

> The second law of thermodynamics says that irreversible processes produce entropy. Energy-converting activities in nature and technique are irreversible. Their entropy production involves heat and particle emissions.

To be more specific, entropy production within a system is given by the integral of entropy production density σ_S over the volume of this system. Since, according to the second law of thermodynamics, this integral is positive for irreversible processes within the volume, no matter how small this volume is, the second law also says that

$$\sigma_S > 0. \tag{3.17}$$

This most general formulation of the second law of thermodynamics can be made more explicit for a nonequilibrium system that contains N different sorts of particles k, which do not undergo chemical reactions.[12] If thermodynamic variables such as temperature and pressure remain well defined *locally* during all changes, nonequilibrium thermodynamics [51] yields the density of (dissipative) entropy production ($\sigma_{S,\mathrm{dis}}$) as

$$\sigma_S = \sigma_{S,\mathrm{dis}} \equiv \mathbf{j}_Q \nabla \frac{1}{T} + \sum_{k=1}^{K} \mathbf{j}_k \left(-\nabla \frac{\mu_k}{T} + \frac{\mathbf{f}_k}{T} \right) > 0. \tag{3.18}$$

---

[12]If chemical reactions occur, one gets $\sigma_S = \sigma_{S,\mathrm{dis}} + \sigma_{S,\mathrm{chem}}$ in (3.18). The entropy production densities $\sigma_{S,\mathrm{dis}}$ and $\sigma_{S,\mathrm{chem}}$ are positive separately because $\sigma_{S,\mathrm{dis}}$ is given by products of vectorial currents and forces, which cannot interfere with the scalar currents and forces that make up $\sigma_{S,\mathrm{chem}}$; see also the Appendix.

The gradient operator ∇ represents the vectorial sum of spatial derivatives, which operate on the absolute temperature T and the chemical potentials μ_k of the particles of type k. The gradients of temperature and chemical potentials, and specific external forces $\mathbf{f}_k$, which act on the particles, represent the generalized forces that drive the heat current density $\mathbf{j}_Q$ and the diffusion current densities $\mathbf{j}_k$. At each point all diffusion current densities cancel, so total mass is conserved everywhere:

$$\sum_{k=1}^{K} \mathbf{j}_k = 0. \tag{3.19}$$

Equation (3.18), which is derived in the Appendix of Chap. 3, shows that the second law of thermodynamics covers the dissipation of energy *and* matter: In irreversible processes, heat currents of density $\mathbf{j}_Q$ carry away degraded energy, and diffusion currents of densities $\mathbf{j}_k$ spread matter in space. There is no need for a "fourth law of thermodynamics" to take care of the dissipation of matter, as *Nicholas Georgescu-Roegen* suggested [13, 14]. Claiming to have discovered a "fourth law," he created some confusion [15], which temporarily slightly obscured his merits of calling the second law to the attention of economists. Since the publication of his seminal book *The Entropy Law and the Economic Process* [16], awareness has grown in the scientific community that thermodynamics *is* relevant for economics [17–20].

Equation (3.18) is important from an ecological point of view. It tells us that entropy production is unavoidable (greater than 0) whenever inhomogeneities (gradients) and forces "make something happen," and that entropy production generates emissions of heat and matter at every point of the system. The currents due to $\mathbf{j}_Q$ and $\mathbf{j}_k$ carry out the mandate of the second law of thermodynamics to distribute energy and matter as evenly as possible in space and over the states of motion.

In the nonequilibrium system of industrialized planet Earth, heat and particle currents emanate from furnaces, reactors, and heat engines. These emissions change the energy flows through and the chemical composition of the biosphere to which the living species and their populations have adapted in the course of evolution. If these changes are so big that they cannot be balanced by the biological and anorganic processes driven by the exergy input from the Sun and the radiation of heat into space, and if they are so rapid that biological, social , and technological adaptation deficits develop, the emissions are perceived as environmental pollution.

3.6.1   Sources and Substances

The principal substances emitted in energy conversion are sulfur oxides (SO_2), nitrogen oxides (NO_x), dust, carbon monoxide (CO), hydrocarbons ($C_m H_n$), radioactive substances, "others" (especially H_2S), and greenhouse gases. They are related to

Table 3.2 Air pollutants from energy carriers

Energy Carrier	SO_2	NO_x	Dust	CO	C_mH_n	Radioactive substances	"Others"	Greenhouse gases
Coal	X	X	X	X	X	X		X
Oil	X	X	X	X	X			X
Natural gas		X		X				X
Uranium						X		
Sun, wind, water								
Biomass	X	X	X	X	X			X
Geothermal heat							X	

Table 3.3 Emissions from energy conversion and their noxious effects, without greenhouse gases

	SO_2	NO_x	Dust	CO	C_mH_n	Radioactive substances
Origin	S content	N content	Fly ash	Incomplete		Fission,
		of coal and oil		combustion		activation
		high-temperature			Direct	
		combustion			liberation	
Conversion	H_2SO_3	HNO_2				
	H_2SO_4	HNO_3			O_3	
Effects on						
Humans		Disease of respiratory organs		Inhibits O_2		Cancer,
	heart,	Reduced diffusion		transport		mutations
	circulation	of CO		in blood		
Ecosystems		Forest damage				
Buildings		Corrosion				

their primary energy sources in Table 3.2. Carbon dioxide (CO_2) and ozone (O_3) are greenhouse gases, originating from the combustion of coal, oil, natural gas, and the products manufactured therefrom, and methane (CH_4) from coal mining and the hauling and distribution of natural gas; biomass growing is associated with nitrous oxide (N_2O) emissions, if nitrate fertilizers are used, and its combustion – if unbalanced by recultivation – creates CO_2. (If one also takes into account the emissions associated with the construction of energy conversion facilities, "sun, wind, water" carry the marks (X in Table 3.2) of the primary energy carrier(s) used in the production of photovoltaic cells or solar-thermal collectors, armaments, windmills, and dams.)

The direct damage potentials of the emissions listed in Table 3.2 have been known for a long time. They are indicated in Table 3.3. Sulfur and nitrogen oxides are irritant gases, which enhance the resistance of the respiratory tract. SO_2 causes constrictions of the bronchi, and NO_x reduces the arterial partial pressure of O_2. In experimental exposure of animals to NO_x, destructive processes in the surface cells of lung vesicles and of the bronchial system, and a reduction of the defensive power of the lungs against pathogenic agents were observed. Respirable dust with a grain size below $10\,\mu m$, which represents about 85% of all dust emissions, can penetrate into the bronchi and lungs, sickening them. During the London smog catastrophes in 1948 and 1952, more than 4,000 additional deaths because of pneumonia and heart diseases were registered. This triggered decisive measures against air pollution. As a result, such smog catastrophes should no longer occur, at least not in western industrialized countries.

The "Waldsterben," that is, the alarming increase of forest damage to which acid rain, caused by SO_2 and NO_x emissions, contributes significantly, was the first pollution effect to shock the German public profoundly. Thus, in the mid-1980s, it became politically possible to introduce low legal emission limits for SO_2 and NO_x. They hold for power stations with a thermal power of more than 300 MW. For instance, a coal-fired power plant with an electrical power output of 750 MW must not emit more than 6,200 t of SO_2 and 4,100 t of NO_2 (amount of NO_x computed as amount of NO_2) annually. In practice, power stations can do much better. Their SO_2 and NO_x (NO_2) emissions decreased from 2×10^6 and 10^6 t in 1980 to 0.3×10^6 and 0.5×10^6 t in 1989 in West Germany. This shows what engineers and industry can do, if they have to.

On the other hand "Freedom of the Sea" has prevented reduction of emissions from ships, which handle 90% of global transportation. Their diesel engines burn cheap, extremely dirty heavy fuel oil. Little progress has been made so far in reducing the resulting emissions by international regulations. Thus, shipping traffic accounts for 9% of all global SO_2 emissions – this is about as much as the emissions from all 800 million cars – and produces as much respirable dust as 300 million cars. Also, 29% of all NO_x emissions in Europe are attributed to ships.[13]

3.6.2   Anthropogenic Greenhouse Effect

A pressing problem of global pollution is the enhancement of the greenhouse effect by emissions. The natural greenhouse effect, described in Chap. 2, is the reason for Earth's average surface temperature being at a comfortable $+15°C$ instead of a chilling $-18°C$. It is caused by trace gases radiatively active in the infrared, water vapor, and clouds in the upper troposphere. The anthropogenic greenhouse effect consists in an increase of the average surface temperature of Earth because of increasing concentrations of infrared-active trace gases in the atmosphere, caused by human activities. These gases are also called greenhouse gases. They are listed in Table 3.4. They absorb about 87% of the infrared radiation from the surface of Earth and radiate nearly all of it back.

---

[13] Süddeutsche Zeitung, 7/8 August 2010, p. 20.

Table 3.4 Contributions of infrared (*IR*) active gases to the natural and the present anthropogenic greenhouse effect, their present (2005 estimate) and preindustrial (1750) concentrations [22], and average time of residence (*ATR*) in the atmosphere [23]. *CFC* chlorofluorocarbons, *ppm* parts per million, *ppb* parts per billion. (The relative molecular greenhouse warming potential is that of 100 years after emission; see Table 3.5)

IR-active gas	Contributions to greenhouse effect		2005/1750 concentration	ATR (years)
	Natural (%)	Anthropogenic (%)		
CO_2	26	76.7	379/277 ppm	5–10
CH_4	2	14.3	1.77/0.72 ppm	10
N_2O	4	7.9	0.32/0.27 ppm	130
CFC	–	≈ 1	0	>55
O_3	8		40/5 ppb	
H_2O	60		2.6%/2.6%	
Others	2.5			

Figure 3.6 shows the radiation balance of Earth. Earth and its atmosphere receive short-wave solar radiation mainly in the wavelength range between 0.3 and 3 μm. This range comprises the spectrum of visible light (between 0.4 and 0.7 μm), the spectrum of near-infrared radiation (with wavelengths longer than 0.7 μm), and ultraviolet radiation (with wavelengths shorter than 0.4 μm). Earth's surface, in turn, emits longer-wave infrared radiation with wavelengths essentially between 3 and 50 μm. In this spectral range absorption by water vapor and the greenhouse gases is high, as indicated by the absorption spectra in the lower part of Fig. 3.6.

The essential point of the anthropogenic greenhouse effect is that the infrared absorption maxima of the greenhouse gases shown in Fig. 3.6 are in the "windows" where H_2O absorbs little. Thus, increases of the atmospheric concentrations of the greenhouse gases block more and more the passage of heat radiation through the "windows" of water vapor.

Energetically most important are the open atmospheric water vapor window between 7 and 13 μm – where the infrared radiation of Earth's surface reaches a maximum – and the spectral region from 13 to 18 μm, where the infrared radiation is only partly absorbed by water vapor. In the central part of the open window from 8 to 12 μm, the quasicontinuous absorption of water vapor is important. It increases with the square of water vapor pressure, so – because of the exponential increase of the partial pressure of H_2O with temperature increase – there is a strong feedback between temperature increase and infrared absorption by water vapor. This feedback is the reason why increasing CO_2 concentrations lead to a substantial overall anthropogenic greenhouse effect, despite the fact that the two main CO_2 absorption bands at 15 and 4.3 μm are already saturated to a large extent.[14] Additional CO_2 causes a small temperature increase δT; the partial

---

[14]The absorption bands of a given gas are saturated if – because of its atmospheric concentration – there is a rather low likelihood that an additional molecule in this spectral region will absorb further radiation. This molecule is overshadowed, so to speak, by the other molecules of the same gas [27].

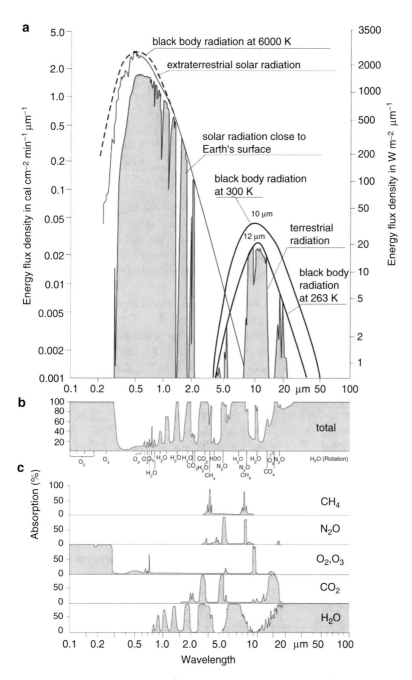

Fig. 3.6 (**a**) Energy flux density [W/(m² μm), *right ordinate*; cal/(min cm² μm), *left ordinate*] versus wavelength of solar insolation and of terrestrial heat emission; (**b**) total absorption spectrum of the principal greenhouse gases in the atmosphere close to Earth's surface; (**c**) individual absorption spectra of the principal greenhouse gases [24]

Table 3.5 Relative molecular greenhouse warming potential (*GWP*) of infrared-active trace gases ("greenhouse gases") for different numbers of years after emission [23].

Trace gas	GWP				
	20 years	50 years	100 years	200 years	500 years
CO_2	1	1	1	1	1
CH_4	35	20	11	7	4
N_2O	260	280	270	240	170
CFC-11	4,500	4,300	3,400	2,400	1,400
CFC-12	7,100	7,600	7,100	6,200	4,100

pressure of Earth's H_2O increases exponentially with δT, and infrared absorption by water vapor increases with the square of water vapor pressure. As a net result, the anthropogenic greenhouse effect caused by additional CO_2 emissions increases with the logarithm of CO_2 concentration, so, for every doubling of the atmospheric CO_2 content, the temperature of Earth's surface increases by the same amount (of approximately 2.5°C).

The situation is different for CH_4, N_2O, and the chlorofluorocarbons, the concentrations of which are much lower than the CO_2 concentration. But one additional molecule of these gases increases the atmospheric infrared absorption by a much greater factor than one additional CO_2 molecule. A measure of the absorption effectiveness of these gases relative to the absorption effectiveness of CO_2 is the relative molecular greenhouse warming potential (GWP). The emission-related GWP of a gas indicates (relative to CO_2) how much radiation is absorbed in the atmosphere in the long run by the emission of 1 kg of that gas. This takes into account the average time of residence of the gas in the atmosphere. Table 3.5 shows the time-dependent GWP (except for ozone, where the situation is unclear).

The GWP and the emission quantities determine the contributions of the greenhouse gases to the anthropogenic greenhouse effect. The combined climate-change effects of all gases with different GWPs is summarized by the concept of "carbon dioxide equivalents" (CO_{2e}). In the year 2000, the emissions of CO_{2e} were 42 Gt. Figure 3.7 shows the sources. Table 3.6 indicates specific life-cycle CO_2 emissions of different energy systems.

Empirically, one finds that the average global surface temperature has risen by about 0.8°C within the last 140 years; see Fig. 3.8

Deterministic climate model simulations on powerful computers, based on equations that describe the physical processes assumed to be relevant for climate evolution, suggest that during industrial times the anthropogenic greenhouse effect is very likely to have caused a temperature increase of about 1 ± 0.5°C. This agrees with the results of multiple statistical regression models that estimate the weight of climate-influence parameters from empirical time series of climate data, and – by simulating changes of these parameters – estimate the resulting climate effects; they indicate a temperature increase of 0.6–0.8°C since 1800 [23, 24]. These statistical studies have been continued and updated, using regression and neural network models. The results are as follows: (1) the decrease of the global mean surface air

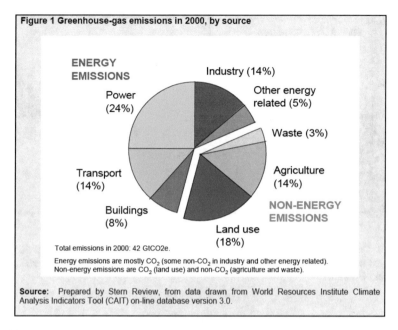

Fig. 3.7 Sources of greenhouse gas emissions in 2000, with greenhouse gas concentration being 430 ppm CO_{2e} [21]

Table 3.6 Specific total life-cycle CO_2 emissions, in grams per kilowatt-hour of electric energy, for different energy systems. Included are emissions from operation, maintenance, and production of the system. For nuclear power plants, emissions from waste disposal are excluded [29]

Lignite-fired power plant	1,200–1,100
Hard-coal-fired power plant	1,000–700
Natural-gas-fired power plant	700–430
Photovoltaic cell	90–190
Gas-powered combined heat and power unit	≈50
Water power plant	10–40
Nuclear power plant	6–32
Wind park	10–20

temperature since its climax in 1998 may be attributed to the fact that simultaneously solar radiative forcing and the influence of the El Niño/Southern Oscillation phenomenon has decreased since 1998, and, in addition, another climate-relevant volcanic eruption occurred; (2) anthropogenic radiative forcing was not interrupted, so, despite this cooling episode, greenhouse-gas-induced warming continued; (3) it is even possible that relatively large cooling due to the anthropogenic sulfate aerosol effect masked an anthropogenic warming as large as about 1.5°C in industrial times since 1860 [25].

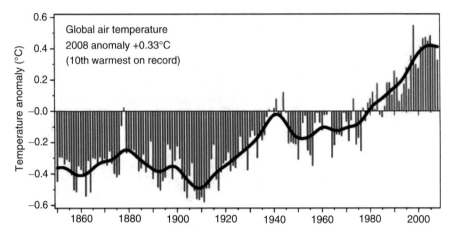

Fig. 3.8 Combined global land and marine surface temperature record from 1850 to 2008. The year 2008 was tenth warmest on record, exceeded by 1998, 2005, 2003, 2002, 2004, 2006, 2001, 2007, and 1997 [30]

Thus, deterministic climate modeling and statistical climate change analyses both reveal the influence of anthropogenic greenhouse gas emissions on the world's climate. They agree with empirical observations within the error margins. Since these different methods also arrive at similar conclusions for the future, one should take their scenarios of future climate evolution seriously.

The United Nations Intergovernmental Panel on Climate Change (IPCC) accumulates knowledge and updates research results from the world's leading experts on climate change. It consists of the three working groups "Science" (WG1), "Impacts" (WG2), and "Mitigation"(WG3). Several hundred authors cooperate in the reports of each working group. Since its foundation in 1988 by a resolution of a United Nations plenary assembly, the IPCC has delivered reports that subsequently have narrowed the still substantial uncertainties of climate-change scenarios. These reports have caught the attention of decision makers worldwide. To achieve this, the complex results of climate research have to be simplified for presentation in easy-to-read executive summaries. This, and some problems with temperature data communication, encouraged critics to pick on leading IPCC experts in Great Britain, after hackers had accessed their e-mail correspondence. In March 2010, after having investigated accusations of scientific misdemeanor put forward in a mischievous campaign, the science committee of the British parliament acquitted the scientists who had been attacked of the accusation of data manipulation. In May 2010, more than 250 members of the US National Academy of Sciences signed a statement entitled "Climate Change and the Integrity of Science" [26] that says:

"We are deeply disturbed by the recent escalation of political assaults on scientists in general and on climate scientists in particular. All citizens should understand some basic scientific facts. There is always some uncertainty associated with scientific conclusions; science never absolutely proves anything. When someone says that society should wait until scientists are absolutely certain before taking any action, it is the same as saying society should never take action. For a problem as potentially catastrophic as climate change, taking no action poses a dangerous risk for our planet. . . .

Many recent assaults on climate science and, more disturbingly, on climate scientists by climate change deniers are typically driven by special interests or dogma, not by an honest effort to provide an alternative theory that credibly satisfies the evidence. The Intergovernmental Panel on Climate Change (IPCC) and other scientific assessments of climate change, which involve thousands of scientists producing massive and comprehensive reports, have, quite expectedly and normally, made some mistakes. When errors are pointed out, they are corrected. But there is nothing remotely identified in the recent events that changes the fundamental conclusions about climate change:

(i)   The planet is warming due to increased concentrations of heat-trapping gases in our atmosphere. A snowy winter in Washington does not alter this fact.
(ii) Most of the increase in the concentration of these gases over the last century is due to human activities, especially the burning of fossil fuels and deforestation.
(iii) Natural causes always play a role in changing Earth's climate, but are now being overwhelmed by human-induced changes.
(iv) Warming the planet will cause many other climatic patterns to change at speeds unprecedented in modern times, including increasing rates of sea-level rise and alterations in the hydrologic cycle. Rising concentrations of carbon dioxide are making the oceans more acidic.
(v)   The combination of these complex climate changes threatens coastal communities and cities, our food and water supplies, marine and freshwater ecosystems, forests, high mountain environments, and far more.

Much more can be, and has been, said by the world's scientific societies, national academies, and individuals, but these conclusions should be enough to indicate why scientists are concerned about what future generations will face from business-as-usual practices. We urge our policy-makers and the public to move forward immediately to address the causes of climate change, including the unrestrained burning of fossil fuels.

We also call for an end to McCarthy-like threats of criminal prosecution against our colleagues based on innuendo and guilt by association, the harassment of scientists by politicians seeking distractions to avoid taking action, and the outright lies being spread about them. Society has two choices: We can ignore the science and hide our heads in the sand and hope we are lucky, or we can act in the public interest to reduce the threat of global climate change quickly and substantively. The good news is that smart and effective actions are possible. But delay must not be an option."

Climate change deniers give talks and interviews, write e-mails, Internet blogs, and books purporting that there is no such thing as an anthropogenic greenhouse effect. Some of them even claim that the natural greenhouse effect would violate the second law of thermodynamics.

The quoted statement of the academy members is in line with the "Stern Review Report on the Economics of Climate Change"[21], published by British government in 2007. The report is based on the available scientific knowledge concerning the consequences of increased greenhouse-gas concentrations. This knowledge has been presented in the "IPCC Third Assessment Report 2001" and the subsequent literature on feedback mechanisms that enhance climate change. The conclusion is that if the present trend of CO_{2e} emissions continues, one has to expect global average temperature increases of 2–3°C , or even more, by 2100.

Climate change affects especially the following areas:

- *Food production*. Substantial crop failures threaten developing countries, and diminished crops are to be expected in the developed regions. Only in higher latitudes is there a chance of higher agricultural yields because of CO_2 fertilization.
- *Earth's water supply*. Glaciers will melt, the water level of rivers will decrease, and drinking water will become scarce for many people. Sea level rise threatens coastal regions and big cities such as New York, London, Tokyo, Shanghai, and Hong Kong.
- *Ecosystems*. Healthy, natural ecosystems can only adapt to temperature changes of not more than 0.1°C per decade. Thus, coral reefs and the Amazonian rain forest may collapse partly or totally, and the forests in the intermediate and higher degrees of latitude, already damaged by acid rain and other pollution effects, will die. Many species are threatened by extinction. On the other hand, microbes adapt easily to changing environmental conditions, because of their rapid change of generations. Thus, viruses, bacteria, and pests will spread to northern latitudes and higher altitudes.
- *Extreme weather events*. Storms, flooding, droughts, heat waves, and forest fires will occur more often. The intensity of hurricanes will increase.
- *Risks of rapid climate change and major irreversible impacts*. Natural carbon sinks may be weakened. Natural methane release may increase, for instance, by the thawing of the permafrost. Irreversible melting of the Greenland ice sheet may start. A collapse of the Atlantic thermohaline circulation and of the West Antarctic Ice Sheet cannot be excluded.
- *Peace*. Hitherto unknown mass migrations of environmental refugees threaten social stability.

Ice core data from the Russian Antarctic Vostok station allow reconstruction of the atmospheric CO_2 concentrations during the 160,000 years prior to the Industrial Revolution; temperature variations relative to the present level have been obtained by means of the deuterium method.[15] The variations of CO_2 concentrations and temperature are strongly correlated [27]. This indicates one of the positive-feedback mechanisms that are activated in the ocean–atmosphere system by temperature

---

[15]Low deuterium levels suggest particularly high temperatures, whereas high deuterium levels indicate low temperatures.

changes: When it gets warmer, as happened between 145,000 and 130,000 years before the present, CO_2 is released from the warming oceans, and its atmospheric concentration increases; this then increases the greenhouse effect, which causes additional warming.

Of course, another change of Earth's orbital parameters, such as the one that triggered the last ice age, can reduce temperature and cause reabsorption of CO_2 by the cooling oceans. But we should not count on another ice age coming to rescue us from the anthropogenic greenhouse effect and its feedback mechanisms.

The near-surface temperature in the Northern Hemisphere has varied by up to 7°C during the past million years (Fig. 3.9). Even in the most recent history there have been temperature changes of 1°C around the 15°C median. It was particulary warm 700–1,000 years ago, and Greenland got its name. Then, between 1400 and 1850, during the so-called Little Ice Age, the global mean temperature dropped well below 15°C. Subsequently, the atmosphere warmed up considerably. But variations have been limited to about 1°C since the Neolithic Revolution 10,000 years ago when temperature stabilized around 15°C. This rendered farming and cattle breeding possible and human civilization evolved.

Temperature changes of 2–3°C, or even more, within one century, as they are to be expected because of the anthropogenic greenhouse effect, have never occurred since the end of the last ice age. Never has human civilization been challenged by climate changes comparable to those that loom in the decades ahead.

Occasionally, people argue that the medieval temperature maximum was quite beneficial for people in Europe and that, therefore, we should not worry about global warming. They overlook the fact that the present global population distribution is the result of (more or less) optimal adaptation to the present climate zones, and that the world population has more than quintupled since 1850 to its 2008 level of 6.7 billion people. If a large fraction of these people have to migrate because of rapid climate changes that cause inundations of densely populated, low-lying areas and desertification of formerly fertile regions, international tensions may grow to a level that will be too dangerous for a world with nuclear weapons.

In summary, we must realize that the combustion of fossil fuels in heat engines and furnaces, and fire clearing of woods, have caused emissions of CO_2 that have increased the atmospheric concentration of CO_2 from its preindustrial level of 280 to 380 ppm by the end of the twentieth century. In 2000 about 8.8×10^9 t of carbon was emitted globally in the form of 32.27×10^9 t of CO_2[16] [28]. Other greenhouse gases such as methane, also originating from nonenergetic sources such as agriculture, add to CO_2 emissions. The concentration change of the infrared-active trace gas CO_2 changes the radiation budget of the earth–atmosphere system sketched in Fig. 2.1 in Chap. 2. The resulting increase of the average surface temperature of Earth is expected to reach at least 2–3°C by the end of the twenty-first century if global CO_2 emissions are not reduced so drastically – by about 75% by 2050 – that the concentration of all greenhouse gases is stabilized at 550 ppm CO_{2e}. If the enormous

---

[16]The complete oxidation of 1 t of carbon results in 3.67 t of CO_2.

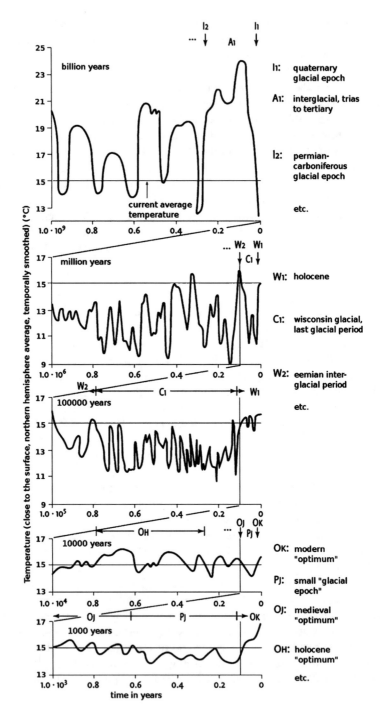

Fig. 3.9 Variations of the average near-surface temperature in the Northern Hemisphere during the past billion, million, 100,000, 10,000, and 1,000 years. (From various sources and [22])

economic-technological challenge of such a drastic reduction of fossil-fuel burning, which requires a fundamental change of the global energy system, cannot be met, severe ecological damage is to be expected. The associated economic losses have been estimated by the "Stern Review" [21].

In 1987 the German Bundestag unanimously decided to establish a study commission (Enquete-Kommission) on "Preventive Measures to Protect the Earth's Atmosphere" to deal with the growing threats to the Earth's atmosphere. This commission held hearings of the world's leading experts on energy, emissions, and climate change. Its final 935-page report was published on May 24, 1990 [31]. Its draft for an International Convention on Climate and Energy was aimed at the forthcoming United Nations Conventions on Climate Change. The principal findings of this report are still valid. Subsequent IPCC studies have reduced uncertainties and added new risks. Figure 3.10 presents the vision of the Bundestag commission of how energy-related CO_2 emissions should decline in order to prevent severe climate changes. But actual global CO_2 emissions due to energy conversion are rising and stay close to the upper "business-as-usual curve." Thus, tremendous efforts will be necessary to reach the generally agreed aim of reducing global emissions by 50% by 2050. There is not much time left to take action. And more than 100 years have passed since the Swedish physical chemist Svante Arrhenius pointed out in 1895 that the world will have to deal with the phenomenon now known as the anthropogenic greenhouse effect.

3.6.3 Pollution Control and Heat Equivalents of NO_x, SO_2, CO_2, and Nuclear Waste

The entropy-production equation (3.18) is helpful for designing strategies for mitigating harmful particle emissions. The options are essentially:

1. Reduce the number and intensity of irreversible wealth-producing processes. Since all these processes are driven by energy conversion, a reduction of primary energy consumption with unchanged energy services by appropriate energy conservation technologies is mandatory. Exploitation of the conservation potentials are supported by thermoeconomic optimization methods [34–36].
2. Use the renewable energies(Sects. 2.5.3 and 2.6.3) nourished by sunshine. The entropy production that is coupled to the generation of solar radiation occurs outside the biosphere.
3. An option for reducing greenhouse gas emissions is energy production in nuclear fusion reactors and inherently safe nuclear fission reactors, discussed in Sects. 2.6.1 and 2.6.2. This option is burdened with development costs and has to solve the problem of radioactive waste.
4. Convert the emissions of particles into the more benign emission of heat. The latter, however, is unavoidable, because entropy production is unavoidable: $\sigma_S > 0$ *always* holds. The conversion of particle emissions into heat emissions

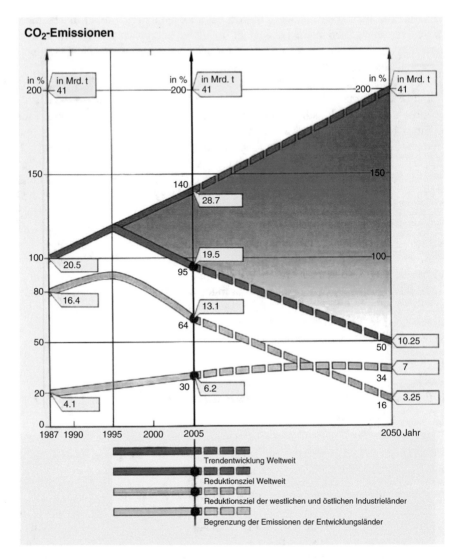

Fig. 3.10 Annual CO$_2$ emissions (Gt) from energy conversion according to the reduction plan of the study commission on "Preventive Measures to Protect the Earth's Atmosphere"of the German Bundestag [31]. The *top curve* represents the global emission trend as it was expected in 1990. The global CO$_2$ emissions of 31.5 Gt in 2008 [32] are pretty much on this trend line. The curve below the trend, which starts declining in 1995, represents global emissions that may be compatible with climate stability. The next curve indicates how much the industrial countries should reduce their emissions, and the *lowest curve* shows how much the developing countries may add, if an emission goal of 10.25 Gt is to be met in 2050. The percentages refer to the base year 1987. This historical document shows that the ideas about the essential requirements for climate stabilization did not change much between 1990 and 2010. In 1990 one saw the need to halve global CO$_2$ emissions by 2050, and in 2010 a "modelling study suggests that the massive mobilization necessary to halve greenhouse-gas emissions by 2050 is 'barely feasible' unless global development patterns shift radically" [33]

requires exergy-driven technological processes. Examples are desulfurization and removal of nitrous oxides from power plant exhaust gases, carbon (dioxide) capture and storage , and the disposal of radioactive waste via electromagnetic mass drivers or rockets in interstellar space [37,38]. The exergy needed should be obtained from sources that emit only small particle current densities themselves.

Presently, people do not worry about heat emissions (except in the cases when power stations must be shut down because the cooling rivers have dried out). But this would change if world energy consumption were to increase by a factor 20 over the 2004 level of 1.34×10^{13} W. Since most of the primary energy consumed eventually ends up in heat, total heat emissions would then approach the *heat barrier* of about 3×10^{14} W. This is roughly 0.2% of the power the Earth receives from the Sun. Anthropogenic heat flows of that magnitude are likely to cause climate changes even without the anthropogenic greenhouse effect [39]. Local temperature increases observed in urban areas, where room heating and traffic increase heat emissions relative to the urban hinterland, alert us to the heat barrier.[17]

To indicate how, in principle, material emissions can be converted into heat emissions, and how this accelerates the approach to the heat barrier, a detailed model calculation of the heat equivalent of noxious substance (HEONS) has been performed for a coal-fired power plant [38]. The following is the summary [40] of the basic results.

The HEONS is defined as the waste heat generated by pollution control of a noxious substance in a production process of given utility output, divided by the primary energy input into a not-pollution-controlled production process of the same utility output.

Pollution control is the process that keeps the injection of the noxious substance into the biosphere below a limiting value by means of suitable technical installations. The limits must be determined by risk assessment and social consensus. (This is the harder part of the problem.) The HEONS aggregates the heterogeneous pollutants in terms of thermal pollution.

The heat emissions from a not-pollution-controlled production process can usually simply be set equal to the heating value, i.e., the enthalpy, of total primary input. This can be determined by the methods of energy analysis [36]. To compute the HEONS for a given pollution control process, a methodology similar to that of energy analysis has been developed. For this one must fix the physical and chemical process boundaries, decide which indirect energies and heat emissions shall be taken into account, and establish rules from which to calculate the energy and waste heat balances.

---

[17]All the factors that determine the climate in the "urban heat islands" are described in [22] (pp. 339–343).

3.6.3.1 Physical and Chemical Boundaries for Pollution Control Processes

Substances may be proper to the biosphere or noxious. Proper substances, by definition, are present in their natural concentrations as a permanent passive background. If they are involved in pollution control processes, their consideration in balances of energy and materials begins and ends with the form they assume under standard environmental conditions. For instance, water vapor, emitted into the environment during a process of pollution control, is taken into account in the energy balances as a waste heat that is given by the energy of condensation of vapor to water at the temperature of the environment. Noxious substances, on the other hand, have to be traced, in principle, all the way from their entering the system, by being mined or produced in chemical or nuclear reactions, until they leave the system, by being properly deposited or transformed into proper substances. In actual calculations this tracing and accounting is stopped if one arrives at emissions and concentrations that are below the limiting values accepted by society.

Pollution control terminates with waste disposal. The following disposal options are admitted in the analysis:

- Open disposal sites for inert and ecologically harmless substances in suitable, nonvolatile forms that are stable for long times under environmental conditions, e.g., calcium sulfate from desulfurization.
- Closed disposal sites for inert but ecologically possibly harmful substances conditioned into nonvolatile forms that are stable for long times under disposal conditions; heavy metal dust from power plant filters is such a substance, for instance.
- Deep space for aggressive, volatile, ecologically dangerous substances if society does not tolerate storage deep in the Earth, e.g., burned-out fuel rods from nuclear power stations.

3.6.3.2 Waste Heat Analysis in First Order

This analysis determines the heat emitted in all processes that provide the necessary materials and energies for pollution control and that remove and dispose of all noxious substances generated during pollution control itself. The heat emissions due to the production of the pollution control installations, e.g., desulfurization plants and disposal sites, are disregarded.

Computation of the HEONS requires the energy balances of all pollution control processes in first-order waste heat analysis. The resulting heat emissions are assigned to the different noxious substances in a unique way according to the following rules:

1. Mechanical and electrical energy is directly converted into waste heat if it is used to surmount friction. It is also counted as waste heat if it is transformed into potential energy that does not perform useful work subsequently, but rather

dissipates more or less slowly. The total waste heat related to the use of mechanical and electrical energy (equivalent to exergy) also includes all waste heat generated during the conversion of primary energy into the considered exergy at known technical conversion efficiencies.

2. Chemical processes with endothermal reactions emit that fraction of process heat as waste heat that is not converted into chemical binding energy. In exothermal chemical processes the reaction enthalpy determines the waste heat. Exceptions are chemical processes that occur during primary energy conversion itself, e.g., in a furnace, because all heat involved here is part of useful heat.

3. The decrease of conversion efficiencies by pollution control shall be compensated for, if possible, by increased primary energy input Q. The additional waste heat ΔQ of the production process is included in the HEONS. It is determined by the initial primary energy or process heat Q_0, the conversion efficiency without pollution control η_0, and the reduced conversion efficiency with pollution control, η, under the condition of constant utility, i.e., $\eta Q = \eta_0 Q_0$:

$$\Delta Q \equiv Q(1 - \eta) - Q_0(1 - \eta_0) = Q_0(\eta_0 - \eta)/\eta. \qquad (3.20)$$

If this compensation for reduced conversion efficiency is not possible, the HEONS must include the waste heat of an additional, pollution-controlled process that makes up for diminished productivity.

4. The energy required for the transportation of materials within the biosphere is completely counted as waste heat, because all of it is dissipated by friction sooner or later.

3.6.3.3   HEONS for NO_x, SO_2, and CO_2 from a Coal Power Plant

We consider a power plant fired with low-sulfur hard coal. One normal cubic meter of its flue gases typically contains 2,000 mg sulfur dioxide (SO_2) and 1,000 mg nitrous oxide (NO_x computed on the basis of NO_2). Carbon dioxide (CO_2) occupies about 15% by volume of the dry flue gas. The German "Large Power Station Regulation Law" of 1985 for power plants of more than 300 $MW_{thermal}$ prescribes limiting values of 200 mg NO_x and 400 mg SO_2 per normal cubic meter after at least 85% desulfurization. For CO_2 there is no law except emission trading. According to recommendations of the German Physical Society and the German Meteorological Society [41], CO_2 emissions should be reduced by at least two thirds.

   NO_x is removed from the flue gases by a catalytic reaction, which involves TiO_2 as a catalyst and ammonia (NH_3) as the reducing agent. The waste heat analysis shows that for 80% NO_x removal, the heat equivalent is 23.71 $MJ/MWh_{thermal}$, which is 0.66 % of the primary energy input into a plant of the same electricity output without NO_x abatement. To a good approximation, the waste heat is equal to the primary energy required for NO_x removal.

Desulfurization by wet limestone scrubbing has the biggest market share in Germany. For 95% SO_2 removal, the waste heat analysis yields the heat equivalent of SO_2 as 158.4 MJ/MWh$_{thermal}$, which is 4.4%.

The removal and disposal of carbon dioxide, now called carbon (dioxide) capture and storage, was first investigated theoretically for postcombustion processes, such as chemical scrubbing with alkanolamine [42] and refrigeration under pressure [43]; physical absorption by Selexol after CO-shifting of the flue gas was considered too [44]; disposal options discussed were storage of CO_2 in the deep sea or in empty oil and gas fields, and carbon recycling by solar carbon technologies and biomass conversion into fuels [45, 46]. Alkanolamine scrubbing with 90% CO_2 removal would lower the power plant efficiency from 38% to 29%. If refrigeration under pressure [47] and (risky) deep sea disposal are to reduce CO_2 emissions by 66% for unchanged electricity output, the power plant efficiency would decrease from 38% to 28%, and the total heat equivalent of CO_2 is 1.38 GJ/MWh$_{thermal}$, which is 38.3% [38].

Actual plans for carbon (dioxide) capture and storage consider separation of CO_2 by the oxyfuel process, where combustion of coal occurs in pure oxygen, and storage of CO_2 is planned in geological formations such as empty gas fields and saline aquifers; a pilot project is presently under development in Germany by the power company Vattenfall [48]. The oxyfuel process with precombustion gas separation requires most of its energy for the separation of the oxygen from the air. The CO_2 heat equivalent should be of the same order of magnitude as that computed for postcombustion gas separation. Whether society will accept the risks associated with large-scale CO_2 storage within the Earth's crust or the deep sea remains to be seen.

3.6.3.4   HEONS for Radioactive Waste from a Nuclear Power Plant

For a variety of reasons, underground disposal sites have not yet been accepted as solutions to the problem of what to do with burned-out fuel rods from nuclear power plants. The problem is one of risk assessment, not of quantity. A 1,300-MW$_{electric}$ boiling water reactor burns about 12,840 fuel rods and produces just 56 t of radioactive waste in 7,000 h of annual operation, whereas a coal power plant of the same electricity output emits 8.5×10^6 t of CO_2 per year. Thus, the mass of CO_2 exceeds that of radioactive waste by a factor of more than 100,000.

The HEONS concept cannot project uncertain risks into a single number. However, it is possible to compute the HEONS for different pollution control scenarios associated with different risk categories, where acceptance of a certain risk is again a matter of social consensus. For nuclear waste one can consider disposal in the Earth's crust, on the one hand, and shooting the radioactive material into interstellar space, where it can do no harm whatsoever, on the other hand.

The HEONS calculation was done for deep space disposal [37, 38], which is certainly much more energy intensive than the German plan of storing nuclear waste in empty rock salt caverns. Rockets and electromagnetic launchers were

Table 3.7 Technical data of nuclear fuel rod disposal in deep space by a mass driver

Projectile	Telephone pole shaped
Mass	1,000 kg
Launch velocity	12.3 km/s
Velocity at top of atmosphere (escape velocity)	11 km/s
Kinetic energy at launch	76 GJ
Ablation loss, SiC shield	3% of mass
Energy loss	20%
Acceleration	1,000g $(g = 9.81\,\text{m/s}^2)$
Launcher length	7.8 km
Launch duration	1.2 s
Average force	9.81×10^6 N
Average power	60 GW
Capacitor charging time	1.5 min

considered as the means of accelerating the well-packed, burned-out fuel rods to escape velocity. For hydrogen rockets, each of which would carry 350 kg of fuel rods into deep space, the heat equivalent of nuclear waste is 0.27%, and the required energy input would be 0.34% of the thermal energy produced by a boiling-water reactor. However, 25,000 rocket takeoffs per year would be necessary to take care of the waste fuel rods of all nuclear power plants that were globally operational in the 1990s. It is hard to believe that this ever can be done.

Mass drivers, operating like the linear motors that power magnetically levitated trains, developed and tested in Germany and Japan, and commercially operational in China, have been proposed by Henry Kolm of MIT [49] for nuclear waste disposal. Prototypes were built by him and Gerard K. O'Neill and coworkers at Princeton University for space manufacturing purposes [50]. The mass driver for nuclear waste disposal in space receives its energy from huge capacitor banks charged by a 1,000-MW nuclear power plant. Its characteristics are given in Table 3.7.

If 40 fuel rods of a total mass of 175 kg constitute the payload of one 1,000-kg steel projectile with silicon carbide shielding, the heat equivalent of nuclear waste is 0.075%, and the energy consumed in deep space disposal is 0.097% of the thermal energy produced by the nuclear reactor. It exceeds the HEONS, because part of it is the energy of the projectile on its trajectory of no return. One may wish to minimize the risks associated with the projectile crashing, caused by a failure during acceleration along the 7.8-km-long track, and put only one fuel rod instead of 40 into one 1,000-kg projectile. Then the heat equivalent increases to 3% and the required specific energy increases to 3.9%. Thus, in the most energy expensive mass driver scenario, the HEONS of nuclear waste is an order of magnitude smaller than that of CO_2.

To get rid of the fuel rods of 160 power plants, each at 1,300 MW$_{\text{electric}}$, whose total generating capacity is 56% of the global 371.7-GW$_{\text{electric}}$ nuclear capacity installed in 2007, one would need more than 50,000 electromagnetic launches per year, each with a 40-fuel-rod payload. One launcher, used continuously every

12 min day and night [49], would make 43,800 launches per year. Each launch generates an explosion-like pressure wave in the atmosphere with an energy content ten times the energy of an average lightning discharge (roughly 0.1 MWh in the principal discharge). This would probably disturb significantly the balance of ions and the chemistry of the atmosphere close to the launching site. The consequences are unknown, but certainly deserve the inclusion of the relatively small amount of energy for transportation to desert launching sites in the waste heat balance.

The HEONS concept does not solve the problem of risk assessment. However, it indicates the minimum risk that is associated with the use of energy sources and technologies. This risk is the approach to the heat barrier, which looms at anthropogenic heat emissions that reach some per mills of the power received from the Sun.

> The heat barrier is the ultimate limit to industrial growth on Earth. Further growth is possible if the emissions coupled to energy conversion are no longer a burden on the biosphere. This may be achieved by solar power satellites and space industrialization [50], discussed in Sect. 2.6.3. Otherwise, one has to give up the growth paradigm.

Appendix: Nonequilibrium Thermodynamics and the Second Law

In this appendix, the important equation (3.18) for the density of entropy production is derived within the formalism of nonequilibrium thermodynamics. This *phenomenological* description of many-body systems is only concerned with macroscopic system parameters and is completely independent of quantum mechanics and statistics. It is introduced in the following as if we knew nothing about entropy. We follow presentations of nonequilibrium thermodynamics by Kluge and Neugebauer [51] and Kammer and Schwabe [52].

Gibbs's Fundamental Equation

Careful analyses of the work a system may exchange with its environment by changing its internal energy U have led physicists to the following conclusions:

1. For the description of thermodynamic systems and the analysis of heat-to-work conversion, one needs, besides the internal energy U, a second state property, called *entropy S*. This state property must be related to the work-performing capability of internal energy U and the heat transfer involved in changes of U.

2. If heat δQ crosses the system boundaries, changing the internal energy from U to $U + \delta Q$, the entropy of the system changes by an amount dS that increases with δQ.
3. In reversible processes, where the system passes infinitely slowly without internal friction or other forms of dissipation through a sequence of equilibrium states, an infinitesimal entropy change dS is proportional to the infinitesimal amount of heat δQ_{rev} the system receives at absolute temperature T. The proportionality factor must be such that its product with the heat δQ_{rev} – which, as a boundary-crossing energy form, is not a state property – becomes a state property. It turns out that this proportionality factor is the inverse of the absolute temperature T. Thus, the entropy change in an infinitely slow *reversible* process is[18]

$$dS = \frac{\delta Q_{rev}}{T}. \qquad (3.21)$$

This is the usual way entropy is introduced in physics textbooks – and from this hardly any student gets a feeling for what entropy *really* is. The statistical definition of (3.16) is more helpful. But the phenomenological definition of entropy provides a powerful tool for analyzing nonequilibrium processes, to which all energy-conversion processes belong.

Let us first recall equilibrium thermodynamics.

Consider a homogeneous system of internal energy U and volume V. Then, if the system receives the heat δQ_{rev} in a reversible process at absolute temperature T, the change of its internal energy, dU, and the mechanical work $p dV$ it performs by volume change dV at pressure p add up to δQ_{rev} according to the first law of thermodynamics:

$$\delta Q_{rev} = dU + p dV. \qquad (3.22)$$

Combination with (3.21) results in

$$dS = \frac{1}{T}(dU + p dV). \qquad (3.23)$$

Let the system contain K different sorts of molecules, and let m_k be the total mass of all molecules of type k. Then, entropy S is a state function of the system variables $U, V, m_1, ..., m_K$, and the total differential of $S(U, V, ...m_k...)$ is

$$dS = \left(\frac{\partial S}{\partial U}\right)_{V,m} dU + \left(\frac{\partial S}{\partial V}\right)_{U,m} dV + \sum_{k=1}^{K} \left(\frac{\partial S}{\partial m_k}\right)_{U,V,m_i \neq m_k} dm_k. \qquad (3.24)$$

---

[18]In (2.54) the index "rev" has been omitted for the sake of simplicity.

If all m_k are fixed, so that $dm_k = 0$, comparison of (3.23) and (3.24) yields

$$\left(\frac{\partial S}{\partial U}\right)_{V,m} = \frac{1}{T}, \qquad \left(\frac{\partial S}{\partial V}\right)_{U,m} = \frac{p}{T}. \tag{3.25}$$

In the general case of variable m_k, the chemical potential per unit mass of the molecules of type k is defined as

$$\mu_k \equiv -T\left(\frac{\partial S}{\partial m_k}\right)_{U,V,m_i \neq m_k}. \tag{3.26}$$

Inserting (3.25) and (3.26) into (3.24) one obtains

$$dS = \frac{1}{T}dU + \frac{p}{T}dV - \sum_{k=1}^{K} \frac{\mu_k}{T}dm_k, \tag{3.27}$$

which, by multiplication with T, becomes Gibbs's fundamental equation

$$T dS = dU + p dV - \sum_{k=1}^{K} \mu_k dm_k. \tag{3.28}$$

Furthermore, the relation

$$T S = U + pV - \sum_{k=1}^{K} \mu_k m_k \tag{3.29}$$

holds, because of Euler's theorem for linearly homogeneous functions, such as U and S.

Starting from these equations of equilibrium thermodynamics, we calculate entropy changes in irreversible *nonequilibrium* processes.

Consider a system in which things happen, i.e., irreversible processes occur. This may be a rod of copper with its ends at different temperatures, the expanding gas in the working cylinder of a diesel engine, or a booming industrial country. The laws of thermodynamics, although not sufficient for a complete description of these systems, are, of course, valid in all of them. The system may exchange energy and matter with its environment through material or imaginary system boundaries. Although the system as a whole is not at all in thermodynamic equilibrium, one can still analyze it in terms of locally defined thermodynamic variables such as temperature and pressure if the irreversible processes are neither too fast nor associated with too strong inhomogeneities in, e.g., mass or energy densities. Then local equilibrium prevails between the atoms, molecules, and energy quanta in volume elements ΔV that are small on a macroscopic scale but still so large on an atomic scale that they contain a huge number of particles. For instance, in copper, a tiny box of volume $\Delta V = 10^{-12}\,\text{cm}^3$ contains 2.6×10^{10} atoms. As long as the temperature variations along an edge of the box are small compared with the average temperature in the box, one has local thermodynamic equilibrium. The same is true for transport processes in gases if the variation of temperature along the mean free path, i.e., the average distance a gas atom travels between two collisions, is small compared with the temperature itself; and for compression and expansion processes, the time during which a macroscopically noticeable change of volume by, say, 1% occurs must be much larger than the relaxation time within which internal equilibrium between the atoms is reestablished after a perturbation. Most irreversible processes associated with human activities can be described with the assumption of local thermodynamic equilibrium.

We subdivide the system of total volume V into macroscopically small and microscopically large boxes of volume ΔV. On the one hand, the volume of each box is supposed to be much smaller than V, $\Delta V \ll V$ and, on the other hand, the edge length of each box is supposed to be much larger than the average distance between the particles of the system and their mean free path. A given box contains the total mass Δm, the mass Δm_k of a molecule of type k, the internal energy ΔU, and the entropy ΔS. These quantities and all other thermodynamic variables may change with the spatial coordinate $\mathbf{r}$ of the box center and time t. We work with *local*, specific quantities, defined by:

$$\rho = \Delta m/\Delta V : \quad \text{total mass density}$$

$$\rho_k = \Delta m_k/\Delta V : \quad \text{mass density of molecule of type } k$$

$$c_k = \Delta m_k/\Delta m = \rho_k/\rho : \quad \text{concentration of molecule of type } k$$

$$u = \Delta U/\Delta m : \quad \text{specific internal energy}$$

$$s = \Delta S/\Delta m : \quad \text{specific entropy}$$

$$v = \Delta V/\Delta m = 1/\rho : \quad \text{specific volume}$$

Replacing U by ΔU, S by ΔS, V by ΔV, m_k by Δm_k in (3.28) and dividing by Δm, one obtains Gibbs's fundamental equation of nonequilibrium thermodynamics:

$$T\,ds = du + p\,dv - \sum_{k=1}^{K} \mu_k dc_k. \tag{3.30}$$

Equation (3.30) relates the entropy field $s(\mathbf{r}, t)$ to the independent field variables $u(\mathbf{r}, t)$, $v(\mathbf{r}, t)$, and $c_k(\mathbf{r}, t)$, where one c_k is determined by the other ones, because

$$\sum_{k=1}^{K} c_k = \frac{1}{\rho}\sum_{k=1}^{K} \rho_k = 1. \tag{3.31}$$

In addition to the state functions that are known from equilibrium thermodynamics, new velocity variables enter the description of nonequilibrium systems, where gradients of pressure, and thus forces, cause mechanical movements. The motions are described by the velocity fields $\mathbf{v}_k(\mathbf{r}, t)$ of the different molecules of type k and the field of the barycentric velocity $\mathbf{v}(\mathbf{r}, t)$.

Balance Equations

General Balance Equation in Local Form

We want to calculate the change of entropy with time, ds/dt, from (3.30) in the form of a balance equation. This requires the balance equations for internal energy u, specific volume v, and concentration c_k.

Balance equations are important in many fields. One needs them, for instance, for the proper management of financial affairs. Suppose there is an amount A_M of money in a bank account. It can change in different ways: Money may flow into the account from one's employer who pays his employees via bank transfers; reversely, money flows out of the account when the account owner pays bills with his credit card; and money is produced within the account when the bank pays interest on the deposit. People usually watch carefully the change with time of money in their account, which is dA_M/dt. The balance equation they have to keep track of in order not to get into trouble is as follows: $dA_M/dt =$ (money paid into the account minus money withdrawn from the account) plus interest.

Physical quantities in a given volume V may change in time correspondingly. To establish their balance equations we define the following:

$$A(t) = \text{any extensive quantity in volume } V.$$

$$\tilde{a}(\mathbf{r}, t) = \Delta A/\Delta V = \text{the density of the extensive quantity } A.$$

$$a(\mathbf{r}, t) = \frac{1}{\rho}\tilde{a}(\mathbf{r}, t) = \Delta A/\Delta m = \text{the mass-specific quantity } A.$$

The time-dependent

$$A(t) = \int_V \tilde{a}(\mathbf{r}, t)dV \tag{3.32}$$

may change in time, like money in a bank account, by inflows minus outflows, which is $d_a A(t)/dt$, plus internal production, which is $d_i A(t)/dt$:

$$\frac{dA(t)}{dt} = \frac{d_a A(t)}{dt} + \frac{d_i A(t)}{dt}. \tag{3.33}$$

The first term on the right-hand side of this equation is

$$\frac{d_a A(t)}{dt} = -\int_O \mathbf{J}_A(\mathbf{r}, t) \cdot d\mathbf{O}, \tag{3.34}$$

where $\mathbf{J}_A(\mathbf{r}, t)$ is the current density of A through the element $d\mathbf{O}$ of the surface O that encloses the volume V. The integral over this surface is the sum of all inflows minus the sum of all outflows; the vector $d\mathbf{O}$ is directed outwardly, which results in the minus sign. The second term is

$$\frac{d_i A(t)}{dt} = \int_V \sigma_A dV, \tag{3.35}$$

where σ_A is the production density of A, that is, the quantity A produced per unit volume and unit time.

Combining (3.32–3.35) and changing the surface integral of $\mathbf{J}_A(\mathbf{r}, t)$ into a volume integral over

$$\mathrm{div}\mathbf{J}_A \equiv \nabla\mathbf{J}_A = \partial J_{Ax}/\partial x + \partial J_{Ay}/\partial y + \partial J_{Az}/\partial z$$

according to the theorem of Gauss, we obtain

$$\frac{d}{dt} \int_V \tilde{a}(\mathbf{r}, t)dV = \int_V [-\mathrm{div}\mathbf{J}_A + \sigma_A] \, dV. \tag{3.36}$$

Since this equation must hold for any volume V, the integrands must be equal too:

$$\frac{\partial}{\partial t}\tilde{a}(\mathbf{r}, t) = -\mathrm{div}\mathbf{J}_A + \sigma_A. \tag{3.37}$$

This is the local balance equation of any extensive quantity A.

Expressing the density of A by the quantity of A per mass element Δm,

$$\tilde{a}(\mathbf{r}, t) = \rho(\mathbf{r}, t)a(\mathbf{r}, t), \tag{3.38}$$

we obtain the general local balance equation for A:

$$\frac{\partial}{\partial t}(\rho a) + \mathrm{div}\mathbf{J}_A = \sigma_A. \tag{3.39}$$

Mass Balance Equation in Local Form

It is assumed that there are no chemical reactions, so there is no mass production density: $\sigma_{m_k} = 0$. We define ρ_k as the mass density of molecules of type k, $\mathbf{J}_{m_k}$ as the mass current density of molecules of type k, and $\mathbf{v}_k(\mathbf{r}, t)$ as the velocity (field) of molecules of type k, defined by $\mathbf{J}_{m_k} \equiv \rho_k \mathbf{v}_k$.
The general mass balance equation (where we have to put $A = m_k$ and $a = \Delta m_k/\Delta m$) then becomes

$$\frac{\partial}{\partial t}\rho_k(\mathbf{r}, t) + \mathrm{div}\mathbf{J}_{m_k} \equiv \frac{\partial}{\partial t}\rho_k(\mathbf{r}, t) + \mathrm{div}\rho_k \mathbf{v}_k = 0. \tag{3.40}$$

The total mass density is given by

$$\rho = \sum_{k=1}^{K} \rho_k. \tag{3.41}$$

The center-of-mass (barycentric) velocity $\mathbf{v}(\mathbf{r}, t)$ in the nonequilibrium system is defined as

$$\rho \mathbf{v} \equiv \sum_{k=1}^{K} \rho_k \mathbf{v}_k , \tag{3.42}$$

so

$$\mathbf{v}(\mathbf{r}, t) \equiv \sum_{k=1}^{K} \frac{\rho_k}{\rho} \mathbf{v}_k = \sum_{k=1}^{K} c_k \mathbf{v}_k. \tag{3.43}$$

Summation of (3.40) over all k yields the local balance equation for total mass, which expresses mass conservation:

$$\frac{\partial}{\partial t} \rho(\mathbf{r}, t) + \mathrm{div}\rho\mathbf{v} = 0. \tag{3.44}$$

General Balance Equation in Substantial Form

Often one is interested in the change with time of a quantity $a(\mathbf{r}, t)$ within a volume element that moves along with the center-of-mass velocity $\mathbf{v}(\mathbf{r}, t)$. This change with time is given by the substantial time derivative

$$\frac{\mathrm{d}}{\mathrm{d}t} \equiv \frac{\partial}{\partial t} + (\mathbf{v} \cdot \nabla) = \frac{\partial}{\partial t} + v_x \frac{\partial}{\partial x} + v_y \frac{\partial}{\partial y} + v_z \frac{\partial}{\partial z}. \tag{3.45}$$

The total time derivative $\frac{\mathrm{d}}{\mathrm{d}t}$ is the change with time noticed by an observer who "sits" in the volume element that moves with velocity $\mathbf{v}(\mathbf{r}, t)$. The partial time derivative $\frac{\partial}{\partial t}$ is the change with time noticed by an observer who sits at a fixed point in space. $(\mathbf{v} \cdot \nabla) = v_x \frac{\partial}{\partial x} + v_y \frac{\partial}{\partial y} + v_z \frac{\partial}{\partial z}$ is the change with time caused solely by the flow with $\mathbf{v}(\mathbf{r}, t)$.

We want to rewrite the local general balance equation (3.39) in such a way that we obtain the change with time of a that is measured by an observer who sits in the volume element that moves with $\mathbf{v}$. For this purpose we consider the following mathematical recasting according to the rules of vector analysis:

$$\frac{\partial}{\partial t}(\rho a) + \mathrm{div}(a\rho\mathbf{v}) = a\frac{\partial\rho}{\partial t} + \rho\frac{\partial a}{\partial t} + a\nabla(\rho\mathbf{v}) + \rho\mathbf{v}(\nabla a)$$

$$= a\left[\frac{\partial\rho}{\partial t} + \mathrm{div}\rho\mathbf{v}\right] + \rho\left[\frac{\partial a}{\partial t} + \mathbf{v}(\nabla a)\right] = \rho\frac{\mathrm{d}a}{\mathrm{d}t}. \tag{3.46}$$

In the second line of (3.46) the first bracket vanishes because of the mass-conservation equation (3.44), and the second bracket is changed into the substantial derivative of a according to definition (3.45). Observing the local balance equation (3.39), we can write (3.46) as

$$\rho\frac{da}{dt} = \sigma_A - \mathrm{div}\mathbf{J}_A + \mathrm{div}(a\rho\mathbf{v}). \tag{3.47}$$

We define

$$\mathbf{j}_A \equiv \mathbf{J}_A - a\rho\mathbf{v}. \tag{3.48}$$

This is the conductive current density of A. It gives the quantity of A that flows per unit time through the surface unit of the volume element dV that moves with the center-of-mass velocity $\mathbf{v}$. Furthermore, $a\rho\mathbf{v}$ is the convective current density of A; it gives the quantity of A that is transported through a *fixed* cross section by the total mass flow.

With these definitions, (3.47) becomes the general balance equation in *substantial* form:

$$\rho\frac{da}{dt} + \mathrm{div}\mathbf{j}_A = \sigma_A. \tag{3.49}$$

Mass Balance Equation in Substantial Form

The diffusion current density $\mathbf{j}_k$, which is the conductive current density of mass m_k, is defined as

$$\mathbf{j}_k \equiv \rho_k(\mathbf{v}_k - \mathbf{v}). \tag{3.50}$$

Identifying $\mathbf{j}_k$ with $\mathbf{j}_A$ and $\mathbf{J}_{m_k} = \rho_k\mathbf{v}_k$ with $\mathbf{J}_A$ in (3.48), we must identify $\rho_k\mathbf{v}$ with $a\rho\mathbf{v}$. Thus, in this special case of $A = m_k$ we have $\rho_k = a\rho$, so

$$a = \frac{\rho_k}{\rho} = c_k, \tag{3.51}$$

and the mass balance equation in substantial form becomes (with $\sigma_{m_k} = 0$)

$$\rho\frac{dc_k}{dt} + \mathrm{div}\mathbf{j}_k = 0. \tag{3.52}$$

According to (3.43)

$$\mathbf{v}(\mathbf{r}, t) = \sum_{k=1}^{K} c_k\mathbf{v}_k. \tag{3.53}$$

Furthermore, $\sum_{k=1}^{K} \rho_k = \rho$. Thus,

$$\sum_{k=1}^{K}\mathbf{j}_k = \sum_{k=1}^{K}\rho_k\mathbf{v}_k - \mathbf{v}\sum_{k=1}^{K}\rho_k = \rho\left[\sum_{k=1}^{K}c_k\mathbf{v}_k - \mathbf{v}\right] = 0. \tag{3.54}$$

The sum of all diffusion current densities vanishes.

Momentum Balance Equation

Let the many-particle system be an "ideal liquid," that is, a liquid or gas where one may disregard internal friction. The pressure is p, and there are external forces which result in

$$\text{specific forces } \mathbf{f}_k$$

per unit mass of particle component k. Then one can show from Euler's equation for ideal liquids that the momentum balance equation in substantial form is [51]

$$\rho \frac{d\mathbf{v}}{dt} + \nabla p = \sum_{k=1}^{K} \rho_k \mathbf{f}_k. \tag{3.55}$$

Energy Balance Equation

External force fields with specific forces $\mathbf{f}_k$ are not considered as part of the system. Their contributions to the locally conserved *specific energy* e are separately taken into account by an energy source term σ_e, which consists of the sum of all power densities of the system components k:

$$\sigma_e = \sum_{k=1}^{K} \rho_k \mathbf{f}_k \cdot \mathbf{v}_k. \tag{3.56}$$

According to the general balance equation in substantial form (3.49), the substantial energy balance equation is

$$\rho \frac{de}{dt} + \text{div} \mathbf{j}_e = \sum_{k=1}^{K} \rho_k \mathbf{f}_k \cdot \mathbf{v}_k. \tag{3.57}$$

$\mathbf{j}_e \equiv \mathbf{J}_e - e\rho\mathbf{v}$ is the conductive current density of the specific energy e, where $\mathbf{J}_e$, the current density of e, satisfies the general local balance equation (3.39) specified to e:

$$\frac{\partial}{\partial t}(\rho e) + \text{div} \mathbf{J}_e = \sigma_e. \tag{3.58}$$

By definition, the specific internal energy u of the system is the difference between the specific energy e and the specific kinetic energy $\frac{1}{\Delta m}(\Delta m \mathbf{v}^2/2) = \mathbf{v}^2/2$:

$$u \equiv e - \frac{1}{2}\mathbf{v}^2. \tag{3.59}$$

With the help of the momentum balance equation (3.55), and after some calculations, one obtains the substantial balance equation for internal energy u as

$$\rho \frac{du}{dt} + \mathrm{div} \mathbf{j}_Q = -p \, \mathrm{div} \, \mathbf{v} + \sum_{k=1}^{K} \mathbf{j}_k \cdot \mathbf{f}_k, \tag{3.60}$$

where

$$\mathbf{j}_Q \equiv \mathbf{j}_e - \mathbf{v} p \tag{3.61}$$

is the conductive current density of internal energy u; this is the heat current density.

Entropy Balance Equation

We could write down the entropy balance equations immediately by inserting the appropriate entropic terms into the general local and substantial balance equations (3.39) and (3.49). However, we prefer to derive the local entropy balance equation explicitly, repeating the reasoning that led to the general local balance equation (3.39). It is hoped is that this makes patent the unrestricted validity of the second law of thermodynamics, and its general form as well, thus clarifying things also for those people who argue that the second law of thermodynamics holds only for closed systems and is not applicable to open systems such as Earth and its economies.[19]

The system considered has the total volume V; O is the surface of that volume and defines the system boundaries. (We may think of a northern industrial country and the atmosphere above its land area.) At a given instant of time t the total system entropy is $S(t)$. During an infinitesimal time interval dt this entropy changes by dS. The change may be due to an exchange $d_a S$ of entropy with the environment (this may be the entropy input from solar radiation during the day or the entropy export into space by infrared radiation, especially at night) and to internal entropy production $d_i S$ because of irreversible processes (occurring, e.g., in the cars, steam turbines, and blast furnaces of the country):

$$dS = d_a S + d_i S. \tag{3.62}$$

(For reversible processes $d_i S$ is zero and $d_a S$ is given by (3.21)). In general, real-life processes $d_a S$ may be positive or negative (e.g., positive on a cloudy day when a warm wind blows from the south and negative during a clear night, when heat is radiated into space). With (3.62), the total change with time of entropy is

$$\frac{dS}{dt} = \frac{d_a S}{dt} + \frac{d_i S}{dt}. \tag{3.63}$$

---

[19]Such statements can be even heard from people with a Ph.D. degree in physics. Furthermore, at an international physics conference on energy and the environment in the early 1990s, an invited speaker from a well-known institution of a country that is a member of the United Nations Security Council gave a talk in which he claimed to have mathematically disproven the second law of thermodynamics. This was printed without any critical comment in the report on the conference, published by a physics journal.

Whether total entropy $S(t)$ increases, decreases, or stays constant depends upon the magnitudes of $d_a S/dt$ and $d_i S/dt$, and the sign of $d_a S/dt$. The *sign* of $d_i S/dt$, however, is known. It is always positive for irreversible processes, i.e., the processes that occur in real life:

$$\frac{d_i S}{dt} > 0. \tag{3.64}$$

Equation (3.64) follows from overwhelming empirical evidence and is the most *general* formulation of the second law of thermodynamics. It holds in all systems, whether they are open or closed. (If the system boundaries are such that neither energy nor matter can cross them, the system is *closed* and $d_a S = 0$. Then entropy increases as long as irreversible processes occur. These processes cease when equilibrium is reached and entropy is maximum.)

The second law of thermodynamics, (3.64), can be cast in a form that shows the relation between entropy production and emissions. For this purpose we define $s(\mathbf{r}, t)$ as mass-specific entropy, $\rho s(\mathbf{r}, t)$ as entropy density, $S = \int_V \rho s dV$, $\mathbf{J}_S(\mathbf{r}, t)$ as entropy current density, and $\sigma_S(\mathbf{r}, t)$ as entropy production density. By definition we have

$$\frac{dS}{dt} = \int_V \frac{\partial \rho s}{\partial t} dV ,$$

$$\frac{d_a S}{dt} = -\int_O \mathbf{J}_S(\mathbf{r}, t) d\mathbf{O}$$

$$= -\int_V \nabla \mathbf{J}_S dV,$$

$$\frac{d_i S}{dt} = \int_V \sigma_S(\mathbf{r}, t) dV.$$

Here

$$\nabla \mathbf{J}_S \equiv \partial J_{Sx}/\partial x + \partial J_{Sy}/\partial y + \partial J_{Sz}/\partial z \quad (\equiv div \mathbf{J}_S).$$

The combination of these definitions with (3.63) results in the entropy balance equation

$$\int_V \left[\frac{\partial \rho s}{\partial t} + \nabla \mathbf{J}_S - \sigma_S \right] dV = 0. \tag{3.65}$$

Since (3.65) is true for any arbitrary volume V, the integrand itself must vanish, and we obtain the *local* entropy balance equation

$$\frac{\partial \rho s}{\partial t} + \nabla \mathbf{J}_S = \sigma_S(\mathbf{r}, t). \tag{3.66}$$

The all-important information added to this equation by the second law of thermodynamics is that the *density* of entropy production is always *positive* for irreversible processes:

$$\sigma_S(\mathbf{r}, t) > 0. \tag{3.67}$$

From the local entropy balance equation (3.66) and the general substantial balance equation (3.47)–(3.49), we obtain the substantial entropy balance equation:

$$\rho \frac{ds}{dt} + \nabla \mathbf{j}_S = \sigma_S(\mathbf{r}, t),\tag{3.68}$$

where

$$\mathbf{j}_S \equiv \mathbf{J}_S - s\rho\mathbf{v}\tag{3.69}$$

is the conductive entropy current density.

Density of Entropy Production

An explicit expression for the density of entropy production σ_S in terms of generalized forces and flows can be derived from Gibbs's fundamental equation (3.30) for local thermodynamic equilibrium and the balance equations for energy and mass.

From (3.30) we obtain an infinitesimal change of specific entropy, ds, as

$$ds = \frac{1}{T}du + \frac{p}{T}dv - \frac{1}{T}\sum_{k=1}^{K}\mu_k dc_k.\tag{3.70}$$

The change with time of specific entropy in a mass element along its trajectory – for instance, CO_2 particles within a cubic micrometer passing through a heat exchanger of a power station – is given by the substantial time derivative

$$\frac{ds}{dt} = \frac{1}{T}\frac{du}{dt} + \frac{p}{T}\frac{dv}{dt} - \frac{1}{T}\sum_{k=1}^{K}\mu_k\frac{dc_k}{dt}.\tag{3.71}$$

This equation is rewritten in such a way that it assumes the form of the substantial entropy balance equation (3.68). In so doing, we use the substantial balance equations (3.60) and (3.52) for internal energy u and concentration c_k. Furthermore, we need the balance equation for the specific volume $v = 1/\rho$. It is obtained from the substantial balance equation for density ρ,

$$\frac{d\rho}{dt} + \rho\,\mathrm{div}\,\mathbf{v} = 0,\tag{3.72}$$

which results from the local mass balance equation (3.44). It is

$$\rho\frac{dv}{dt} - \mathrm{div}\,\mathbf{v} = 0.\tag{3.73}$$

The combination of the three balance equations (3.52), (3.60), and (3.73) with (3.71) multiplied by ρ is evaluated, using mathematical manipulations outlined in [51,52]. This finally yields

$$\rho \frac{ds}{dt} + \text{div}\left(\frac{\mathbf{j}_Q}{T} - \sum_{k=1}^{K} \mu_k \frac{\mathbf{j}_k}{T}\right) = \mathbf{j}_Q \cdot \nabla\frac{1}{T} + \sum_{k=1}^{K} \mathbf{j}_k \cdot \left(-\nabla\frac{\mu_k}{T} + \frac{\mathbf{f}_k}{T}\right). \quad (3.74)$$

Comparison with the substantial entropy balance equation (3.68),

$$\rho \frac{ds}{dt} + \text{div}\mathbf{j}_S = \sigma_S(\mathbf{r}, t), \quad (3.75)$$

suggests the identification of the conductive entropy current density with

$$\mathbf{j}_S = \frac{\mathbf{j}_Q}{T} - \sum_{k=1}^{K} \mu_k \frac{\mathbf{j}_k}{T}, \quad (3.76)$$

and of the entropy production density with[20]

$$\sigma_S = \sigma_{S,\text{dis}} \equiv \mathbf{j}_Q \cdot \nabla\frac{1}{T} + \sum_{k=1}^{K} \mathbf{j}_k \cdot \left(-\nabla\frac{\mu_k}{T} + \frac{\mathbf{f}_k}{T}\right). \quad (3.77)$$

We have obtained the important result that entropy production by irreversible processes in a many-particle system of N different types of molecules is given by the heat current density $\mathbf{j}_Q$, driven by the gradient of temperature T, and diffusion current densities $\mathbf{j}_k$ of the different types of molecules k, $1 \leq k \leq K$, driven by the gradients of chemical potentials μ_k and temperature, as well as by specific external forces $\mathbf{f}_k$. The heat and particle current densities, and the gradients and forces driving them, determine the "dissipative" entropy production density $\sigma_{S,\text{dis}}$. There may also be chemical (and physical) reactions that change the molecular composition of the system. The entropy production density of them, σ_{chem}, is the sum of products of scalar flows and forces which cannot interfere with the vectorial flows $\mathbf{j}_Q$, $\mathbf{j}_k$ and the vectorial forces acting on them [52], so both $\sigma_{S,\text{dis}}$ and σ_{chem} must be positive individually, because of the second law of thermodynamics. (Entropy production by friction-tension may be added to $\sigma_{S,\text{chem}}$.)

In summary, the second law of thermodynamics says that in nonequilibrium systems, where usual thermodynamic variables such as temperature, pressure, internal energy, and entropy can still be defined *locally*, total entropy production density is positive,

$$\sigma_S = \sigma_{S,\text{dis}} + \sigma_{\text{chem}} > 0, \quad (3.78)$$

---

[20]These identifications are mathematically not unique, but they are the ones that satisfy physical criteria such as Galilean invariance.

and its components "dissipative" entropy production density, $\sigma_{S,\text{dis}}$, and entropy production density due to physical and chemical reactions, σ_{chem}, are separately positive as well:

$$\sigma_{S,\text{dis}} = \mathbf{j}_Q \cdot \nabla \frac{1}{T} + \sum_{k=1}^{K} \mathbf{j}_k \cdot \left(-\nabla \frac{\mu_k}{T} + \frac{\mathbf{f}_k}{T} \right) > 0 \text{ and } \sigma_{\text{chem}} > 0. \qquad (3.79)$$

Note that at each point all diffusion current densities cancel according to (3.54) (so total mass is conserved everywhere):

$$\sum_{k=1}^{K} \mathbf{j}_k = 0. \qquad (3.80)$$

Therefore, one can always express one $\mathbf{j}_k$ by the negative sum of the others.

References

1. Bergmann, L., Schäfer, C.: Lehrbuch der Experimentalphysik, Bd. 7, Erde und Planeten, 2nd Edn., (W. Raith ed.). Walter de Gruyter, Berlin (2001)
2. Hamburger Bildungsserver: Ewärmung des Weltozeans 1955–2003. http://lbs.hh.schule.de/welcome.phtml?unten=/klima/klimafolgen/meeresspiegel/sterisch.html
3. Hägele, P., Evers, P.: Freche Verse, physikalisch. Vieweg + Teubner, Wiesbaden (1995)
4. Reif, F.: Fundamentals of statistical and thermal physics. McGraw-Hill, New York (1965)
5. Bardeen, J., Cooper, L.N., Schrieffer, J.R.: Theory of Superconductivity. Phys. Rev. **108**, 1175–1204 (1957)
6. Landau, L. D., Lifschitz, E.M.: Lehrbuch der Theoretischen Physik V, Statistische Physik Teil 1, 6th Edn. Akademie-Verlag, Berlin (1984)
7. Ruelle, D.: Zufall und Chaos. Springer, Berlin, Heidelberg (1992)
8. Stahl, A.: Entropiebilanzen und Rohstoffverbrauch. Naturwissenschaften **83**, 459, 1996
9. Lebowitz, J. L.: Emergent Phenomena—Entropy and phase transitions in macroscopic systems. Physik Journal **6**, 41–46 (2007)
10. Zeh, H.D.: The Direction of Time (2nd Ed.). Springer, Berlin, Heidelberg, New York (1992)
11. Kraus, K.: Über die Richtung der Zeit. Phys. Blätter, **29**, 1–19 (1973); see also: Becker, H.: Theorie der Wärme, pp. 104–106. Springer, Berlin (1975)
12. Kohelet (The Preacher, Ecclesiastes), 1,2–1,11
13. Georgescu-Roegen, N.: Energy and Economic Myths. Pergamon, New York (1976)
14. Georgescu-Roegen, N.: The entropy law and the economic process in retrospect. East. Econ. J. **12**, 3–23 (1986)
15. Letters to the Editor: Recycling of Matter. Ecol. Econ. **9**, 191–196 (1994)
16. Georgescu-Roegen, N.: The Entropy Law and the Economic Process. Harvard University Press, Cambridge (1971)
17. van Gool, W., Bruggink J.J.C. (eds.): Energy and Time in the Economic and Physical Sciences. North-Holland, Amsterdam (1985)
18. Faber, M., Niemes, H., Stephan, G.: Entropy, Environment, and Resources. Springer, Berlin (1987)
19. Faber, M., Proops, J.: Evolution, Time, Production, and the Environment (2nd Ed.). Springer, Berlin (1994)

20. Daly, H. E.: On Nicholas Georgescu-Roegen's contributions to economics: an obituary essay. Ecol. Econ. **13**, 149–154 (1995)
21. Stern Review Report on the Economics of Climate Change, http://www.hm-treasury.gov.uk/ independent_reviews/stern_review_economics_climate_change/stern_review_report.cfm
22. Schönwiese, C.-D.: Klimatologie. Ulmer UTB, 3rd Edn., Stuttgart (2008)
23. Schönwiese, C.-D., Rapp, J., Meyhöfer, S., Denhard, M., Beine, S.: Das "Treibhaus"–Problem: Emissionen und Klimaeffekte. Eine aktuelle wissenschaftliche Bestandsaufnahme. Berichte des Instituts für Meteorologie und Geophysik der Universität Frankfurt. No. 96, (1994)
24. Schönwiese, C.-D.: Klimatologie. Ulmer UTB, 2nd Edn., Stuttgart (2003)
25. Schönwiese, C.-D., Walter, A., Brinckmann, S.: Statistical assessments of anthropogenic and natural global climate forcing. An update. Meteorol. Z. **19** (1), 003–010 (2010)
26. Science 7 May 2010, Vol. 328. no 5979, pp. 689-690; http://www.sciencemag.org/cgi/content/full/328/5979/689
27. German Bundestag (Ed.): Protecting the Earth's Atmosphere. An International Challenge. Deutscher Bundestag, Referat Öffentlichkeitsarbeit, Bonn (1989)
28. Bundesanstalt für Geowissenschaften und Rohstoffe: Emissionsdatenbank, http://www.bgr. bund.de/cln_006/nn_333908/DE/Themen/Klimaentwicklung/Bilder/co2_emiss1850_2000_g. html
29. Mauch, W.: Kumulierter Energieaufwand—Instrument für nachhaltige Energieversorgung. Forschungsstelle-für-Energiewirtschaft Schriftenreihe, Band 23 (1999); combined with data from Öko-Institut Darmstadt, 2006.
30. Brohan, P., Kennedy, J.J., Harris, I., Tett, S.F.B., Jones, P.D.: Uncertainty estimates in regional and global observed temperature changes: a new dataset from 1850. J. Geophys. Res. **111**, D12106 (2006) doi: 10.1029/2005/JD006548 with updates (until 2009) from http://www.cru. uea.ac.uk/cru/data/temperature/
31. Dritter Bericht der Enquete Kommission Vorsorge zum Schutz der Erdatmosphäre, p. 855. Deutscher Bundestag, Drucksache 11/8030, Bonn (1990)
32. International Economic Platform for Renewable Energies (IWR), Press release of 10 August 2009; http://www.renewable-energy-industry.com/business/press-releases/newsdetail. php?changeLang=de_DE&newsid=3185
33. Tollefson, J.: Missed 2050 climate targets will reduce long-term options. Nature, 11 January 2010, doi:10.1038/news.2010.6
34. Kenney, W. F.: Energy Conservation in the Process Industries. Academic Press, Orlando (1984)
35. Groscurth, H.-M., Kümmel, R.: Thermoeconomics and CO_2-Emissions. Energy—Intntl. J. **15**, 73–80 (1990)
36. Blok, K.: Introduction to Energy Analysis. Techne Press, Amsterdam (2006)
37. Schüssler, U., Kümmel, R.: Schadstoff-Wärmeäquivalente. ENERGIE, **42**, 40–49 (1990)
38. Kümmel, R., Schüssler, U.: Heat equivalents of noxious substances: a pollution indicator for environmental accounting. Ecol. Econ. **3**, 139–156 (1991)
39. von Buttlar, H.: Umweltprobleme. Phys. Blätter **31**, 145–155 (1975)
40. Kümmel, R., Schüssler, U.: Valuation of Environmental Cost by Heat Emissions from Pollution Control. In: Hohmeyer, O., Ottinger, R.L. (eds.) External Environmental Costs of Electric Power, pp. 147–158. Springer, Berlin (1991). (The misprints that sneaked into this article during final production are absent in [38].)
41. German Physical Society and German Meteorological Society: The threat of man made global changes in climate. Phys. Blätter **43**, 347–349 (1989).
42. Steinberg, M., Cheng, H.C., Horn, F.: A system study for the removal, recovery and disposal of carbon dioxide from fossil fuel power plants in the US. BNL-35666 Informal Report, Brookhaven National Laboratory, Upton (1984)
43. Fricke, J., Schüssler, U., Kümmel, R.: CO_2–Entsorgung. Phys. Unserer Zeit **20**, 56–81 (1989), and references therein.
44. Hendriks, C. A., Blok, K., Turkenburg, W.C.: The Recovery of Carbon Dioxide from Power Plants. In: Okken, P.A., Swart, R.J., Zwerver, S. (eds.) Climate and Energy, pp. 125–142. Kluwer, Dordrecht (1989)
45. Okken, P.A., Swart, R.J., Zwerver, S. (eds.): Climate and Energy. Kluwer, Dordrecht (1989)

46. Kümmel, R., Groscurth, H.-M., Schüssler, U.: Thermoeconomic Analysis of Technical Green-house Warming Mitigation. Int. J. Hydrogen Energy **17**, 293–298 (1992), and references therein.
47. Schüssler, U., Kümmel, R.: Carbon Dioxide Removal from Fossil Fuel Power Plants by Re-frigeration under Pressure. In: Jackson, W.D. (ed.) Proc. 24th Intersociety Energy Conversion Engineering Conference, pp. 1789–1794. IEEE, New York (1989)
48. http://www.vattenfall.de/www/vf/vf_de/225583xberx/228407klima/228587co2-f/index.jsp
49. Kolm, H.: Mass driver up-date. L-5 News **5**, 10–12 (1980).
50. O'Neill, G. K.: The High Frontier. William Morrow & Co., New York (1977)
51. Kluge, G., Neugebauer, G: Grundlagen der Thermodynamik. Spektrum Fachverlag, Heidelberg (1993)
52. Kammer, H.-W., Schwabe, K.: Thermodynamik irreversibler Prozesse. Physik-Verlag Weinheim (1985)

Chapter 4
Economy

How long will researchers working in adjoining fields ...
abstain from expressing serious concern about the splendid
isolation in which academic economics now finds itself?

Wassily Leontief, 1982 [1]

4.1 Complementary Perspectives on the Economy

Economic policies influence our lives profoundly. To be sound, they should be based on a thorough understanding of the economic system. Consisting of people, machines, and natural resources, the economy is arguably one of the most complex systems that science tries to understand.

The behavior of humans is the subject of the social sciences. Its modeling is influenced by the modeler's philosophical view of people, available knowledge of behavioral patterns, and quantitative methods at hand. Economics, as far as it understands itself exclusively as a social science, has concentrated on that. The energy–matter world, which is also part of the economic system, has been given little, if any, attention. Making good on that, and thus complementing the social-science perspective with that of the natural sciences, is the purpose of this chapter and, in fact, of this book. It is part of the effort of physicists in the growing econophysics community to contribute to a better understanding of the economy.

The natural science method of gaining knowledge is observation of the world, systematic ordering of the information obtained, and the building of models that quantitatively describe those parts of the world one is interested in. It is hoped that, like pieces of a puzzle, the partial descriptions will join in forming a complete picture, eventually. The models must be free from internal contradictions, be open to falsification by quantitative comparison of their results with empirical facts, and allow predictions under well-defined assumptions. Let us start with observation.

An extraterrestrial observer who had watched the evolution of life and human civilization unfold on Earth during the last four billion years would have noted during all that time no physical inputs into the biosphere other than solar energy, cosmic radiation, rocks from outer space, emissions from volcanic eruptions, and gravitational influences from celestial bodies. He would have seen in detail how life thrives on the energy from the Sun, whereas ionizing particles, meteorites, and volcanic emissions cause mutations and extinctions. The observer would conclude

R. Kümmel, *The Second Law of Economics: Energy, Entropy, and the Origins of Wealth*,
The Frontiers Collection, DOI 10.1007/978-1-4419-9365-6_4,
© Springer Science+Business Media, LLC 2011

that varying energy flows and energetic interactions are the drivers of evolution. Historians who record technological and economic development since the Industrial Revolution [2] note the pivotal role of heat engines. To understand heat engines, thermodynamics, the science of energy conversion, has been developed. It reigns in those parts of the economy where wealth is produced by machines. Therefore, its insights should be incorporated into economic theory. This requires mathematics.

Mathematics, however, has become controversial in economics. An example is a press release[1] of October 23, 2009, titled: "The Financial Crisis: How Economists Went Astray. Two Nobel Laureates and over 2000 Signatories Uphold That Economists Have Mistaken Mathematical Beauty for Economic Truth." The signatories signed a Web petition in support of an article[2] by the Nobel laureate Paul Krugman, saying: "Few economists saw our current crisis coming, but this predictive failure was the least of the field's problems. More important was the profession's blindness to the very possibility of catastrophic failures in a market economy ... the economics profession went astray because economists, as a group, mistook beauty, clad in impressive-looking mathematics, for truth ...."

Krugman criticizes the mathematical models of human behavior that are used by mainstream economists as the basis for recommendations on deregulation, taxation, and the like. Even without considering their failures as evidence for deficiencies, one may doubt whether economic models have come up so far to the complexity of the system of "humans interacting on markets in a technical world with limited natural resources." The interacting many-particle systems of physics are much simpler. And yet the statistical and field-theoretical apparatus required for their description is considerably more sophisticated and involved than any mathematics used in economic theory.

Then, if Krugman's criticism is legitimate, can we expect that mathematics will make any sense at all in economics? The answer is simple: We do not intend to model complex human behavior. We only examine the traditional mathematical modeling of standard behavioral assumptions – maximization of profit and overall welfare – and point out that even these plain models are flawed, because they disregard technological constraints on the freedom of choice between production factors. This opens the gate for a descent to the productive physical basis of modern industrial economies, which is subject to the laws of physics. Here mathematical methods that have proven successful in thermodynamics are appropriate for the description of how energy conversion contributes to the generation of wealth.

The wealth of nations is traditionally measured by the gross domestic product (GDP). This is the monetary value added of all goods and services produced in a country in 1 year (whereas the value of imported intermediate goods and services is not included in the GDP). The growth of GDP is the aim of economic policy everywhere. There has been some criticism of considering GDP also as a measure of

---

[1]From Geoffrey M. Hodgson, The Business School, University of Hertfordshire, Hatfield, Hertfordshire AL10 9AB, UK; http://www.geoffrey-hodgson.info.

[2]New York Times, September 2, 2009.

the quality of life. It does not include unpaid housekeeping, upbringing of children, and voluntary social activities. On the other hand, GDP in an industrial country comprises all services and goods required to mitigate the effects of traffic accidents, or cure diseases from pollution, whereas in a tropical island republic it does not include the joy of moving leisurely in a warm breeze among lush flowers and colorful fish. Nevertheless, people like to live in countries where a high GDP per capita indicates that the average individual commands a rich consumer basket of material goods and services. The increasing migration from the warm "South" of low GDP per capita to the cold "North" of high GDP per capita demonstrates this. People agree in general that the "North" is better off than the "South," because it uses technology and natural resources more intensively for the production of material wealth. It is important to understand in detail how this happens, and how the basic conditions for future wealth creation may change.

The program of complementing the social-science view of the economy by that of thermodynamics is as follows. We first confront the paradigm of perpetual exponential growth of GDP with the facts on energy conversion and entropy production. Then we derive the conditions for economic equilibrium if factor combinations are constrained technologically. It results that both profit and overall welfare optimization no longer support a basic theorem of orthodox economics, namely, that a production factor's economic weight is equal to its cost share. Being invalid, this theorem is useless for the construction of production functions which describe the evolution of GDP, or parts thereof. As a remedy, the capital–labor–energy–creativity (KLEC) model [3] is developed. It is depicted in Fig. 4.1. This model considers work performance and information processing by the production factors capital, labor, and energy as the fundamental processes of wealth creation, whereas the space available for the evolution of the production system provides natural resources and establishes restrictions. The working of human creativity via ideas, inventions, and value decisions is coupled to the arrow of time. The model estimates the importance of the productions factors for the generation of value added on the basis of standard empirical data. Thus, the theoretical findings can be checked by comparison with experience.

The sections in this chapter contain material from [4].

4.2 Energy, Entropy, and the Growth Paradigm

The Nobel laureate in economics in 1972, Paul A. Samuelson, answers the question "Why study economics?" in his textbook *Economics* [5] (pp. 14–15), with the following quote from *The General Theory of Employment, Interest and Money* by Lord Keynes: "The ideas of economists and political philosophers, both when they are right and when they are wrong, are more powerful than it is commonly understood. Indeed the world is ruled by little else. Practical men, who believe themselves to be quite exempt from any intellectual influences, are usually the slaves of some defunct economist. Madmen in authority, who hear voices in the air, are

Fig. 4.1 The capital–labor–
energy–creativity (KLEC)
model of wealth production
in the physical basis of the
economy [3]

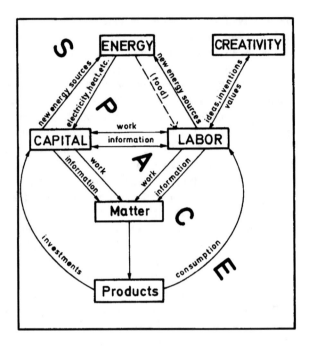

destilling their frenzy from some academic scribbler of a few years back. I am sure
that the power of vested interests is vastly exaggerated compared with the gradual
encroachment of ideas. Not, indeed, immediately, but after a certain interval; for in
the field of economic and political philosophy there are not many who are influenced
by new theories after they are 25–30 years of age, so that the ideas which civil
servants and politicians and even agitators apply to current events are not likely to
be the newest. But, soon or late, it is ideas, not vested interests, which are dangerous
for good or evil."

The necessity of continuous exponential economic growth is such an idea. The
"Stern Review Report on the Economics of Climate Change" assumes an annual
global growth rate of 1.9% for 200 years into the future [6, 7]. Consequently,
this report, published by the British government [6], explicitly states (on p. 160):
"…even with climate change the world will be richer in the future as a result of
economic growth." On the other hand, the theory of economic growth has been
a frontier of research for quite some time without having been completed so far,
despite recent attempts to formulate an "endogeneous" growth theory [8–10].

Politicians in OECD countries aspire after annual growth rates not below 3%,
because under the present framework of the market, only sufficiently high growth
rates can provide new jobs to replace the ones that have fallen prey to increasing
automation and globalization. The developing countries strive for even higher rates
of economic growth in order to catch up with the standard of living in the highly

industrialized countries. Furthermore, strong economic growth shall reduce the growth of state indebtedness, which is a major problem these days.

Unfortunately, perpetual exponential economic growth, desirable as it may be from a social point of view, collides with the energetic and environmental restrictions established in a finite world by the first two laws of thermodynamics, according to which nothing happens in the world without energy conversion and entropy production. The first warning of such a collision was given in 1972 with the publication of *The Limits to Growth* [11]. Although this report commissioned by the Club of Rome shocked natural scientists and engineers into a new awareness of the relevance of thermodynamics for economic and environmental stability, most economists condemned it fervently. One of the principal objections was that technological progress, which has seemingly blasted all limits to growth since Thomas R. Malthus predicted doom at the beginning of the nineteenth century, has not been taken into account sufficiently. Apparently, mainstream economists have nearly unlimited confidence in the creativity of scientists and engineers. But none of those who actually advance technological progress will ever try to beat the first two laws of thermodynamics. The presently increasing worries about "Peak Oil" [12, 13] and climate change indicate that the problems pointed out by *The Limits to Growth* may become virulent earlier than expected, and that the hope that the Club of Rome report would fulfill its purpose of serving as a self-destroying prophecy may have been in vain.

There are many formulations of the first law and the second law of thermodynamics. Here we use the ones from Chaps. 2 and 3, which elucidate their economic and environmental importance.[3]

Quantitatively, the first law of thermodynamics states:

> Energy, including the energy equivalent of mass, is a conserved quantity. It consists of valuable *exergy* and useless *anergy*.

Useful physical work can only be obtained from exergy conversion. The principal carriers of primary energy – the fossil and nuclear fuels, and solar radiation as well – are practically 100% exergy. Heat at the temperature of the environment is an example of anergy. Later, in the context of the second law of thermodynamics, we will see that in all energy conversion processes useless anergy increases at the expense of valuable exergy. This is what is meant by "energy consumption."

From an engineering point of view, it is obvious that there is no industrial production without the performance of physical work on matter in conjunction with information processing. Therefore, energy conversion, or more precisely, exergy

---

[3]It should also be obvious from these formulations that the first law and the second law of thermodynamics apply to *all* systems, no matter,whether they are closed or open.

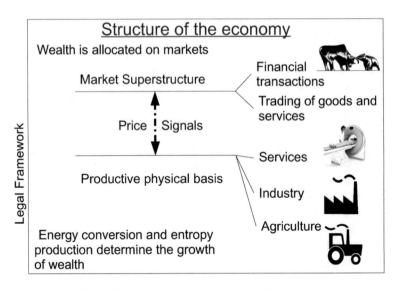

Fig. 4.2 Productive physical basis and market superstructure of the economy

conversion, is a prime mover of modern industrial economies. But, as discussed later, mainstream economic theory has quite a different view on energy.

An economic system – a national market economy or a subsector of it – consists, roughly speaking, of a physical basis that produces goods and services, and a market superstructure, where economic actors trade the products of the basis. Price signals from supply and demand provide the feedback between the productive physical basis and the market superstructure. The scheme of the economy is sketched in Fig. 4.2.

Whereas in bygone "socialist" economies the market was replaced by the planning bureaucracy, in recent decades capitalist economies have added a financial superstructure, which does not exchange real goods or services, but rather exchanges more and more complex derivatives, starting with equity shares and debt instruments, "rights" to buy or sell in the future, insurance policies, and increasingly sophisticated and nontransparent packages of these things such as mortgage-backed securities, collateralized debt obligations, and credit default swaps. The present financial crisis that started with the burst of the US mortgage bubble in 2007, and which has thrown the world into the worst recession since the 1929 crash, indicates that this additional financial superstructure is a permanently endangered, unstable house of cards [14].

Since economics understands itself as a social science, in fact as the queen of social sciences [5], it is mainly concerned with the behavior of the actors in the market superstructure, where everybody is striving for wealth. The main focus of economic research has been based so far on observations whose summary may be called the "first law of economics": "Wealth is allocated on markets, and the legal framework determines the outcome." In this chapter the focus is widened so that it includes the engineering mechanisms of production in the realm of the productive physical basis. This will lead to the second law of economics: "Energy conversion and entropy production determine the growth of wealth."

Three production factors, also called "inputs," are busy in the productive physical basis of an industrial economy:

1. Energy-converting devices and information processors together with all buildings and installations necessary for their protection and operation. They represent the production factor capital K.
2. The capital stock K is manipulated and supervised by people, who constitute the production factor labor L.
3. The machines of the capital stock are activated by energy (more precisely exergy), which is the production factor E.

As a rule, the professional qualification of labor has to increase with the energy flows through the capital stock it controls.

The energy-converting and information-processing machines with their energetic inputs are open thermodynamic systems, subject to the laws of nature when they produce the output Y. The output is the sum of all goods and services produced within an economic system. Its measure is the GDP, or parts thereof. The natural environment, in which all economic systems are embedded, serves as the reservoir of temperature and pressure for the heat engines, transistors, and all other energy conversion devices of the capital stock. It also contains the energetic and material resources.

In preindustrial agrarian economies the productive basis was the photosynthetic collection of solar energy by plants, which provided food, fuel, and timber. Since plants grow on land, land has been considered as a basic factor of production. In feudal agrarian societies it gave economic and political power to its owners. But this power was due to photosynthesis and the people and animals tilling the soil. The production site itself is not a production factor. Sure, nutrients in the soil and climatic conditions matter, but modern hydrocultures show that food can be grown in all places and spaces where energy, minerals, and water are available. Fortunate land owners may become rich because of valuable energetic or mineral deposits in the ground below their land. But their land is just the passive top of a container, so to speak. It is not an active factor that performs work or processes information. Therefore, "land," or its three-dimensional extension "space," is rather part of natural restrictions. The areas of agricultural and industrial production sites, and the tops of nature's treasure chests, are limited by the surface of Earth. Furthermore, the quantity of tolerable emissions of pollutants is limited by the absorption capacity of the biosphere, which encompasses the land.

This takes us to the second law of thermodynamics, which is behind the collateral damage caused by production processes in the physical basis. It summarizes all experiences that people have accumulated from processes of change in the energy–matter world. It states that all real-life processes, which are irreversible, produce entropy. Therefore, the density $\sigma_S(\mathbf{r}, t)$ of the entropy that is produced per unit time at point $\mathbf{r}$ and at time t by an irreversible process is inevitably larger than zero.

Quantitative details of entropy production were derived in the Appendix of in Chap. 3. Considering nonequilibrium systems containing N different sorts of molecules k that are *locally* in thermodynamic equilibrium, and which do not undergo chemical reactions,[4] one finds that the entropy production density σ_S is equal to the "dissipative" entropy production density $\sigma_{S,\mathrm{dis}}$ [16]:

$$\sigma_S(\mathbf{r}, t) = \sigma_{S,\mathrm{dis}}(\mathbf{r}, t) = \mathbf{j}_Q \nabla \frac{1}{T} + \sum_{k=1}^{N} \mathbf{j}_k \left[-\nabla \frac{\mu_k}{T} + \frac{\mathbf{f}_k}{T} \right] > 0. \qquad (4.1)$$

In words: Positive entropy production density at the space–time point $(\mathbf{r}, t)$ consists of the heat current density $\mathbf{j}_Q$, driven by the gradient (∇) of temperature T, and diffusion current densities $\mathbf{j}_k$, driven by gradients of chemical potentials μ_k divided by T and by specific external forces $\mathbf{f}_k$. Thus, (4.1) describes entropy production in the macroscopically small and microscopically large volume elements of many-particle nonequilibrium systems with mass and energy flows.

The irreversible industrial production processes generate emissions of heat and particles according to (4.1). Once the heat emitted has assumed the temperature of the environment, it is all useless anergy. Then, part of the exergy contained in the energy that drives the production process has been consumed. Furthermore, the emissions of particles and heat changes the composition of and the energy flows through the biosphere to which the living species and their populations have adapted in the course of evolution. If these changes are so big that they cannot be balanced by the biological and anorganic processes that are driven by the exergy input from the Sun and the radiation of heat into space, and if they occur so rapidly that biological, social, and technological adaptation deficits develop, the emissions are perceived as environmental pollution. (CO_2 is a typical example of how quantities and emission rates of a sort of molecule determine its environmental impact. Practically nobody worried about CO_2 emissions before the 1970s – except Svante Arrhenius – and now they are feared as a driver of climate change.) As long as heat emissions are considered as environmentally more benign than particle emissions, one can transform the latter into the former by appropriate technologies such as desulfurization, denitrification, and perhaps, carbon (dioxide) capture and storage. The heat equivalents of noxious substances (HEONS), described in Sect. 3.6.3, are a measure of the exergy consumed in pollution abatement and the resulting additional

---

[4]If chemical reactions occur too, there is also the entropy production density $\sigma_{S,\mathrm{chem}}$. Then, the total entropy production density consists of two components, $\sigma_{S,\mathrm{dis}}$ and $\sigma_{S,\mathrm{chem}}$, where the latter involves products of scalar currents and forces, which cannot interfere with the vectorial products in $\sigma_{S,\mathrm{dis}}$ in (4.1). Thus, both $\sigma_{S,\mathrm{dis}}$ and $\sigma_{S,\mathrm{chem}}$ are positive individually [15].

heat burden on the environment. They have been calculated for a coal power plant in percentages of the primary energy consumed by such a power plant of the same electricity output without pollution abatement. They amount to less than 5% for SO_2 and NO_x abatement and about 39% for carbon (dioxide) capture and storage [17]. Should, however, present global waste heat emissions of about 1.4×10^{14} W increase by a factor of 20 so that they approach the so-called heat barrier of 3×10^{14} W – which is roughly 0.2% of the power the Earth receives from the Sun – global climate changes are to be expected even without the anthropogenic greenhouse effect [19]. This is inferred from observed local climate changes in areas where heat emissions reach a few per mill of solar insolation. In this sense the second law of thermodynamics draws the ultimate limits to growth on Earth.

Our economies face a threefold challenge: (1) prepare for declining oil reserves; (2) reduce carbon dioxide emissions drastically – an 80% reduction by 2050 is recommended for the highly industrialized countries [6, 18]; and (3) invest heavily in non-fossil-fuel technologies and energy conservation. Energy conservation should have high priority, as it also minimizes emissions for unchanged energy services. Assessments of how this will affect economic growth require economic methods and models that correctly incorporate the physical laws on energy and entropy [20].

There are economists who realized that energy and entropy matter in economics. For instance, as early as 1927, Tryon [21] stated: "Anything as important in industrial life as power deserves more attention than it has yet received from economists.... A theory of production that will really explain how wealth is produced must analyze the contribution of the element energy." And in 1974, Binswanger and Ledergerber [22] flatly declared: "The decisive mistake of traditional economics... is the disregard of energy as a factor of production." The economic importance of entropy was highlighted in 1971 by Georgescu-Roegen's book *The Entropy Law and the Economic Process* [23]. (Unfortunately, Georgescu-Roegen created a bit of confusion later by postulating the existence of a fourth law of thermodynamics that would describe the dissipation of matter. But the second law of thermodynamics already takes care of that, as (4.1) shows.) Since the 1972 publication of *The Limits to Growth* and the first oil-price shock in 1973–1975, a growing number of natural scientists and economists have ventured into interdisciplinary research on the relevance of energy conversion and entropy production for economic evolution [3, 20, 24–40].

On the other hand, mainstream economics is unconcerned about the first law and the second law of thermodynamics, despite their governing all energetic and material processes of wealth production. Of course, one realizes the problem of pollution and climate change [6], but does not really believe in limits to growth in finite systems such as planet Earth. For instance, when discussing resource problems, the Nobel laureate in economics Robert A. Solow stated: "The world can, in effect, get along without natural resources." However, he noted that "if real output per unit of resources is effectively bounded – cannot exceed some upper limit of productivity which in turn is not far from where we are now – then catastrophe is unavoidable" [41]. Since entropy production decreases exergy, the output of goods and services per energy unit *is* bounded. Whether catastrophe is unavoidable depends on how society meets the economic and environmental challenges that rise from the laws of physics. The option of expanding the economic system beyond the biosphere, e.g., by solar power satellites and space industrialization, as discussed in [42–44] and Sect. 2.6.3, is taken seriously by few economists only.

When it comes to the question of how wealth is produced, standard production theory usually takes only capital K and labor L into account. The focus widened somewhat when the oil price shocks of 1973–1975 and 1979–1981, shown in Fig. 4.7, and the accompanying recessions, known as the first and the second energy crises, prompted investigations that considered energy E, sometimes in combination with materials M, as an additional factor of production [45–52].

But in most (*KLE* or *KLEM*) models energy was given a tiny economic weight only. The marginal role attributed to energy in standard economic theory was very clearly described by the econometrician Denison. In a controversial discussion on whether the first oil price explosion in 1973–1975 could have been related to the simultaneous worldwide recession, he argued: "Energy gets about 5 percent of the total input weight in the business sector ... the value of primary energy used by nonresidential business can be put at $ 42 billion in 1975, which was 4.6 percent of a $ 916 billion nonresidential business national income. ... If ... the weight of energy is 5 percent, a 1-percent reduction in energy consumption with no change in labor and capital would reduce output by 0.05 percent" [53]. As said above, "output" means GDP, or parts thereof, if subsectors of a national economy are considered, and "input" means production factor. "Input weight" is synonymous with "output elasticity." The *output elasticity* of a production factor measures the *productive power* of the factor in the sense that (roughly speaking) it gives the percentage of output change when the factor changes by 1%, while the other factors stay constant. The exact definition of output elasticities is given by (4.3).

The quoted input weight corresponds to the cost share of energy in total factor cost, which is roughly 5% on an OECD average. The cost share of capital has been about 25% and that of labor roughly 70% during recent decades. For reasons outlined in Sect. 4.4, orthodox economics considers the output elasticity of a production factor as equal to its cost share. We call this the "cost-share theorem." On the basis of the cost-share theorem, orthodox economics makes a Legendre transformation, given by (4.10), from factor quantities to factor prices. This may have been seen as a justification for restricting growth and production analyses to the market superstructure without observing the thermodynamic laws that reign in the physical basis.

According to the cost-share theorem, reductions of energy inputs by up to 7%, observed during the first energy crisis in 1973–1975, could have only caused output reductions of 0.35%, whereas the observed reductions of output in industrial economies were up to an order of magnitude larger. Thus, from this perspective, the recessions of the energy crises are hard to understand. In addition, cost-share weighting of production factors has the problem of the Solow residual. The Solow residual accounts for that part of output growth that cannot be explained by the input growth rates weighted by the factor cost shares. It amounts to more than 50% of total growth in many cases. Standard neoclassical economics attributes this difference formally to what is being called "technological progress" or, sometimes, "manna from Heaven." Learned papers on the measurement of technological progress abound. In empirical studies this progress is usually modeled by a time-dependent exponential function that multiplies a (Cobb–Douglas) production function [58]. The dominating role of technological progress "has led to a criticism of the neoclassical model: it is a theory of growth that leaves the main factor in economic growth unexplained" [59], as the founder of neoclassical growth theory, Robert A. Solow, admitted himself.[5] Recent endogeneous growth theories [10] and quantitative economic climate-change assessments [61] also employ cost-share weighting of production factors.

The cost-share theorem follows from the maximization of profit or overall welfare if one assumes that there are no constraints whatsoever on the combination and substitution of production factors.[6] The following qualitative view on wealth production and its relevant factors may help us understand why technological constraints on automation and capacity utilization must be taken into account in the optimization calculus of Sect. 4.4.

4.3   Preanalytic Vision: The Law of Diminishing Returns

Figure 4.1 depicts our preanalytic vision of wealth creation in the productive physical basis of modern economies. We consider capital, labor, and energy as the fundamental physical factors of production and creativity as a nonquantifiable input from human ingenuity, whose economic impact can only be detected ex post facto. Space contains natural resources, accommodates the production system, and absorbs its emissions.

Introducing energy as a third factor of production on an equal footing with capital and labor has been motivated by the observation that long-term economic growth in

---

[5]Nevertheless, one even tries to estimate "technological progress" for the future. In 2007 the investment bank Goldman Sachs projected economic growth of the Gulf Cooperation Council countries (Saudi Arabia, United Arab Emirates, Kuwait, Qatar, Oman, and Bahrain) until 2050 with the help of the neoclassical Cobb–Douglas function $Y = AK^\alpha L^{1-\alpha}$ of capital K and labor L, disregarding energy altogether; A is the level of "technical progress" and α "is the share of income that accrues to capital"[60].

[6]During an international conference on natural resources, a young economist gave a talk on energy in the economy. He explained that, because of the first law of thermodynamics, it is impossible to substitute capital for energy completely. A world-famous mathematical economist jumped up, interrupted him, and shouted with a reddening face: "You must never say that! There is always a way for substitution!"

industrialized countries has been accompanied by considerable capital deepening, i.e., increase of the capital/labor quotient, *without* a significant increase of the capital coefficient, which is the capital/output quotient.[7] This would contradict one of the most famous laws of economics [5], the law of diminishing returns, if standard economics did not call "technological progress" to the rescue.

The law of diminishing returns says:

> At a given state of technology, the additional input of a factor, at constant inputs of the other factors, results in an increase of output. Beyond a certain point, however, the additional return from an additional unit of the variable factor will decrease. This decrease is because one unit of the increasing factor is combined with smaller and smaller quantities of the fixed factors.

Since the coal-powered steam engine started the Industrial Revolution, heat engines and other energy conversion devices have become the core of the industrial capital stock, and growing energy inputs into the capital stock have gone hand in hand with economic growth. Therefore, physical and engineering logic suggests we consider "capital divided by the sum of labor and energy" as decisive for the law of diminishing returns. Then, there is no need to assume that technological progress camouflages the law of diminishing returns [5]. Rather, the evolution of the quotient capital/(labor + energy) does not deviate too much from that of the quotient capital/output, see, e.g., Figs. 4.3–4.6. This also suggests splitting technological progress into a quantitative part, measured by energy conversion, and a remaining qualitative part, called creativity. This facilitates (1) checking quantitatively the surmise that a substantial share of growth attributed to technological progress by standard economics is due to increasing energy conversion, and (2) computing the residual that is left for creativity. These considerations lead to the mathematical form of capital's output elasticity given by (4.34) [54]. Here, "labor" refers to routine labor, which is measured in man-hours worked per year, and energy is measured by energy (exergy) quantities "consumed" per year. In this sense one may consider labor and energy as flows which are combined with the stock of capital.

There are researchers who see an important difference between funds and flows, and who have a different concept of labor. They conceive "capital and labor ... as funds or agents that transform the flow of natural resources into a flow of products. The dominant relation between funds and flows is complementary. Substitutability between fund and flow is strictly marginal, limited to reducing process waste" [35]. This view is based on (1) identifying labor with the total workforce and not with man-hours actually worked per year, and (2) observations on short timescales of

---

[7]This is shown for the USA between 1900 and 1975 by Fig. 37.3 in Samuelson's *Volkswirtschafts-lehre II*, the German translation of the second part of [5], and for later years by Figs. 4.3–4.6.

days or months, during which capital goods and people do not change appreciably, except for accidents and war-time destruction.[8] Things look different on a timescale that is comparable to the lifetime of humans and machines – let us say some decades. Capital goods and workers enter the production process new and vigorous and leave it depreciated and aged. They are transformed flows too, just as energy. In fact, energy's lifetime of usefulness can be prolonged by thermoeconomic optimization. This involves heat recovery via heat exchanger networks, heat pumps, cogeneration, and heat storage in appropriate media [55–57]. Exergy optimization can also draw on stocks of electric energy stored in batteries, current-carrying (super)conductors [56], and pumped-up water of hydropower plants. Thus, on a long timescale, where all factors deteriorate, the difference between funds and flows is blurred.

The pivotal role of energy is shown clearly if we imagine a future where an inexhaustible source of energy, e.g., nuclear fusion in the Sun or on Earth, powers a mighty, totally automated, self-repairing and self-replicating factory. This production system would be initially endowed with an amount of materials so huge that it cannot be exhausted before the economic lifetime of the products has expired; in the long run, materials are almost exclusively provided by the recycling of all depreciated products. There is a continuous output of consumer and investment goods and only one input: energy, or rather, and more precisely, exergy (and some materials to replace the atoms lost in recycling). Labor is completely substituted by energy-driven machines and has disappeared as a factor of production. Capital, in the course of time, becomes more and more abundant; and an observer in some distant future would conclude that energy is the main (or even only) factor of production and that all the self-repairing and self-replicating factories are just the product of this one agent. The observer's awareness of energy as *a* or *the* factor of production will be even stronger if the cost of this indispensable factor dominates in the business accounts. Furthermore, watching materials being permanently recycled and knowing that, except for nuclear reactions, no atom is really changed in industrial production but is only given different neighbors, it will be obvious to the observer that materials, shaped into products by the production factor(s), are only passive partners of the production process.

---

[8]On short timescales one observes, e.g., how workers in an iron and steal works put iron ore and coke on a conveyor belt, which transports the material into a blast furnace. The molten iron comes out, and is then further formed, pressed, or squeezed by man-operated or computer-controlled machines into steel blocks, plates, and wires. These, in turn, are transported by people in heat-engine-powered trucks to steel-processing factories, where they, e.g., are turned into shining passenger cars. During the time of this production process the people, machines, and installations involved do not deteriorate noticeably, but remain intact, as a rule. On the other hand, the energy, flowing in as coal, gasoline, and electricity and dumped as heat into the environment, although being conserved quantitatively, has lost all its quality, that is, exergy. The iron ore involved is also transformed into dust, slag, and cars.

Thus, prolonged timescales and technological extrapolations show that physically *and* economically capital, labor, and energy fall into one and the same category of production factors. Their limitational and substitutional relations are determined by technological constraints.

Finally, the above scenario of an energetic fool's paradise cannot develop within the space of Earth's biosphere, because the coupling of energy conversion to entropy production and its associated thermal and material pollution establishes the ultimate law of diminishing returns [54]:

> If technological progress is limited to Earth-bound technology, the additional input of the production factor energy at constant value of the factor space causes an increase of production. Beyond a certain point, however, the additional return (of traditional goods and services) from an additional energy unit will decrease. This decrease is because the additional energy unit is being combined with a decreasing magnitude of space still capable of absorbing pollution. Therefore, an increasing share of energy must be dedicated to pollution abatement within the limits drawn by thermodynamics.

This preanalytic vision of industrial production will be formalized by the KLEC model in Sect. 4.5, which is based on the physical aggregation of output and factors outlined in Appendix 3 of Chap. 4. The ultimate law of diminishing returns is cast in a rudimentary mathematical model in Sect. 4.5.8. Before that, however, we show that the KLEC model and its results are not at variance with economic equilibrium, as calculated from standard behavioral assumptions, if the search for equilibrium observes stringent technological constraints.

4.4   How Technological Constraints Change Economic Equilibrium

The output of an economic system, i.e., the GDP or parts thereof, is generated by the production factors of the physical basis. Economic actors select the combinations of production factors, observing price signals from the market superstructure. The market acts as a control loop whose elements are all interconnected with each other. In such a structure, signals are processed in parallel, and decisions are rapid. In a planned economy, on the other hand, each element of the control loop is only connected to its two neighbors, so signals are processed in series, and decisions are slow. Therefore, the market is *the* system for the efficient allocation of goods and services according to the principles of supply and demand. The wealth *distribution* that results from such an allocation depends very much on the legal framework of the market. This is provided by lawmakers. Taxation of production factors is part of the legal framework. Societies with preferences for laissez-faire usually have lower taxes and levies and higher income inequalities than social market economies.[9] These facts are summarized by:

> Wealth is allocated on markets,
> and the legal framework determines the outcome.

We call this the "first law of economics."

When modeling the behavior of market actors, economics uses extremum principles, which have proven so successful in physics. If the variables of a system adjust within given constraints so that a system-specific objective becomes an extremum, the system is said to be in equilibrium. There are time-independent equilibrium conditions and others that depend on time t. We call them static and dynamic. In this sense the equilibrium condition "Gibbs free energy must be minimum" for a thermodynamic system in contact with a temperature and pressure reservoir is static, whereas Hamilton's principle of least action, from which the Lagrange equations of motion can be derived in classical mechanics, is dynamic. In *formal* one-to-one correspondence to these physical examples, economic equilibria are defined by the maximum of either profit or time-integrated utility (overall welfare). Needless to say, this involves assumptions about the behavior of the economic actors. Thus, the economic equilibrium conditions result from optimization postulates. The consequence of rejecting them is discussed below.

The mathematical derivation of both types of economic equilibrium starts from the assumption that the output Y of an economic system is produced by three production factors X_1, X_2, and X_3, the combinations of which are subject to technological constraints. Then, identifying the three factors with capital K, labor L, and energy (exergy) E, we will show that their combinations are constrained technologically by limits to the degree of automation and to the degree of capacity utilization.

---

[9]The simplest measure of income inequality is the Lorenz curve. It indicates what fraction, $y\%$, of total national income is owned by the poorest $x\%$ of households whose income is less than the income of those households that pertain to the range on the abscissa above x. The numerical values on both the y ordinate and the x abscissa go from 1 to 100. If all incomes were equal, the lowest 10% would also receive 10% of total income and would be "down" only in the way of counting. In that case the Lorenz curve would coincide with the bisecting line between the x-axis and the y-axis. But in reality the poorest 10%, 20% ... have less than 10%, 20% ... of total income, and the Lorenz curve lies below the bisecting line. The Gini coefficient of a country is equal to the area between the Lorenz curve of that country and the bisecting line expressed as a proportion of the whole triangle below the bisecting line. In the mid-1980s the Gini coefficient of Finland, Sweden, and Norway was close to 20%, whereas it exceeded 30% for Switzerland, Ireland, and the USA [63]. Since then, inequality of wealth distribution has increased globally and within countries, as Sect. 4.6 shows.

The corresponding shadow prices, which translate technological constraints into monetary terms, are derived. This indicates *why* realistic models of production must replace the schemes of mainstream economics. *The reader who is only interested in the derivation of such models from first principles can go from the end of Sect. 4.4.1 directly to Sect. 4.5.*

4.4.1 Output Elasticities

We follow standard economic theory and assume that output Y is uniquely determined by the inputs. Then, it is a state function of the economic system (in the same sense as internal energy (2.49), enthalpy (2.59) and the free energies (2.62) and (2.65) are state functions of physical systems). As a state function, Y must be twice-differentiable with respect to the (time-dependent) production factors within the accessible "space" spanned by the factors X_1, X_2, and X_3. Furthermore, Y may also depend explicitly on time t. Thus, we assume that a macroeconomic production function $Y(X_1, X_2, X_3; t)$ describes economic evolution. In Appendix 3 of Chap. 4 we discuss objections against the concept of the macroeconomic production function and respond to them by aggregating output, capital, labor, and energy in physical terms.

Infinitesimal changes of twice-differentiable functions are total differentials. Dividing the total differential dY of the production function by Y, one obtains the growth equation (which does not yet include pollution and scarcity of materials). This equation says that the growth rate dY/Y of output is given by the sum of the weighted growth rates dX_i/X_i of the inputs, and by the change of time since an initial time t_0:

$$\frac{dY}{Y} = \varepsilon_1 \frac{dX_1}{X_1} + \varepsilon_2 \frac{dX_2}{X_2} + \varepsilon_3 \frac{dX_3}{X_3} + \delta \frac{dt}{t - t_0}. \tag{4.2}$$

The weight ε_i, with which the growth rate of the factor X_i contributes to the growth of output, is what economists call the "output elasticity of X_i." It measures the productive power of the production factor and is given by the change of output Y due to the change of X_i, which is $\partial Y/\partial X_i$, divided by Y/X_i:

$$\varepsilon_i \equiv \frac{X_i}{Y} \frac{\partial Y}{\partial X_i}, \quad i = 1, 2, 3. \tag{4.3}$$

Human contributions to growth that cannot be captured by changes of the X_i may manifest themselves in an explicit time dependence of the production function and give rise to

$$\delta \equiv \frac{t - t_0}{Y} \frac{\partial Y}{\partial t}. \tag{4.4}$$

At any *fixed* time t the contributions from the growth rates of all factors to the growth of output must add up to 100%, so the output elasticities must satisfy the the so-called constant returns to scale relation

$$\sum_{i=1}^{3} \varepsilon_i = 1. \tag{4.5}$$

This relation characterizes linearly homogeneous production functions, whose value increases by a certain factor, say, λ if *all* inputs increase by the same factor λ. It is fundamental for the cost-share theorem.[10]

Standard economics often uses linearly homogeneous production functions such as Cobb–Douglas and constant-elasticity-of-substitution functions. Occasionally, one also considers increasing or decreasing returns to scale. The corresponding production functions have the property $Y(\lambda X_1, \lambda X_2, \lambda X_3) = \lambda^{\nu} Y(X_1, X_2, X_3)$, with $\nu > 1(<1)$ for increasing (decreasing) returns to scale. This, however, implies that changes of inputs by λ are associated with alterations of the state of technology. The state of technology changes, for instance, when the thermodynamic effectiveness [25] of the production process changes. Such time-related changes are excluded by the condition "at fixed time t" in the derivation of (4.5). Linearly homogeneous production functions can take care of alterations of the state of technology by an explicit time dependence. This may manifest itself in time-changing technology parameters like those of the LinEx function (4.35).

4.4.2   Shadow Prices

Economic equilibrium conditions relate factor inputs to factor prices. The standard assumptions for computing macroeconomic equilibrium are that the actions of all economic agents result in the maximization of either profit or overall welfare. Profit is output Y minus factor cost, and overall welfare is the time integral of a utility function. One may question these assumptions, e.g., because of game-theoretical findings and the experiences of the 2008 financial market crash. If one rejects them, one also rejects the cost-share theorem, and there is no reason to believe in the tiny output elasticity of energy. If one accepts them, we will show that the cost-share

---

[10] An increase of all inputs by λ must increase output by λ, because at the fixed state of technology that exists at the given time t a, say, doubling of the production system doubles output; in other words, two identical factories with identical inputs of capital, labor, and energy produce twice as much output as one factory. Thus, the production function must be linearly homogeneous in (X_1, X_2, X_3) so that $Y(\lambda X_1, \lambda X_2, \lambda X_3) = \lambda Y(X_1, X_2, X_3)$ for all $\lambda > 0$ and all possible factor combinations. Differentiating this equation with respect to λ according to the chain rule and then putting $\lambda = 1$, one obtains the Euler relation $X_1(\partial Y/\partial X_1) + X_2(\partial Y/\partial X_2) + X_3(\partial Y/\partial X_3) = Y$. Dividing this by Y yields $(X_1/Y)(\partial Y/\partial X_1) + (X_2/Y)(\partial Y/\partial X_2) + (X_3/Y)(\partial Y/\partial X_3) = 1$. With (4.3), this becomes (4.5).

theorem is killed by hitherto disregarded technological constraints. In either case, alternative methods of computing output elasticities are required. The results of one method are presented in Sect. 4.5.3.

4.4.2.1 Profit Maximization

We assume that the three production factors $(X_1, X_2, X_3) \equiv \mathbf{X}$ have the exogeneously given prices per factor unit $(p_1, p_2, p_3) \equiv \mathbf{p}$, so total factor cost is $\mathbf{p(t)} \cdot \mathbf{X(t)} = \sum_{i=1}^{3} p_i(t)X_i(t)$. The factors can vary independently within technological constraints until profit

$$G(\mathbf{X}, \mathbf{p}, t) \equiv Y(\mathbf{X}, t) - \mathbf{p} \cdot \mathbf{X} \tag{4.6}$$

becomes maximum.

The technological constraints are labeled by a. They can be brought into the form of equations,

$$f_a(\mathbf{X}, t) = 0, \tag{4.7}$$

with the help of slack variables. Slack variables change inequalities into equalities. They define the range in factor space within which the factors can vary independently at time t. They are given explicitly in Appendix 4 of Chap. 4 for the factors capital, labor, and energy. There are two technological constraints on these factors. Thus, a is either, say, A or B.

In the maximum of profit G, infinitesimally small changes of the X_i do not change G. The same is true for $G + \sum_a \mu_a f_a(\mathbf{X}, t)$, if constraints exists as the ones described by (4.7). The constants μ_a are called *Lagrange multipliers*. They may depend on the equilibrium values of the inputs X_i. The necessary condition for a maximum of profit in the presence of constraints is derived in Appendix 1 of Chap. 4. It yields the three equilibrium conditions

$$\varepsilon_i = \frac{X_i[p_i + s_i]}{\sum_{i=1}^{3} X_i[p_i + s_i]}; \quad i = 1, 2, 3. \tag{4.8}$$

Here, s_i, defined as

$$s_i \equiv -\mu_A \frac{\partial f_A}{\partial X_i} - \mu_B \frac{\partial f_B}{\partial X_i} \tag{4.9}$$

is called the *shadow* price of the production factor X_i in economics.

In the absence of technological constraints, the Lagrange multipliers μ_a and the shadow prices s_i would be zero, and the equilibrium conditions (4.8) would turn into the cost-share theorem: On the right-hand side of (4.8) the numerator would be the cost $p_i X_i$ of the factor X_i, the denominator would be the sum of all factor costs, and the quotient, which is equal to the output elasticity ε_i, would represent the cost share of X_i in total factor cost. This would also justify the neoclassical duality of production factors and factor prices, which is often used in orthodox growth analyses. This duality is a consequence of the Legendre transformation that results from the requirement that profit $G(\mathbf{X}, \mathbf{p})$ is maximum *without* any constraints on X_1, X_2, and X_3. Then (4.65) in Appendix 1 of Chap. 4 would hold with $\mu_a = 0$ and yield equilibrium values $X_{1M}(\mathbf{p})$, $X_{2M}(\mathbf{p})$, and $X_{3M}(\mathbf{p})$. With $\mathbf{X}_M(\mathbf{p})$, the profit function would turn into the price function

$$G(\mathbf{X}_M(\mathbf{p}), \mathbf{p}) = Y(\mathbf{X}_M(\mathbf{p})) - \mathbf{p} \cdot \mathbf{X}_M(\mathbf{p}) \equiv g(\mathbf{p}). \qquad (4.10)$$

The essential information on production would be contained in the price function $g(\mathbf{p})$, which is the Legendre transform of the production function $Y(\mathbf{X})$. (This is in formal analogy to the Hamilton function being the Legendre transform of the Lagrange function in classical mechanics, or to enthalpy and free energy being Legendre transforms of internal energy in thermodynamics.) However, because of the technological constraints and the resulting shadow prices, the cost-share theorem and (4.10) are not valid. For an understanding of the economy, prices are not enough.

4.4.2.2 Intertemporal Welfare Maximization

Maximization of intertemporal (overall) welfare is an alternative to the derivation of equilibrium conditions from profit maximization. It tests the sensitivity of the equilibrium relations between output elasticities and factor prices to modified behavioral assumptions.

The starting point is the optimization formalism described by Samuelson and Solow [64] stating: "... society maximizes the (undiscounted) integral of all future utilities of consumption subject to the fact that the sum of current consumption and of current capital formation is limited by what the current capital stock can produce."[11] We follow the optimization procedure of Samuelson and Solow [64]

---

[11] No or low discounting of future benefits and losses is rejected by many economists. On the other hand, leading economists such as Ramsey [65], Arrow [66], Solow [41], and others question time preferences and discounting for ethical reasons. For instance, Solow [41] says "we ought to act as if the social rate of time preference were zero (though we would simultaneously discount future consumption if we expected the future to be richer than the present)." Stern [6, 7], following them, reasons that the only sound ethical basis for placing less value on the utility of future generations is the uncertainty whether or not the world will exist, or whether those generations will all be present. He uses a pure time discount rate of 0.1%. This corresponds to a probability of 90.5% that humanity will not have perished within 100 years. Stern defends his low discount rate against criticism from

with three modifications: (1) there is not one variable production factor but three –
X_1, X_2, and X_3; (2) there are constraints on magnitudes and combinations of these
factors; (3) as in Hellwig et al. [67], optimization of time-integrated utility is done
within finite time horizons.

Thus, in overall welfare optimization we assume that society maximizes the
(undiscounted)[12] integral W of utility U between the times t_0 and t_1, where the
factors evolve along a curve symbolized by $[s]$. We limit the calculation to the
simplest case that the utility function U only depends on consumption C: $U =
U[C]$. Consumption is the difference between output and capital formation. As
discussed in Appendix 2 of Chap. 4, capital depreciation is included in the capital
price, so it is sufficient to consider only the actual increase of capital, $dX_1/dt \equiv \dot{X}_1$,
in capital formation. Then, consumption is

$$C = Y(\mathbf{X};t) - \dot{X}_1. \tag{4.11}$$

The sum of consumption and capital formation at any time t between t_0 and
t_1, which is the output $Y(\mathbf{X};t)$, is limited by what the capital stock $X_1(t)$ in
combination with the two other production factors $X_2(t)$ and $X_3(t)$ can produce
with due regard of the constraints on these factors. Thus, the optimization problem
is as follows:

Maximize overall welfare

$$W[s] = \int_{t_0}^{t_1} U[C]dt, \tag{4.12}$$

subject to the same technological constraints as in profit maximization: $f_A(\mathbf{X},t) = 0$,
$f_B(\mathbf{X},t) = 0$. In addition, there is an economic constraint: the total cost $\mathbf{p} \cdot \mathbf{X}$ of
producing consumption C by means of the factors $(X_1, X_2, X_3) \equiv \mathbf{X}$ must not
diverge. Rather, its magnitude $c_f(t)$ must be finite at all times t, where each price
per factor unit, p_i, is exogeneously given:

$$c_f(t) - \sum_{i=1}^{3} p_i(t)X_i(t) = 0. \tag{4.13}$$

The variational formalism of intertemporal welfare optimization includes an additional Lagrange multiplier μ, which takes care of the constraint (4.13). The
formalism, presented in Appendix 2 of Chap. 4, is the same as that used to derive the
Lagrange equations of motion from Hamilton's principle of least action in classical

---

a number of authors [7]. Given the diverging views on discounting, and for the sake of simplicity,
we disregard discounting. We will come back to discounting of the future in Sect. 4.6.4.

[12]If one multiplied $U[C]$ in the integrand in (4.12) by $\exp(-\delta t)$, δ being the pure time discount
rate, one would have to subtract $\delta\frac{dU}{dC}$ from $\frac{d}{dt}(\frac{dU}{dC})$ in (4.86) and (4.87).

mechanics. The resulting equilibrium conditions can again be written as a relation between output elasticities, factor prices, and shadow prices in the form of (4.8). But the shadow prices are now

$$s_i \equiv -\sum_{a=A}^{B} \frac{\mu_a}{\mu} \frac{\partial f_a}{\partial X_i} - \delta_{i,1} \frac{1}{\mu} \frac{d}{dt} \left(\frac{dU}{dC} \right), \quad i = 1, 2, 3; \qquad (4.14)$$

here $\delta_{i,1}$, is 1 for $i = 1$ and 0 otherwise. The shadow prices in (4.14) differ from the shadow prices in (4.9) in two aspects. First, there is the term $\delta_{i,1} \frac{1}{\mu} \frac{d}{dt} (\frac{dU}{dC})$. This term originates from (4.11) and is due to taking capital formation into account in intertemporal utility optimization, whereas capital formation is not an issue in profit optimization. Second, ratios μ_a/μ of Lagrange multipliers take the positions of the Lagrange multipliers μ_a in (4.9). If one does profit maximization subject to the additional constraint $\mathbf{p} \cdot \mathbf{X} = c_f$, which fixes factor cost, one gets μ_a/μ instead of μ_a in the equations that then replace (4.8) and (4.9). Thus, the second difference is rather a formal one.

The first difference vanishes if one can disregard decreasing marginal utility and approximate the utility function $U[C]$ by a linear function in C. For instance, if the function of decreasing marginal utility [6] is $U[C] = C_0 \ln \frac{C}{C_0} + U_0$, and if it can be approximated by its Taylor expansion up to first order in $\frac{C}{C_0} - 1$, one has

$$U[C] \approx C - C_0 + U_0, \qquad (4.15)$$

$\frac{d}{dt} \left(\frac{dU}{dC} \right) = 0$, and the shadow prices are practically the same.[13]

In summary, the standard assumptions about the behavior of economic actors lead to the result:

If entrepreneurs are not free to choose the optimum set of production factors at will, but rather have to observe technological constraints on factor combinations that limit access to certain regions of factor space, equilibrium conditions derived from profit or welfare optimization no longer support the (cost-share) theorem that output elasticities should be equal to factor cost shares. Neither do they support the Legendre transformation from factor quantities to factor prices. Therefore, realistic output elasticities that reflect the true productive power of production factors cannot be obtained from models that only consider the market. A description of how wealth is created in the productive physical basis must complement the perception of an industrial economy.

---

[13] A linear approximation of $\ln x$ is acceptable for $x < 4$.

4.5   The Second Law of Economics

Equations (4.8), (4.9), and (4.14) show that output elasticities cannot be obtained from the conditions for economic equilibrium if there are technological constraints on the factor inputs. We must look for additional criteria that facilitate calculation of output elasticities and production functions. Therefore, we descend to the physical basis of the economy. Here we find the technological constraints on capital, labor, and energy, and are guided to a new method of computing output elasticities, production functions, and economic growth.

4.5.1   Levels of Wealth Creation

Technological constraints matter in economic systems where energy conversion is an essential part of production. This is the case in modern industrial economies like the ones of the so-called G7 countries of Canada, France, Germany, Italy, Japan, the UK, and the USA. They consist of the sectors agriculture(including forestry and fishery), goods-producing industry (manufacturing, mining and quarrying, construction, electricity, gas and water), and services. Services include trade, transport, and communications, the service industries, the public sector, and private households.

Although the wealth of preindustrial societies was mostly created in agrarian primary production, and most of the population worked there, this has changed fundamentally in the course of industrialization and the accompanying rural exodus. For example, in the Federal Republic of Germany in 1950, only five million people, 25% of all employed people, were still working on farms and generated 11% of gross value added. Mechanization of agriculture, which started in the USA between the two world wars and in Europe after World War II, changed the situation dramatically again. Tables 4.1 and 4.2 show that agricultural production in the G7 countries has declined to about 2% of GDP, or less, within 40 years, and employment was about 4%, or less, of total employment in 2009. What has happened in agriculture is also going on in goods-producing industry: energy-powered machines are replacing routine human labor. More and more people are working in the service sector.

Since the share of labor in total factor cost is between 65% and 75%, the products of the service sector are expensive, and their share in gross value added is the biggest. This does not mean, however, that the physical production in agriculture and industry has declined. To the contrary, despite decreasing numbers of employees, the quantities of food and industrial goods produced have stayed constant or even increase. Productivity grows mostly because of increasing automation. As a consequence, the products of the agrarian and industrial sectors have become cheaper and cheaper, so their contribution to value added has decreased. Table 4.3 shows how the relative values of basic goods changed in Germany (western states) between 1960 and 2008.

Table 4.1 Gross value added (GVA; GDP) in agriculture (*A*), goods-producing industry (*I*), and services-producing sectors (*S*) as percentage of total GVA (GDP). The sums differ from 100 because of rounding. (Sources: Institut der deutschen Wirtschaft, Cologne, for the 1970 and 1992 data; the CIA World Fact Book for the 2009 data)

Country	A			I			S		
	1970	1992	2009	1970	1992	2009	1970	1992	2009
Canada	4.2	2.7	2	36.3	31.5	28.4	59.4	65.9	69.6
France	6.9	2.9	2.1	41.5	29.7	19.0	51.6	67.4	78.9
Germany[a]	3.4	1.2	0.9	51.7	39.6	27.1	44.9	59.2	72.0
Italy	8.1	3.2	2.1	42.6	32.3	25.0	49.4	64.6	72.9
Japan	5.9	2.1	1.6	45.1	39.4	23.1	49.1	58.5	75.4
UK	2.8	1.7	1.2	42.5	31.7	23.8	54.6	66.6	75
USA	2.7	1.9	1.2	34.1	28.5	21.9	61.8	68.3	76.9

[a]Until 1990 only the part that was the old Federal Republic of Germany (i.e., West Germany) before reunification

Table 4.2 Structure of employment as percentage of total civilian employment in agriculture (*A*), goods-producing industry (*I*), and services-producing sectors (*S*). (Sources: Institut der deutschen Wirtschaft, Cologne, for the 1970 and 1992 data; the CIA World Fact Book for the 2009 data)

Country	A			I			S		
	1970	1992	2009	1970	1992	2009	1970	1992	2009
Canada	7.6	4.4	2	30.9	22.7	22	61.4	73.0	76
France	13.5	5.2	3.8	39.2	28.9	24.3	47.2	65.9	71.8
Germany	8.6	3.1	2.4	49.3	38.3	29.7	42.0	58.5	67.8
Italy	20.2	8.2	4.2	39.5	32.2	30.7	40.3	59.6	65.1
Japan	17.4	6.4	4	35.7	34.6	28	46.9	59.0	68
UK	3.2	2.2	1.4	44.7	25.6	18.2	52.0	71.3	80.4
USA	4.5	2.9	0.6	34.4	24.6	22.6	61.1	72.5	76.8

Table 4.3 Buying power of the work minute in Germany (western states) in the years 1960, 1991, and 2008. The average time an industrial employee has to work for the purchase of the consumer goods listed is given in minutes. (Source: Institut der deutschen Wirtschaft, Cologne)

Good	1960	1991	2008
Bread, 1 kg	20	11	11
Butter, 250 g	39	6	5
Sugar, 1 kg	30	6	5
Milk, 1 L	11	4	4
Beef, 1 kg	124	32	32
Potatoes, 2.5 kg	17	10	14
Beer, 0.5 L	15	3	3
Gasoline, 1 L	14	4	6
Electricity, 200 kWh	607	191	201
Refrigerator	9,390	1,827	1,432
Washing machine	13,470	3,207	2,008

Thus, wealth is created on three levels of an economy's productive physical basis. The fundamental level, which is absolutely indispensable for the physical existence of humans, is that of agriculture. The middle level, without whose products modern people can hardly imagine living, is that of goods-producing industry. On top is the level of human interactions in the services-producing sectors. Thanks to creative energy utilization it has been possible to produce so much of the vital goods on the two lower levels that their prices, which, as a rule, measure scarcity, are low in the industrially advanced countries. Therefore, the economic importance of a level of wealth creation is seemingly smaller the greater its importance is for life.

Obviously, without the growing industrial production of material goods, the service sector could not have expanded. It rests on the pillars of agriculture and goods-producing industries. If they suffer, the service sector suffers even more. This was the painful experience of the former "socialist" countries, when their not sufficiently competitive industries collapsed after the fall of the Iron Curtain. Considering the service sector itself, we note the growing importance of energy. Shopping centers, offices, classrooms, and hospitals are air-conditioned by energy conversion. Computers are taking over more and more information processing in banking, insurance, and administration. They run on electricity provided by central power stations and decentralized generators. Hotels, restaurants, building maintenance, and housekeeping can hardly do without electric appliances these days. Telecommunication is by electromagnetic waves. The field of medicine uses these waves in X-ray apparatus and computer tomography, for irradiation of cancer cells, and laser surgery. Transportation by car, ship, and plane vitally depends on energy conversion in heat engines. And so on. Thus, capital, labor, and energy are the fundamental factors of production in all economic sectors of highly industrialized countries.

4.5.2 The Technological Constraints on Capital, Labor, and Energy

Constraints pertain to production systems. In a world where the capital stock consisted of simple tools such as hammers, pliers, shovels, pick axes, sickles, plows, and winches, the buildings to house them, and transportation equipment such as boats and carts, wealth resulted essentially from the combination of human and animal muscle power with these devices. In such a world, biological constraints dominated, and disregarding technological constraints in a description of its economy may not have been unreasonable.[14]

On the other hand, and in contrast to the preindustrial situation, the production system of an industrialized country *is* subject to binding technological constraints. To illustrate this we consider the capital–labor–energy–creativity (KLEC) model

---

[14]The constraints on the use of wind and water power by sailing ships and ancient mills had little effect on factor substitution.

depicted in Fig. 4.1. The capital stock $K(t)$, as it is reported by the national accounts, consists of all energy-converting and information-processing machines together with all buildings and installations necessary for their protection and operation. Output Y results from work performance and information processing by the combination of such capital with (routine) labor $L(t)$ and energy $E(t)$.[15]

Capital in the absence of energy is functionally inert. Nothing happens. To be productive the machines of the capital stock must be activated by the exergy component of energy. Primary energy in the form of coal, oil, gas, nuclear fuels, or solar radiation, which can be converted to nearly 100% into physical work under appropriate conditions, is practically all exergy and an input ultimately supplied by nature. Furthermore, to be economically productive, capital must also be allocated, organized, and supervised by (human) labor. Economic activities of humans can be subdivided into two components: (1) routine labor, which (by definition) can be substituted by some combination of capital and energy and (2) a residual that gives rise to the output elasticity δ in (4.4) and which is called *creativity*. Creativity, in this sense, is the specific human contribution to production and growth that cannot be provided by any machine, even a sophisticated computer capable of learning from experience. It includes ideas, inventions, valuations, and interactive decisions depending on human reactions and characteristics.

The ultimate lower limit of routine labor inputs is probably unknowable, because it depends to some degree on the limits of artificial intelligence. But we need not concern ourselves with the ultimate limit. At any given time, with a given technology and state of automation, there is a limit to the extent that routine labor can increase output. In other words, the model postulates the possibility of a combination of capital and exergy such that adding one more unskilled worker adds nothing to gross economic output. (In some manufacturing sectors of industrialized countries this point does not seem to be far away.) In such a state of maximally automated production, the output elasticity of routine labor would be vanishingly small.

There is another fairly obvious technological constraint on the combinations of factors. In brief, machines are designed and built for specific exergy inputs. In some cases (e.g., for some electric motors) there is a modest overload capability. Buildings can be overheated or overcooled, to be sure, but this does not contribute to productivity. On average the maximum exergy input is fixed by design. Both energy convertors and energy users have built-in limits. In other words, the ratio of exergy to capital must not exceed a definite upper limit.

---

[15]Some models, such as those in [45, 49], take materials into account as a fourth factor of production. Since materials are passive partners in the production process, and do not contribute actively to output – their atoms and molecules are merely arranged in orderly patterns by capital, labor, and energy when value added is created – we do not include them in the model. Like other models, the present model also disregards land as a production factor. Land area matters mainly as a site for production facilities and for photosynthetic conversion of solar energy into the chemical energy of glucose in agriculture. It does not contribute actively to work performance and information processing.

The bottom line of the above considerations is that the use of capital, labor, and energy in industrial systems is subject to technological constraints that are the consequence of limits to capacity utilization and to the substitution of capital and energy for labor. This substitution changes what we call the "degree of automation."

For the derivation of the constraint equations, output and inputs must be specified by measurement prescriptions. In Appendix 3 of Chap. 4 technological measurement prescriptions for output Y and capital K are proposed in terms of work performance and information processing. It is shown there how the technological measuring units relate to the deflated monetary units in which traditionally output and capital are reported by the national accounts. Routine labor L is measured in man-hours worked per year, as given by the national labor statistics, and energy E is measured in petajoules (or tons of oil equivalents, or quads) per year, as shown by the national energy balances.

Of course, the theory must be independent of the choice of units. Therefore, in our three-factor model, with $X_1 \equiv K$, $X_2 \equiv L$, and $X_3 \equiv E$, it is convenient to introduce new, dimensionless variables, for which we use lowercase letters, by writing inputs and output as multiples of their quantities K_0, L_0, E_0, and Y_0 in a base year t_0. The transformation to the dimensionless time series of capital, $k(t)$, labor, $l(t)$, and energy, $e(t)$, is given by

$$k(t) \equiv \frac{K(t)}{K_0}, \quad l(t) \equiv \frac{L(t)}{L_0}, \quad e(t) \equiv \frac{E(t)}{E_0}, \tag{4.16}$$

and the dimensionless production function is

$$y[k, l, e; t] \equiv \frac{Y(kK_0, lL_0, eE_0; t)}{Y_0}; \tag{4.17}$$

for the sake of notational simplicity we do not always indicate the time dependence of k, l, and e explicitly. From here on we work in the "space" of the dimensionless inputs and outputs, defined by (4.16) and (4.17).

Quantitatively, the degree of automation ρ of a production system is proportional to the actual capital stock k of the system divided by the capital stock $k_m(y)$ that would be required for the maximally automated production of the actual output y. The proportionality factor η is the degree of capacity utilization of the capital stock. Thus, the degree of automation is given by [3]

$$\rho = \eta \frac{k}{k_m(y)}. \tag{4.18}$$

Entrepreneurial decisions, aiming at producing a certain quantity of output y with existing technology, determine the absolute magnitude of the total capital stock k, its degree of capacity utilization η, and its degree of automation ρ. Obviously, ρ and η are functions of capital k, labor l, and energy e. They are definitely constrained by $\rho(k, l, e) \leq 1$ and $\eta(k, l, e) \leq 1$, i.e., the maximum degree of automation cannot be exceeded, and a production system cannot operate above design capacity.[16] Here it is important that (productive) energy input into machines and other capital equipment is always limited by their technical design.

However, there is a technical limit to the degree of automation at time t that lies below 1. We call it $\rho_T(t)$. It depends on mass, volume, and exergy requirements of the machines, especially information processors, in the capital stock. Imagine the vacuum-tube computers of the 1960s, when the tiny transistor, invented in 1947 by Bardeen, Brattain, and Shockley, had not yet diffused into the capital stock. A vacuum-tube computer with the computing power of a 2010 notebook computer would have had a volume of many thousands of cubic meters. In 1960 a degree of automation, that was standard 40 years later in the highly industrialized countries, would have resulted in factories many orders of magnitude bigger than those of today, probably exceeding the available land area.[17]

In the course of time, the technical limit to automation, $\rho_T(t)$, moves toward the theoretical limit 1. This is facilitated by the density increase of information processors (transistors) on a microchip. In accordance with "Moore's law," transistor density has doubled every 18 months during the last four decades. It may continue like that for a while, thanks to nanotechnological progress. But there is a thermo-dynamic limit to transistor density, because the electricity required for information processing eventually ends up in heat. If this heat can no longer escape sufficiently rapidly out of the microchip because of too densely packed transistors, it will melt the conducting elements and destroy the chip. We do not know exactly how far the technical limit to automation can be pushed. For our purposes, however, it is sufficient to know that at any time t such a limit $\rho_T(t)$ exists.

Since the technical properties of the capital stock do not change with η, the constraint on automation applies to the situation of maximum capacity utilization. With $\eta = 1$ in (4.18), the (inequality) formulation of an upper limit to automation is $k/k_m(y) \leq \rho_T(t)$. It is brought into the form of a constraint *equation*, required by the method of the Lagrange multipliers in profit and welfare optimization, with the help of the slack variable k_ρ :

$$f_A(K, L, E, t) \equiv \frac{k + k_\rho}{k_m(y)} - \rho_T(t) = 0; \qquad (4.19)$$

---

[16]Strictly speaking, the limit 1 for η is a sharp technological limit only when "working at full capacity" means working 24 h/day and 365 days/year. There are branches of business where machines have to run for less time per day and per year in order to be considered as working at full capacity. To keep things simple, we disregard these "soft" limits to capacity utilization.

[17]Despite all the advances in computer technology, and despite the fact that the textile industry was the first to become mechanized at the beginning of the Industrial Revolution, it is still not possible to replace the human hand and brain altogether in the sewing of clothing. The sweatshops in developing countries are testimony to that.

k_ρ is the capital that has to be added to k so that the total capital stock $k + k_\rho$, working at full capacity, exhausts the technologically possible automation potential $\rho_T(t)$.

Similarly, the formulation of an upper limit to capacity utilization, $\eta(k, l, e) \leq 1$, is brought into the required form of a constraint equation with the help of the slack variables $e_\eta(t)$ and $l_\eta(t)$:

$$f_B(K, L, E, t) \equiv \eta(k, l + l_\eta, e + e_\eta) - 1 = 0; \qquad (4.20)$$

$l + l_\eta$ and $e + e_\eta$ are the quantities of labor and energy that are needed by the capital stock k so that it can work at full capacity at time t. In a rough approximation, one may assume that $e_\eta(t)$ and $l_\eta(t)$ are related by $e_\eta(t) = d(t) \times l_\eta(t)$. Here, the technological state of the capital stock determines the labor–energy-coupling parameter $d(t)$. The explicit relations between the factors k, l, and e and the slack variables k_ρ, l_η, and and e_η are given by (4.118), (4.119), and (4.126) in Appendix 4 of Chap. 4.

We need an explicit functional form for the degree of capacity utilization η. Since η does not change if k, l, and e all change by the same factor, it is a homogeneous function of degree zero: $\eta = \eta(l/k, e/k)$. A trial form can be derived from a Taylor expansion of $\ln \eta[\ln(l/k), \ln(e/k)]$ around some point "0"$\equiv (\ln(l/k)_0, \ln(e/k)_0)$, up to first order in $\ln(l/k) - \ln(l/k)_0$ and $\ln(e/k) - \ln(e/k)_0$. This approximation yields

$$\eta = \eta_0 \left(\frac{l}{k}\right)^\lambda \left(\frac{e}{k}\right)^\nu, \qquad (4.21)$$

where λ and ν are the derivatives of $\ln \eta$ with respect to $\ln l/k$ and $\ln e/k$ at the point "0." The parameters η_0, λ, and ν can be determined from empirical data on capacity utilization. Then, combining (4.20) with (4.21), one has the complete equation describing constrained capacity utilization. This equation and (4.19) enter the shadow prices (4.9) and (4.14), where one has to identify $X_1 = K_0 k$, $X_2 = L_0 l$, and $X_3 = E_0 e$ and replace subscripts 1, 2, and 3 by K, L, and E.

For instance, in the case of profit maximization, the shadow prices of capital, s_K, labor, s_L, and energy s_E become

$$s_K = -\frac{1}{K_0}\left[\mu_A \frac{\partial f_A}{\partial k} + \mu_B \frac{\partial f_B}{\partial k}\right], \quad s_L = -\frac{1}{L_0}\left[\mu_A \frac{\partial f_A}{\partial l} + \mu_B \frac{\partial f_B}{\partial l}\right], \quad (4.22)$$

$$s_E = -\frac{1}{E_0}\left[\mu_A \frac{\partial f_A}{\partial e} + \mu_B \frac{\partial f_B}{\partial e}\right]. \qquad (4.23)$$

The explicit constraint equations, their partial derivatives, and the slack variables are given by (4.113)–(4.126). Equations (4.72) and (4.73) present the Lagrange multipliers μ_A and μ_B. The inputs are those in equilibrium.

4.5.3 Modeling Production

The production function $y[k, l, e; t]$ must have output elasticities of capital, labor, and energy that are independent of equilibrium conditions and consistent with the technological constraints. The task is to determine $y[k, l, e; t]$ as a mapping of the KLEC model sketched in Fig. 4.1. The explicit time dependence of $y[k, l, e; t]$ takes care of human ideas, inventions, and value decisions, summarized by the concept of creativity. The KLEC model specifies neoclassical "technological progress" in terms of energy and creativity. The quantitative results obtained with it can be compared with empirical data so that the model is open to falsification. The data for and the concepts of capital, labor, energy, and output are those of the national accounts, energy balances, and labor statistics. (Modern tendencies of widening – or blurring – the concept of "capital", e.g., by regarding natural resources as part of total capital inputs [62], are not followed for engineering reasons. As a comment on the "Cambridge Controversy" the relation between physically and monetarily aggregated capital is established in Appendix 3 of Chap. 4.)

In terms of the output y and the inputs k, l, and e, which are the value added and the production factors normalized to their quantities in a base year according to (4.16) and (4.17), the growth equation (4.2) becomes

$$\frac{dy}{y} = \alpha \frac{dk}{k} + \beta \frac{dl}{l} + \gamma \frac{de}{e} + \delta \frac{dt}{t - t_0}. \tag{4.24}$$

The output elasticities ε_i, defined in (4.3), are specified for capital k, labor l, and energy e by α, β, and γ. They, and the contribution of creativity to growth, δ, are given by

$$\alpha \equiv \frac{k}{y} \frac{\partial y}{\partial k}, \quad \beta \equiv \frac{l}{y} \frac{\partial y}{\partial l}, \quad \gamma \equiv \frac{e}{y} \frac{\partial y}{\partial e}, \quad \delta \equiv \frac{t - t_0}{y} \frac{\partial y}{\partial t}. \tag{4.25}$$

We call δ "the output elasticity of creativity.". *The output elasticities measure the productive power of production factors. Roughly speaking, they indicate by what fraction of 1% the output of an economic system changes if a factor changes by 1%. They are the crucial quantities of this chapter.* Deviating from mainstream economics, we calculate them from the standard mathematical requirement on production functions, already observed in the derivation of the growth equation, namely, that $y[k, l, e; t]$ must be a twice-differentiable function of its variables k, l, and e.

A function is twice differentiable if its second-order mixed derivatives are equal. The requirement of twice differentiability, when applied to the thermodynamic potentials internal energy, enthalpy, and free (Helmholtz and Gibbs) energy, leads to the Maxwell relations (2.58), (2.61), (2.64), and (2.67). When applied to the production function $y[k, l, e; t]$, it leads to the partial differential equations

$$l\frac{\partial \alpha}{\partial l} = k\frac{\partial \beta}{\partial k}, \quad e\frac{\partial \beta}{\partial e} = l\frac{\partial \gamma}{\partial l}, \quad k\frac{\partial \gamma}{\partial k} = e\frac{\partial \alpha}{\partial e}. \tag{4.26}$$

If the production function is linearly homogeneous, so that according to (4.5) one can write $\gamma = 1 - \alpha - \beta$, these equations become [3, 24, 28]

$$l\frac{\partial \alpha}{\partial l} = k\frac{\partial \beta}{\partial k}, \quad k\frac{\partial \alpha}{\partial k} + l\frac{\partial \alpha}{\partial l} + e\frac{\partial \alpha}{\partial e} = 0, \quad k\frac{\partial \beta}{\partial k} + l\frac{\partial \beta}{\partial l} + e\frac{\partial \beta}{\partial e} = 0. \tag{4.27}$$

The most general solutions of the second and the third differential equation in (4.27) are

$$\alpha = A\left(\frac{l}{k}, \frac{e}{k}\right), \qquad \beta = B\left(\frac{l}{k}, \frac{e}{k}\right), \tag{4.28}$$

where A and B are any differentiable functions of their arguments. Because of the first equation in (4.27), they are coupled together by

$$\beta = \int^k \frac{l}{k'}\frac{\partial A}{\partial l}dk' + J\left(\frac{l}{e}\right); \tag{4.29}$$

here $J(l/e)$ is any differentiable function of l/e.

The output elasticities, and the combinations of k, l, and e, must satisfy the restrictions

$$\alpha \geq 0, \quad \beta \geq 0, \quad \gamma = 1 - \alpha - \beta \geq 0, \tag{4.30}$$

which result from the technical-economic requirement that all output elasticities must be nonnegative. Otherwise the increase of an input would result in a decrease of output – a situation the economic actors will avoid.

The general form of the twice-differentiable, linearly homogeneous production function with the output elasticities (4.28) is

$$y = e\mathcal{F}\left(\frac{l}{k}, \frac{e}{k}\right). \tag{4.31}$$

According to the theory of partial differential equations, the output elasticities, and thus the production function, could uniquely be determined if β were known on a boundary *surface* in k, l, e-space and if one knew α on a boundary *curve* in that space [3]; see also Appendix 6 of Chap. 4. However, it is practically impossible to obtain this technical-economic information on α and β so that one could exactly compute the output elasticities at a given time t. Therefore, one must replace the unknown exact boundary conditions by less stringent, but technically and economically reasonable *asymptotic* ones.

Before going into this, we consider the trivial solutions of (4.27), which are constants: $\alpha = \alpha_0$, $\beta = \beta_0$. If one inserts them into (4.24) at fixed t, observes $\gamma_0 = 1 - \alpha_0 - \beta_0$, and integrates y from y_0 to y_{CDE} and the factors from $(1, 1, 1)$ to (k, l, e), one obtains the energy-dependent Cobb–Douglas function (CDE)

$$y_{\text{CDE}} = y_0 k^{\alpha_0} l^{\beta_0} e^{1-\alpha_0-\beta_0}. \tag{4.32}$$

This function is often used in quantitative analyses of standard economics,[18] where α_0, β_0, and $1 - \alpha_0 - \beta_0$ are set equal to the cost shares of capital, labor, and energy; these shares have happened to be approximately constant until recently.

The conceptual problem with the Cobb–Douglas function is that it is a production function with complete substitutability of the production factors. Therefore, its use for computing scenarios of the future, for instance in [60], is problematical. Things are different when using energy-dependent Cobb–Douglas functions for the past, when, of course, the empirical inputs of capital, labor, and energy trivially stayed within the range of the physically possible. We will show below that past economic growth can be described approximately by the CDE if the constant output elasticities are for labor much smaller and for energy much larger than the cost shares.

The simplest nontrivial solutions of the partial differential equations (4.27) are obtained if one demands that α reflects the law of diminishing returns and that β

---

[18]Analyses of the economic impacts of climate change such as the DICE model of Nordhaus assume that "Output is produced by a Cobb-Douglas production function in capital, labor, and energy" [61].

vanishes if the system approaches the state of maximum automation. The asymptotic boundary conditions

$$\alpha \to 0, \text{ if } (l + e)/k \to 0, \quad \beta \to 0, \text{ if } k \to k_m(y), \quad \text{and} \quad e \to e_m \equiv ck_m(y)$$
$$(4.33)$$

ascertain the required properties of α and β:

- Since machines do not run without energy and (still) require people for handling them, the additional output due to an additional unit of capital must decrease as the ratio of labor and energy to capital decreases. This is the case if α decreases with $(l + e)/k$.
- The capital stock $k_m(y)$, which enters the definition (4.18) of the degree of automation and the constraint equation (4.19), needs the energy quantity $e_m = ck_m(y)$ when working at full capacity in the state of maximally automated production of the output quantity y. The constant c is the energy-demand parameter of the fully employed capital stock. It depends on the state of technology. The vanishing of β in (4.33) is the definition of the state of maximum automation.

The simplest output elasticities that satisfy the differential equations (4.27), constant returns to scale , and the asymptotic boundary conditions (4.33) are [24]

$$\alpha = a\frac{l + e}{k}, \quad \beta = a\left(\frac{cl}{e} - \frac{l}{k}\right), \quad \gamma = 1 - \alpha - \beta = 1 - a\frac{e}{k} - ac\frac{l}{e}. \quad (4.34)$$

The parameter a indicates the effectiveness with which energy activates and labor handles the capital stock. The negative β term is a direct consequence of the choice of α, as can be seen from the integral in (4.29). The positive β term is a special choice of the function $J\left(\frac{l}{e}\right)$ in (4.29) so that the asymptotic boundary condition (4.33) for β is fulfilled.

The restrictions (4.30) are important for understanding the economic meaning of the the output elasticities (4.34) and of the LinEx function (4.35), which follows from them. They imply $\beta \le 1$ and thus require that l goes to zero as e does. This is consistent with the fact that workers lose their jobs if production ceases because of the lack of energy. Furthermore, the asymptotic boundary condition $\beta \to 0$ if $k \to k_m(y)$ and $e \to e_m \equiv ck_m(y)$ in the approach to the state of maximum automation also implies that $\beta \to 0$ if $e \to ck$ for all $k < k_m$. Then $\alpha \to al/k + ac$, and $\gamma \to 1 - ac - al/k \equiv \gamma_1$. If one interprets $\beta \to 0$ for $e \to ck < ck_m$ as a state of 100% capacity utilization, i.e. $\eta = 1$, in any state of automation, then the nonnegative γ_1 is the output elasticity of e in this state when the capital stock is fully employed and the variation of the energy input is associated with the variation of the capital's degree of automation $\rho < \rho_T$. If one knew the $exact$ output elasticities (4.28) and (4.29) which would satisfy the $exact$ boundary conditions for the partial differential equations (4.27), it is to be expected that the technological constraints $\eta(k, l, e) \le 1$ and $\rho(k, l, e) \le \rho_T \le 1$, on the one hand, and the restrictions $\alpha(k, l, e) \ge 0$,

$\beta(k, l, e) \geq 0$ and $\gamma = 1 - \alpha - \beta \geq 0$, on the other hand, would be equivalent in the sense that they define the same volume of (k, l, e)-space within which the factors can vary independently. In the state of maximum automation of the fully employed capital stock, where $k = k_m$, $e = ck_m$, $\rho = 1$, $\eta = 1$, and $\beta = 0$, the *exact* output elasticity of energy should vanish too. This would correspond to the usual case of a limitational production function in three factors when only the factor capital can be varied independently by entrepreneurial decisions and determines growth completely.

Inserting the output elasticities (4.34) into (4.24) at fixed t, and integrating y from y_0 to y_{L1} and the factors from $(1, 1, 1)$ to (k, l, e) along any convenient path, one obtains the (first) LinEx production function

$$y_{L1} = y_0 e \exp\left[a\left(2 - \frac{l + e}{k}\right) + ac\left(\frac{l}{e} - 1\right)\right], \qquad (4.35)$$

which depends *lin*early on energy and *ex*ponentially on quotients of capital, labor, and energy [24]. In contrast to the CDE the restrictions (4.30) *do* constrain the combinations of factors in the LinEx function. These restrictions are written down explicitly in (4.44).

The LinEx function (4.35) contains the technology parameters a (capital effectiveness), c (energy demand of the capital stock), and y_0. It is a phenomenological function that describes the output of an economy *approximately*. Its deviations from the exact production function correspond to the deviations of the asymptotic boundary conditions (4.33) from the exact boundary conditions indicated below (4.31) and in Appendix 6 of Chap. 4. One cannot expect that the LinEx approximation maps all details of production. What matters is the overall picture, and details may be blurred or distorted. This is as inevitable as the incomplete description of the physical world by the natural sciences and their model approximations. In the end, comparison of the model results with experience will eliminate those models whose approximations are too crude.

The phenomenological LinEx parameters a, c, and y_0 become time-dependent, separately or altogether, when creativity acts and the LinEx function acquires an explicit time dependence: $y_{L1} = y_{L1}[k, l, e; t]$. For instance, $c(t)$ decreases when investments in energy conservation measures improve the energy efficiency of the capital stock. This occurred quite noticeably in response to the oil price shocks and is an example for the (thermodynamically limited) substitution of capital for energy. Structural changes by outsourcing energy-intensive industries may also decrease $c(t)$. The energy imported indirectly via imported, energy-intensive (semifinished) goods does not show in the national energy balances in such cases. Thus, the real energy dependence of an economy may be somewhat underestimated.

The LinEx function has the mathematical property that its l–k and e–k isoquants of constant output are convex, and the l–e isoquant is concave within the factor space whose limits are defined by the restrictions (4.30). This concavity reflects the technology of substituting energy (and capital) for labor within the following scheme. Initially, a fixed quantity of capital k of low degree of automation is combined with much labor and little energy. Then, the degree of automation and the energy input increase. At first, energy takes over the performance of hard physical work, such as lifting and transporting cargos, deforming matter, and digging holes. New fields of production open up. With increasing degree of automation at a constant numerical value of k, the activity of workers is more and more restricted to information processing, which mainly consists of the handling of machines. In the production of a unit of output, the necessary amount of information processing requires less energy than the complementary amount of physical work on matter (see also Appendix 3 of Chap. 4). Therefore, less and less additional energy is needed to substitute for the more and more exclusively information processing workers. Consequently, the l–e isoquant is concave. In this context it is important to remember that the process of substituting energy for labor is one of long duration, because increasing automation is possible only according to the technological progress that increases the density of the information processors within the limited volume of the manufacturing units. Therefore, the concavity of the l–e isoquant does not indicate that the LinEx function describes inefficient technologies but rather reflects the technological constraint that inhibits the rapid realization of the state of maximum automation [28].

Another LinEx-type approximation of the exact production function is obtained if one assumes that an economy depends so much on energy conversion that capital cannot produce without energy. Then, capital's output elasticity must vanish for vanishing energy input, and one has the additional asymptotic boundary condition

$$\alpha \to 0, \quad \text{if} \quad e/k \to 0. \tag{4.36}$$

The simplest output elasticities that satisfy (4.27), (4.33), and (4.36) are

$$\alpha = 2a\frac{e}{k}\left(\frac{l}{k}+1\right), \quad \beta = a\frac{l}{e}\left(c^2 - \frac{e^2}{k^2}\right), \quad \gamma = 1 - \alpha - \beta. \tag{4.37}$$

They result in the second LinEx function

$$y_{L2} = y_0 e \exp\left[a\left(3 - 2\frac{e}{k} - \frac{l}{k}\frac{e}{k}\right) + ac^2\left(\frac{l}{e} - 1\right)\right]. \tag{4.38}$$

Higher LinEx functions, which depend in more complicated ways on quotients of k, l, and e, were calculated in [69]. So far, the first LinEx function y_{L1} has been sufficient for describing economic growth. Its relation to the energy-dependent translog function is given in [28].

4.5.4 *Economic Growth in Germany, Japan, and the USA*

The technology parameters a, c, and y_0 of the LinEx function are integration constants of the differential equations (4.27) for the output elasticities and the growth equation (4.24). They must be determined from the empirical time series of output $y_{\text{empirical}}(t)$. The simplest method of approximating them by constants and fitting the LinEx function to $y_{\text{empirical}}(t)$ already reproduces the observed economic growth in Germany, Japan, and the USA with small residuals [20, 28, 29, 37, 38, 69], and the output elasticities obtained for labor and energy differ significantly from the cost shares. However, the Durbin–Watson coefficients d_{W} of autocorrelation have been mostly below 1. The best d_{W} value, indicating the absence of autocorrelations, is 2. The closer one comes to $d_{\text{W}} = 2$, the more confident one can be that all relevant factors have been taken into account.

To see whether a reduction of autocorrelation has a significant impact on the output elasticities of capital, labor, and energy, we allow for time dependencies of the technology parameters and model them by logistic functions, which are typical for growth in complex systems and innovation diffusion. Let $p(t)$ represent either the capital-effectiveness parameter $a(t)$ or the energy-demand parameter $c(t)$. Its logistic differential equation

$$\frac{d}{dt}\left(p(t) - p_2\right) = p_3\left(p(t) - p_2\right)\left(1 - \frac{p(t) - p_2}{p_1 - p_2}\right) \tag{4.39}$$

has the solution [70]

$$p(t) = \frac{p_1 - p_2}{1 + \exp\left[-p_3\left(t - t_0 - p_4\right)\right]} + p_2, \tag{4.40}$$

with the free (characteristic) coefficients $p_1, \ldots, p_4 \geq 0$; the variable t is dimensionless and given by "time interval divided by one year." As an alternative to logistics, we have also looked into Taylor expansions of $a(t)$ and $c(t)$ in terms of $t - t_0$ with a minimum of free coefficients. With that the output elasticity of creativity δ, defined in (4.25), is given by

$$\delta = \frac{(t - t_0)}{y_{\text{L1}}}\left[\frac{\partial y_{\text{L1}}}{\partial a}\frac{da}{dt} + \frac{\partial y_{\text{L1}}}{\partial c}\frac{dc}{dt} + \frac{\partial y_{\text{L1}}}{\partial y_0}\frac{dy_0}{dt}\right]. \tag{4.41}$$

The free coefficients of the logistic functions, or of the Taylor expansions, are determined by minimizing the sum of squared errors (SSE),

$$\text{SSE} = \sum_i\left[y_{\text{empirical}}(t_i) - y_{\text{L1}}(t_i)\right]^2. \tag{4.42}$$

The sum goes over all years t_i between the initial and the final observation time. It contains the empirical time series of output $y_{\text{empirical}}(t_i)$, and the LinEx function

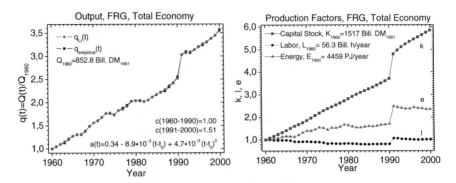

Fig. 4.3 *Left*: Empirical growth *(squares)* and theoretical growth *(circles)* of the dimensionless output $y \equiv q = Q/Q_{1960}$ of the total economy of the Federal Republic of Germany *(FRG)* between 1960 and 2000. *Right*: Empirical time series of the dimensionless factors capital $k = K/K_{1960}$, labor $l = L/L_{1960}$, and energy $e = E/E_{1960}$ [71]. The five coefficients that model the time dependence of the LinEx function parameters a and c and reproduce the drastic structural break at German reunification in 1990 are shown below the output curve

$y_{Lt}(t_i)$ with the empirical time series of k, l, and e as inputs at times t_i. Minimization is subject to the constraints (4.30) that the output elasticities of (4.34) must be nonnegative:

$$\alpha = a \frac{l + e}{k} \geq 0, \quad \beta = a \frac{l}{e}\left(c - \frac{e}{k}\right) \geq 0, \quad \gamma = 1 - a\frac{e}{k} - ac\frac{l}{e} \geq 0. \quad (4.43)$$

These constraints turn into restrictions on $a(t)$ and $c(t)$, or on k, l, and e for given a and c:

$$0 \leq a(t) \leq a_{\max}(t) \equiv \frac{k(t)}{l(t) + e(t)}, \quad \frac{e(t)}{k(t)} \equiv c_{\min}(t) \leq c(t),$$

$$0 \leq a(t)\left[\frac{e}{k} + c(t)\frac{l}{e}\right] \leq 1. \quad (4.44)$$

SSE minimization has been done with the Levenberg–Marquardt method of nonlinear optimization [68] in combination with a new, self-consistent iteration procedure that helps avoid divergences in the fitting procedure or convergence in a side minimum [71]. Details are given in Appendix 6 of Chap. 4.

Data consistency is essential for SSE minimization. It was relatively easy to get consistent sets of data for the German systems – Figs. 4.3, 4.4, 4.13, and 4.14 and Tables 4.11 and 4.12 in Appendix 5 of Chap. 4 – from economic research institutions such as the ifo Institut für Wirtschaftsforschung (Munich), the Statistische Bundesamt (Wiesbaden), and the Institut für Arbeitsmarkt und Berufsforschung (Nuremberg). The Japanese data – Figs. 4.5 and 4.15 and Table 4.13 – were collected during personal visits to Japan with the kind help of Shigeru Yasukawa and Osamu

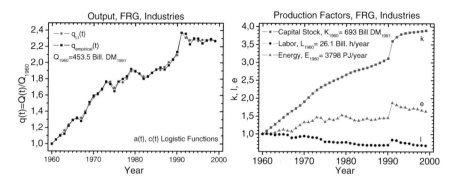

Fig. 4.4 *Left*: Empirical growth (*squares*) and theoretical growth (*circles*) of the dimensionless output $y \equiv q = Q/Q_{1960}$ of the German industrial sector "goods-producing industries" between 1960 and 1999. *Right*: Empirical time series of the dimensionless factors capital $k = K/K_{1960}$, labor $l = L/L_{1960}$, and energy $e = E/E_{1960}$ [71]

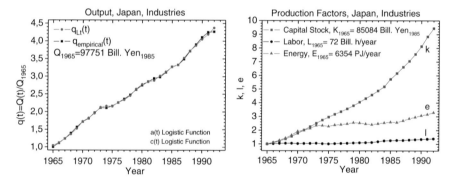

Fig. 4.5 *Left:* Empirical growth (*squares*) and theoretical growth (*circles*) of the dimensionless output $y \equiv q = Q/Q_{1965}$ of the Japanese sector "industries" between 1965 and 1992. *Right*: Empirical time series of the dimensionless factors capital $k = K/K_{1965}$, labor $l = L/L_{1965}$, and energy $e = E/E_{1965}$ [71]

Sato from the Energy System Assessment Laboratory, Tokay Mura, and Kokichi Ito from the Institute of Energy Economics, Tokyo. Buildup of the US database – Figs. 4.6 and 4.16 and Table 4.14 – started in the Department of Economics at Harvard University. Dale W. Jorgenson was very helpful in this. Updates were made during stays at the Institute of Energy Analysis, Oak Ridge, and the University of California at Berkeley.

   German reunification on October 3, 1990 provides an interesting test of the model. The sudden merger of the planned economy of the German Democratic Republic with the market economy of the Federal Republic of Germany was a result of political, social, and economic decisions with far-reaching consequences. It turns out that it is possible to model this working of "creativity" phenomenologically by

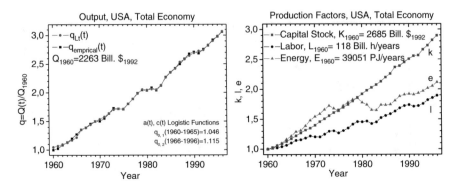

Fig. 4.6 *Left*: Empirical growth (*squares*) and theoretical growth (*circles*) of the dimensionless output $y \equiv q = Q/Q_{1960}$ of the total US economy between 1960 and 1996. *Right*: Empirical time series of the dimensionless factors capital $k = K/K_{1960}$, labor $l = L/L_{1960}$, and energy $e = E/E_{1960}$ [71]

Table 4.4 The free coefficients of $a(t)$ and $c(t)$ in the logistics for the Federal Republic of Germany's industrial sector "goods-producing industries" (*FRG I*), the Japanese sector "industries" (*Japan I*), and the total economy of the USA (*USA TE*)

System	a_1	a_2	a_3	a_4	c_1	c_2	c_3	c_4
FRG I	0.33	0.67	0.19	32	1	1.46	19.1	31
Japan I	0.16	0.2	1.87	20.1	2.75	0.45	0.86	14.61
USA TE	0.21	0.49	0.97	22.64	2.63	0.81	0.81	17.24

just five free coefficients that enter the Taylor series expansion for $a(t)$ and the combination of step functions[19] for $c(t)$ in the model for Germany's total economy: $a(t) = 0.34 - 8.9 \times 10^{-3}(t - t_0) + 4.7 \times 10^{-3}(t - t_0)^2$, $c(t) = 1 \times \theta(1990 - t) + 1.51 \times \theta(t - 1991)$, where the step function $\theta(x)$ is 1 for $x \geq 0$ and 0 otherwise. For the other systems considered a and c are modeled by the logistics function (4.40), with $p_j \equiv a_j$ for $a(t)$ and $p_j \equiv c_j$ for $c(t)$, $j = 1, 2, 3, 4$. The $\{a_j\}$ and $\{c_j\}$ listed in Table 4.4 result from SSE minimization with an iteration procedure that observes that the proper starting values for the numerical iteration (with up to 32,000 iteration steps) are crucial for convergence in the global minimum.

The growth curves obtained from the LinEx function with the $a(t)$ and $c(t)$ are shown in Figs. 4.3–4.6. In these figures dimensionless output $y \equiv Y(K, L, E; t)/Y_0$ is abbreviated by $q(t) \equiv Q(t)/Q_{\text{baseyear}}$ and the LinEx function $y_{L1}(k, l, e; t)$ is abbreviated by q_{Lt}. We note that the theoretical outputs $y_{L1} \equiv q_{Lt}$ closely follow the empirical ones. They also reflect the ups and downs of the energy inputs during

---

[19]Step function results from the logistic (4.40) for $p_3 \to \infty$.

the energy crises of 1973–1975 and 1979–1981. The Solow residual is absent. The sudden enlargement of the system "Federal Republic of Germany" at reunification in 1990 is satisfactorily reproduced too.

The empirical time series of inputs differ significantly in Figs. 4.3–4.6, especially for routine labor l. The number of hours worked per year decreased almost continuously for 40 years in Germany's total economy and in its industrial sector "goods-producing industries," except for the small jumps after reunification on October 3, 1990. It stayed nearly constant in Japan, and almost doubled in the USA. In Germany the introduction of the 35-h week in the 1980s had the effect of a "rationalization whip", as the boss of the German labor unions, Michael Sommer, once admitted himself. This has reenforced the trend to increasing automation, which had already existed before, when Germany's capital stock, destroyed in World War II, was being rebuilt and expanded with the most modern machinery. (There was no social trouble, as long as people who lost their jobs to automation found new jobs in the expanding economy, or fell into the tightly knit social net.) Japan has a similar postwar history. But here part of the culture of firms was not to fire employees in times of recession but to maintain lifelong employment. This produced the enormous loyalty of Japanese workers toward their firms. In the USA, the labor-intensive service sector expanded early and vigorously. Furthermore, the number of "working poor," who must hold more than one poorly paid job to make a decent living, has grown in the USA more rapidly than elsewhere; it reached about 25% of the total US workforce by 2004 [73]. Despite the rather different evolutions of the factor labor, which is considered as the most important production factor by standard growth theory, between 1960 and 1989 real GDP in the Federal Republic of Germany and the USA grew by about the same factors, 2.5 and 2.7, respectively. Apparently, growth of capital stock and growth of energy input are more important than growth of routine labor. (The output elasticities in Table 4.5 are consistent with this observation.) The data on energy inputs indicate that after the first oil price shock, incentives for energy conservation were stronger and worked better in Germany and Japan than in the USA; the former two countries do not have domestic oil, whereas the USA does.[20]

As Fig. 4.7 shows, the inflation-corrected oil price dropped to a near-historic minimum between 1945 and 1970. It "exploded" between 1973 and 1981, and then returned to low levels until the turn of the century. Abundant cheap oil has been one of the principal bases of economic growth since the end of World War II. The oil price shocks of 1973–1975 and 1979–1981 were accompanied by the recessions visible in the growth curves in Figs. 4.3–4.6, which closely follow the curves of energy input. This strong coupling of energy and output had several reasons. One was the worry of producers that energy, without which capital lies idle, may become scarce, so they invested less. For instance, during the first oil price shock, investment

---

[20]In addition, in all highly industrialized countries structural changes away from energy-intensive industries have occurred. More and more energy-intensive goods are being imported from the sectors iron, mining, and quarrying of countries with emerging economies such as China.

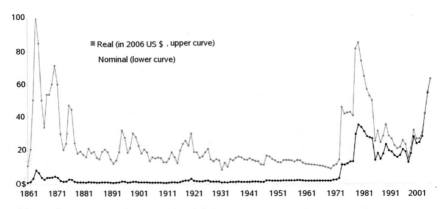

Fig. 4.7 Development of the oil price since 1861 in real 2006 US dollars (*upper curve*) and in nominal dollars (*lower curve*): 1861–1944 US average, 1945–1985 Arabian Light, 1986–2006 Brent spot. See also Fig. 2.5. (Source: Energy Information Administration. Updates at http://en.wikipedia.org/wiki/File%3AOil_Prices_1861_2007.svg)

in the West German industrial sector "goods-producing industries" fell from about DM90 billion in 1973 to less than DM50 billion in 1975 [28]. Another reason was that the buying power of domestic consumers was transferred to the oil producers, so people could buy less. Consequently, reductions in the production of investment and consumer goods reduced the degree of capacity utilization (4.21), with energy inputs going down immediately. Subsequently, labor inputs were reduced, unemployment rose, and consumer confidence collapsed. All this resulted in economic recessions. Whether the rise of the oil price that started in 2004, and is shown until 2009 in Fig. 2.5, was one reason for the collapse of the US housing market, because increasing gasoline bills of commuters contributed to their financial problems, is not yet clear.

4.5.5 The Productive Powers of Capital, Labor, Energy, and Creativity

The output elasticities of capital, labor, energy, and creativity measure the productive powers of these factors. They are computed by inserting the technology parameters $a(t)$ and $c(t)$ and the time series of k, l, and e into α, β, and γ in (4.34) and δ in (4.41). (The law of error propagation produces the largest errors for δ according to (4.41).) Figures 4.8 and 4.9 exhibit the temporal variations of α, β, γ, and δ.

The relatively smooth temporal variations of γ in Figs. 4.8 and 4.9 (with the exception of the 1990 "reunification drop" in the German systems) indicate that the high output elasticity of energy is not a consequence of variations in capacity utilization, as they occurred during the energy crises. It is rather due to the technological dynamics in the evolution of work performance and information processing. These

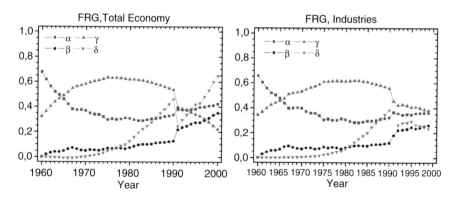

Fig. 4.8 Time-dependent output elasticities α *(squares)*, β *(circles)*, γ *(upright triangles)*, and δ *(inverted triangles)* in the total economy of the Federal Republic of Germany *(left)* and in its industrial sector "goods-producing industries" *(right)* [71]

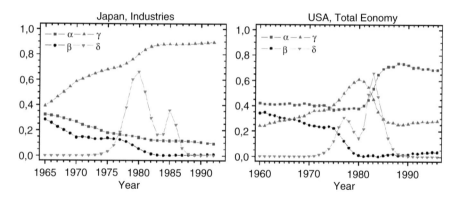

Fig. 4.9 Time-dependent output elasticities in the Japanese sector "industries" *(left)* and in the total US economy *(right)* [71]

elementary processes of production are executed with growing intensity by the capital stock and the energy that activates it. This steadily enhances the realm of energy utilization and constitutes the major part of Solow's "technological progress." Of course, efficiency changes, as indicated by the variations of δ, also contribute somewhat to economic growth, especially during the times of the first and the second oil price shock, when δ has two peaks in Japan and the USA. As shown in [71], the capital efficiency parameter $a(t)$ increases in all systems considered, and the energy demand parameter $c(t)$ decreases for Japan and the USA; if one considers Germany only from 1960 to 1989, $c(t)$ decreases too. However, as indicated in the inset in Fig. 4.3, $c(t)$ rises steeply in 1991 for the German systems, reflecting that the energetically rather inefficient capital stock of the former German Democratic Republic was added to the capital stock of the Federal Republic of Germany. (The corresponding sharp negative peaks of δ in 1991 are not shown in Fig. 4.8.) The

Table 4.5 Time-averaged LinEx output elasticities and statistical quality measures. *FRG TE* total economy of the Federal Republic of Germany before and after reunification, *FRG I* German industrial sector "goods-producing industries," *Japan I* Japanese sector "industries," which produces about 90% of Japanese GDP, *USA TE* total economy of the USA, R^2 the adjusted coefficient of determination, d_W Durbin–Watson coefficient

	FRG TE	FRG I	Japan I	USA TE
System	1960–2000	1960–1999	1965–1992	1960–1996
$\bar{\alpha}$	0.38 ± 0.09	0.37 ± 0.09	0.18 ± 0.07	0.51 ± 0.15
$\bar{\beta}$	0.15 ± 0.05	0.11 ± 0.07	0.09 ± 0.09	0.14 ± 0.14
$\bar{\gamma}$	0.47 ± 0.1	0.52 ± 0.09	0.73 ± 0.16	0.35 ± 0.11
$\bar{\delta}$	0.19 ± 0.2	0.12 ± 0.13	0.14 ± 0.19	0.10 ± 0.17
R^2	>0.999	0.996	0.999	0.999
d_W	1.64	1.9	1.71	1.46

economic system of what was and is the Federal Republic of Germany underwent a dramatic change on reunification on October 3, 1990. Therefore, the contribution of human-decision-determined "creativity" to German economic evolution was relatively large in the last decade of the twentieth century.

In the KLEC model primary energy input, as listed by the national energy balances, is the energy variable. Changes of the capital stock's energy conversion efficiency contribute to the change with time of the technology parameters a and c. Furthermore, outsourcing the production of energy-intensive intermediate goods, and limiting the generation of value added to importing and upgrading them, simulates improvements of energy efficiency.

The time-averaged output elasticities of capital, $\bar{\alpha}$, labor, $\bar{\beta}$, energy, $\bar{\gamma}$, and creativity, $\bar{\delta}$ are presented in Table 4.5 and Figs. 4.10–4.12 for the following economic systems: total economy of the Federal Republic of Germany before and after reunification, German industrial sector "goods-producing industries," Japanese sector "industries," which produces about 90% of Japanese GDP, and total economy of the USA. R^2 is the adjusted coefficient of determination, and d_W is the Durbin–Watson coefficient. Both statistical quality measures turn out to be quite good.[21]

Five free coefficients have been fitted for the total economy of the Federal Republic of Germany before and after reunification, and eight for the German industrial sector 'goods-producing industries" and the Japanese sector "industries." For the total economy of the USA, where k, l, and e are nearly parallel between 1960 and 1965, the fitting algorithm is unstable with respect to minimal changes of the starting values if one fits from 1960 to 1996. This is a well-known problem of

---

[21]The neoclassical macroeconometric multicountry model of capital and labor of the German Bundesbank [58] has R^2 between 0.997 and 0.97 and d_W between 0.65 and 0.24. Estimations of GDP were made for the USA, Japan, Germany, the UK, France, Italy, Canada, Netherlands, and Belgium from the beginning of 1974 to the end of 1995.

Fig. 4.10 Time-averaged output elasticities (productive powers) in the total economy of the Federal Republic of Germany (*top*) and in Germany's industrial sector "goods-producing industries" (*bottom*) [72]

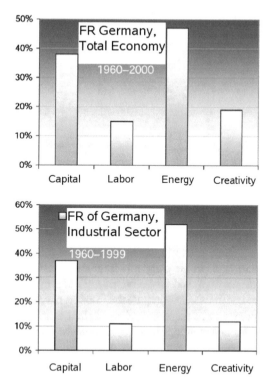

Fig. 4.11 Time-averaged output elasticities (productive powers) in the Japanese sector "industries" [72]

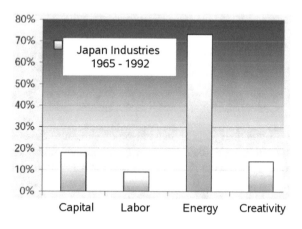

multicollinearity. If one excludes the first 5 years and fits the eight free coefficients in the logistic functions for $a(t)$ and $c(t)$, one obtains a nearly perfect US fit with no autocorrelations ($R^2 = 0.999$ and $d_W = 2.05$), and the time-averaged output elasticities are close to those of the total economy of the USA in Table 4.5 [71]. The latter ones were obtained with the two different values of $q_0(= y_0)$ before and after 1965, which are indicated in Fig. 4.6.

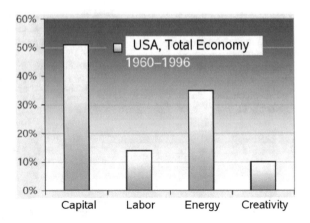

Fig. 4.12 Time-averaged output elasticities (productive powers) in the total US economy [72]

There is an old saying that with a sufficient number of fitting parameters one can even fit an elephant. Because of this we emphasize that the main purpose of the improved modeling of $a(t)$ and $c(t)$ by logistic functions or Taylor series expansions is to compute the maximum contribution to output that may result from creativity in the sense of an explicit time dependence of the production function. This contribution does not significantly reduce the relative weight of energy: the output elasticities in Table 4.5 do not deviate much from those that were obtained from simpler fitting procedures with constant a and c, except for one recalibration in 1978 [20, 28, 29, 38, 74]. This is also confirmed in the next section by cointegration analysis.

> The overall conclusion is that the output elasticity of energy is always much larger than energy's cost share of roughly 5%, and the output elasticity of routine labor is much smaller than labor's cost share of about 70%.

Lindenberger [75], in a reasoning similar to the one that leads to the LinEx function (4.35), calculated the service production function $y_{S1} = y_0 l (e/l)^{ac_m} \exp\{a[2 - (l + e)/k]\}$ as the integral of the growth equation (4.24) with the output elasticities $\alpha = a(l + e)/k$, $\gamma = a(c_m - e/k)$ and $\beta = 1 - \alpha - \gamma$. These elasticities satisfy the set of partial differential equations (4.27) too. The production function y_{S1} reproduces the evolution of German market-determined services between 1960 and 1989 satisfactorily; its time-averaged output elasticities are for labor 0.31 until 1977 and 0.26 after 1978, and for (final) energy they are 0.15 and 0.21, respectively. Thus, even in the labor-dominated sector of the economy, and during times of much less computerized information processing than at present, energy's economic weight substantially exceeds energy's cost share.

4.5.6   Cointegration

How sensitive are the heterodox output elasticities of energy and labor to the mathematical methods of computing them? To answer this question, the output elasticities have been checked by a method that is independent of the concept of the LinEx function. This method is cointegration analysis. The following is a brief summary of the cointegration studies [76, 77].

Simply speaking, cointegration analysis checks whether a linear combination of a number of nonstationary time series is a stationary time series itself. If this is the case, the time series variables are said to be cointegrated, meaning that they are statistically significantly connected. In other words, there is no accidental correlation between the variables, as was the case when the number of babies in Sweden decreased like the number of storks in that country. The literature on cointegration analysis of time series involving output and energy is reviewed in [76]. It also shows explicitly that the time series of output, capital, labor, and energy that are plotted in Figs. 4.13–4.16 are nonstationary and so they can be submitted to cointegration testing.

Cointegration analysis also provides information on output elasticities if one considers the variables

$$\tilde{y}_t \equiv \ln y(t), \ \tilde{k}_t \equiv \ln k(t), \ \tilde{l}_t \equiv \ln l(t), \ \tilde{e}_t \equiv \ln e(t) \tag{4.45}$$

and tests for which values of the constants α_0, β_0, γ_0, and $\tilde{y}_0$ the linear combination

$$u_t = \tilde{y}_t - \tilde{y}_0 - \alpha_0 \tilde{k}_t - \beta_0 \tilde{l}_t - \gamma_0 \tilde{e}_t \tag{4.46}$$

is stationary and likely to represent a cointegration relation. If a set of constants can be found for which (4.46) is indeed stationary, one calls $(1, -\alpha_0, -\beta_0, -\gamma_0)$ the cointegration vector. The right-hand side of (4.46), which can be written as the scalar product of that vector with the vector $(\tilde{y}_t - \tilde{y}_0, \tilde{k}_t, \tilde{l}_t, \tilde{e}_t)$, results from the logarithm of the equation that gives the energy-dependent Cobb–Douglas function (CDE)

$$y_{CDE} \equiv y(t) = y_0 k(t)^{\alpha_0} l(t)^{\beta_0} e(t)^{\gamma_0}, \quad \gamma_0 = 1 - \alpha_0 - \beta_0. \tag{4.47}$$

The CDE was derived earlier as the simplest possible integral of the growth equation (4.24). Stationarity of u_t allows the following conclusion: If the hypothesis of cointegration can be accepted and one identifies the cointegration vector components α_0, β_0, and $\gamma_0 = 1 - \alpha_0 - \beta_0$ with the Cobb–Douglas output elasticities for capital, labor, and energy, then the CDE (4.32, 4.47) is a valid description of output as a function of capital, labor, and energy.

Following Engle and Granger [78], cointegration analysis is done with the standard augmented Dickey–Fuller tests [79, 80], which model the time dependence of the variable u_t in (4.46) by the stochastic process

$$u_t = \rho u_{t-1} + \sum_{i=0}^{p} \theta_i \Delta u_{t-i} + \varepsilon_t. \tag{4.48}$$

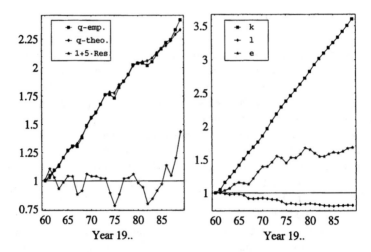

Fig. 4.13 *Left*: Empirical (*squares*) and theoretical (*diamonds*) growth of output of the West German total economy for 1960–1989; $y_0 = 0.998$, and the output elasticities of the energy-dependent Cobb–Douglas function, as obtained from cointegration analysis, are $\alpha_0 = 0.522(\pm 0.029)$, $\beta_0 = 0.094(\pm 0.026)$, and $\gamma_0 = 0.384(\pm 0.039)$; the residuals below the output curves are enlarged by a factor of 5 and shifted upward by 1; $d_W = 0.848$. *Right*: Empirical time series of capital, energy, and labor, *from top to bottom* [77]

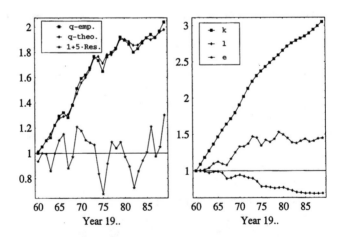

Fig. 4.14 *Left*: Empirical (*squares*) and theoretical (*diamonds*) growth of output of the German industrial sector "goods-producing industries" for 1960–1989; $y_0 = 1.013$, and the Cobb–Douglas output elasticities are $\alpha_0 = 0.442(\pm 0.050)$, $\beta_0 = 0.041(\pm 0.040)$, and $\gamma_0 = 0.517(\pm 0.064)$; residuals below the output curves are enlarged and shifted as in Fig. 4.13; $d_W = 1.165$. *Right*: Empirical time series of capital, energy, and labor, *from top to bottom* [77]

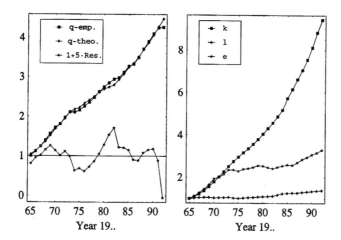

Fig. 4.15 *Left*: Empirical (*squares*) and theoretical (*diamonds*) growth of output of the Japanese sector "industries" for 1965–1992; $y_0 = 1.039$, and the Cobb–Douglas output elasticities are $\alpha_0 = 0.433(\pm0.041)$, $\beta_0 = 0.217(\pm0.024)$, and $\gamma_0 = 0.350(\pm0.027)$; residuals below the output curves are enlarged and shifted as in Fig. 4.13; $d_W = 0.562$. *Right*: Empirical time series of capital, energy, and labor, *from top to bottom* [77]

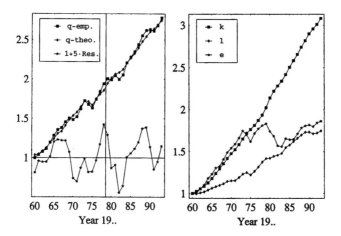

Fig. 4.16 *Left*: Empirical (*squares*) and theoretical (*diamonds*) growth of output of the US sector "industries," which produces about 85% of US GDP; the Cobb–Douglas output elasticities for 1960–1978 are $\alpha_0 = 0.239(\pm0.142)$, $\beta_0 = 0.098(\pm0.057)$, and $\gamma_0 = 0.663(\pm0.153)$; $y_0 = 1.037$; and for 1979–1993 are $\alpha_0 = 0.81(\pm0.151)$, $\beta_0 \approx 0(\pm0.252)$, and $\gamma_0 = 0.19(\pm0.294)$; $y_0 = 0.981$; residuals below the output curves are enlarged and shifted as in Fig. 4.13; for case 1 $d_W = 0.570$ and for case 2 $d_W = 0.852$. *Right*: Empirical time series of capital, energy, and labor, *from top to bottom* [77]

$\Delta u_{t-i} \equiv u_{t-i} - u_{t-i-1}$ is the so-called first difference of u_{t-i}; ε_t is supposed to be independent and identically distributed, following a zero-mean, constant-variance normal distribution, in short: $\varepsilon_t \sim$ i.i.d.N$(0, \sigma^2)$. Our tests involved only $p = 0, 1, 2$, with $\theta_0 \equiv 0$.

According to the theory of stochastic processes, a time series like the linear combination (4.46) is stationary if its mean value, variance, and autocovariance exist, are finite, and are time-independent [81,82]. This, in turn, is the case if $\rho < 1$ in (4.48). On the other hand, if $\rho = 1$, the stochastic process is nonstationary like the "random walk," for which the variance of u_t diverges; then one says that the time series exhibits a "unit root." (The cases $\rho > 1$ and $\rho < 0$, i.e., exploding and alternating time series, do not apply to macroeconomic data, as a rule.)

The regression coefficients ρ and θ_i are estimated by the method of ordinary least squares. The ordinary least squares estimate $\hat{\rho}$ and its standard error $\hat{\sigma}_\rho$ define the quantity

$$\tau = \frac{\hat{\rho} - 1}{\hat{\sigma}_\rho}. \tag{4.49}$$

The distribution of τ for time series with a unit root, i.e., $\rho = 1$, is known from Monte Carlo simulations of Dickey and Fuller [79] and MacKinnon [83]. From this one obtains the likelihood that time series with known $\hat{\rho}$ and $\hat{\sigma}_\rho$ do *not* have a unit root, so they are stationary. The statistical criteria that have been applied to the time series (4.46) with data for Germany, Japan, and the USA, and from which the likelihood of stationarity can be inferred, are presented in detail in [76]. The standard testing methods have been supplemented by a new graphical method that involves the criteria that only those cointegration vectors are acceptable whose α_0, β_0, and γ_0 are nonnegative, so that they make economic sense as Cobb–Douglas output elasticities; minimum residual sum of squares and maximum Durbin–Watson coefficients d_W (below 2) have served as additional selection criteria.

The time series (4.46) turns out to be stationary at the 90–95% confidence levels for the data and cointegration vectors of the economic systems identified in Figs. 4.13–4.16.

The time series for 1960–1996 for the total US economy are only cointegrated under the constraints of nonnegative output elasticities if one includes a time trend, which results in a CDE with a time-dependent exponential; its output elasticities agree with the time-averaged LinEx output elasticities of capital, labor, and energy for "USA TE" in Table 4.5 within the error margins [76], and the Durbin–Watson coefficient is $d_W = 0.584$. In the German, Japanese, and US economic systems of Figs. 4.13–4.16, for which the CDE is an acceptable cointegration relation, the output elasticities deviate somewhat from those in Table 4.5, but the main LinEx finding is confirmed: they are for labor much smaller and for energy much larger than the respective cost shares. The residual sum of squares exceeds 0.99 for all systems, but the Durbin–Watson coefficients are much below those of the LinEx function. In contrast to the LinEx function, the Cobb–Douglas function with its constant output elasticities cannot reproduce German economic growth after reunification. Therefore, cointegration analysis (without structural breaks) is

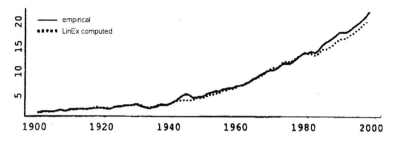

Fig. 4.17 Empirical (*solid line*) and LinEx-computed (*dotted line*) growth of GDP in the USA between 1900 and 1998. The number 20 on the ordinate corresponds to about US $\$_{1990}7$ trillion [87]

possible for Germany only between 1960 and 1989. The CDE reproduces the recessions of the energy crises weakly for Germany, but averages them out for the Japanese sector "industries." For the US sector "industries" the first, and only the first, energy crisis can be reproduced if one splits the total observation time into the two intervals: 1960–1978 and 1979–1993. In the second interval the CDE essentially follows the growth of the capital stock; it does the same during the whole observation time if one foregoes recalibration [77].

In summary, cointegration analysis and CDEs confirm that energy is a powerful factor of production, and (routine) labor is a weak one. Furthermore, CDEs, with output elasticities that are much closer to the time-averaged LinEx output elasticities than to the cost shares, can roughly describe past economic growth.

4.5.7   The "Useful Work" Approach

Ayres and Warr replaced primary energy by "useful work" in the LinEx function. "Useful work" is defined as exergy, multiplied by appropriate conversion efficiencies, plus physical work by animals. The "useful work" data [84] already include most of the efficiency improvements that occurred in energy-converting systems during the twentieth century. In this case, two constant technology parameters suffice to reproduce well the GDP of the US economy between 1900 and 1998 [85, 86]; see Fig. 4.17. The corresponding output elasticities of capital, labor, and energy are shown in Fig. 4.18. The time averages of the output elasticities are similar to the ones in Table 4.5.

Ayres and Warr also checked how things work out if one uses the CDE (4.32) with "useful work" as an energy variable. Figures 4.19 and 4.20 show the results for the USA and Japan between 1900 and 2005. The constant Cobb–Douglas output elasticities of capital, labor, and useful work are $\alpha_0 = 0.33$, $\beta_0 = 0.31$, and $1 - \alpha_0 - \beta_0 = 0.35$ for prewar USA, $\alpha_0 \geq 0.75$, $\beta_0 \approx 0$, and $1 - \alpha_0 - \beta_0 \approx 0.25$ for postwar USA, $\alpha_0 = 0.29$, $\beta_0 = 0.31$, and $1 - \alpha_0 - \beta_0 = 0.40$ for prewar Japan, and $\alpha_0 = 0.39$, $\beta_0 = 0.27$, and $1 - \alpha_0 - \beta_0 = 0.34$ for postwar Japan [86]. During the postwar

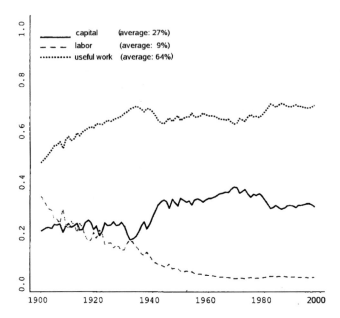

Fig. 4.18 Output elasticities of capital (*solid line*), labor (*dashed line*), and useful work (*dotted line*) for the USA between 1900 and 1998 [87]

Fig. 4.19 Empirical GDP (*solid line*) and Cobb–Douglas-estimated GDP (*squares*) in the USA between 1900 and 2005, excluding 1941–1948 (with rounding errors). (Source: Robert U. Ayres and Benjamin Warr)

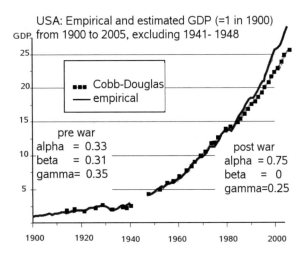

period, the CDE describes growth only to a rough approximation. (Reproduction of GDP between 1900 and 2005 by the LinEx function, not shown here, is better [86].) Nevertheless, the CDE output elasticities are much closer to the LinEx elasticities than to the factor cost shares.

Fig. 4.20 Empirical GDP (*solid line*) and Cobb–Douglas-estimated GDP (*circles*) in Japan between 1900 and 2005, excluding 1941–1948 (Source: Robert U. Ayres and Benjamin Warr)

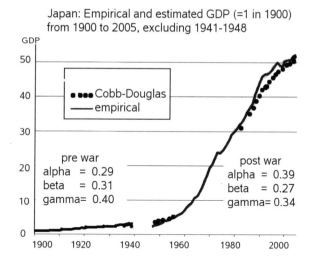

Whatever the details of modeling energy's working as a factor of production, one result is common to all models:

If one foregoes cost-share weighting and determines energy's output elasticity econometrically, one gets numbers up to 10 times larger than the cost share, and the Solow residual disappears. The production factor energy accounts for most of the growth that neoclassical economics attributes to "technological progress."

4.5.8   Pollution, Recycling, and Perspectives on Growth

Economic evolution of Germany, Japan, and the USA was described in the previous sections without considering any environmental and resource limits. This should be more or less justified, because in the past the fraction of capital, labor, and energy that went into pollution abatement such as desulfurization, denitrification, and dedusting, on the one hand, and into recycling, on the other hand, was relatively small. Things are likely to change in the future, when an increasing share of all factors will have to be dedicated to emission mitigation, and to the recovery of materials whose natural deposits are being depleted. This share will be missing for the production of the traditional basket of goods and services people enjoyed in the past. Examples of energy requirements for pollution abatement in electricity generation are given in Sect. 3.6.3. Of course, one could continue to include all work performance and information processing, no matter for what purpose, in GDP. But then GDP would lose much of its significance as a measure of material wealth.

The following describes a crude model for the evolution of traditional value added during the approach to environmental and material limits to growth [3]. It is quite likely that, as in the past, the national statistics on capital, labor, and energy will just report the input quantities without indicating the purposes for which they are used. Therefore, the fact that an additional factor unit will no longer fully contribute to the growth of traditional value added, if part of it is dedicated to emission mitigation and intricate recovery of scarce materials from complex worn-out products, has to be modeled explicitly. This is done by pollution and recycling functions that multiply the output elasticities of capital, labor, and energy in the growth equation (4.24). These functions should gradually go to zero as entropy production and material restrictions force the economy on a path of diminishing wealth creation and increasing environmental protection and resource conservation.

To keep the modeling simple, we consider only the global economy in interaction with the natural environment. The construction of the pollution function is done explicitly. Construction of the recycling function is analogous.

The starting point for the pollution function is the entropy production density $\sigma_S(\mathbf{r}, t) = \sigma_{S,\mathrm{dis}}(\mathbf{r}, t)$, given by (4.1). Integration of it over the system volume V, which is essentially that of the biosphere, yields the entropy production rate at time t as

$$\frac{d_i S}{dt} = \int_V \sigma_S(\mathbf{r}, t) dV. \tag{4.50}$$

From this one can define the *average* entropy production density within V as

$$\frac{1}{V} \frac{d_i S}{dt} = \frac{1}{V} \int_V \sigma_S(\mathbf{r}, t) dV. \tag{4.51}$$

Inserting the right-hand side of (4.1) into the integral yields

$$\frac{1}{V} \frac{d_i S}{dt} = \frac{1}{V} \int_V \{\mathbf{j}_Q \nabla(1/T) + \sum_{k=1}^{N} \mathbf{j}_k [-\nabla(\mu_k/T) + \mathbf{f}_k/T]\} dV. \tag{4.52}$$

With the definitions

- *Thermal* pollution

$$p_{\mathrm{th}} \equiv \frac{1}{V} \int_V \mathbf{j}_Q \nabla \frac{1}{T} dV, \tag{4.53}$$

- *Chemical* pollution by particles of type k

$$p_k \equiv \frac{1}{V} \int_V \mathbf{j}_k [-\nabla(\mu_k/T) + \mathbf{f}_k/T] dV, \tag{4.54}$$

- *Total* pollution

$$p \equiv \frac{1}{V} \frac{d_i S}{dt}, \tag{4.55}$$

(4.52) becomes

$$p = p_{th} + \sum_{k=1}^{N} p_k. \tag{4.56}$$

Thus, total pollution in a given volume V is formally defined as the sum of thermal and material entropy production densities averaged over the volume. The identification of entropy production with pollution is justified by the consideration presented in Sect. 4.2: The heat and particle current densities $\mathbf{j}_Q$ and $\mathbf{j}_k$, which determine p_{th} and p_k, execute the command of the second law of thermodynamics to spread energy and matter as evenly as possible. In the nonequilibrium system of planet Earth, this is done by the heat and particle current densities that emanate from the furnaces, heat engines, and reactors of the energy conversion facilities. These emissions change the energy flows through and the chemical composition of the biosphere to which the living species and their populations have adapted (more or less) optimally in the course of evolution. If these changes are so strong that they cannot be undone by the biological and anorganic processes that are driven by exergy insolation from the Sun and by heat radiation from Earth into space, and if they occur so rapidly that biological, social, and technological adaptation deficits develop, the emissions are perceived as environmental pollution. (An example of a technological deficit of adaptation to rising temperatures is the cooling problem of power plants when the rivers that supply the cooling water become too hot, or even dry up. Another one is the breakdown of air conditioning in the German high-speed ICE trains during the heat spell of July 2010, when outside air temperatures rose to more than 35°C. The systems were designed for temperatures below 33°C. When the systems ceased to operate, lots of passengers collapsed at temperatures of 50°C inside the airtight trains.)

This understanding of pollution involves critical limits and natural purification rates. Therefore, the numerical value of p, as it can be computed in principle from the above equations, cannot capture alone the repercussions of entropy production on ecological and social systems. At least two classes of phenomenological parameters are required in addition. The first class contains the critical limits p_{Cth} for thermal pollution, and p_{Ck} $(k = 1, \dots, N)$ for the different types of polluting molecules. In the end, society has to fix these limits on the basis of medical, chemical, biological, and climatological risk assessments. (The negotiations on the United Nations Framework Conventions on Climate Change show that agreement on such limits is often only reached by tricky diplomatic bargaining and horse trading.) The other class contains the purification rates p_{0th} and p_{0k} by which nature, that is, the Sun and space, mitigates pollution and slows down the approach to the critical limits.

A simple combination of these elements yields the pollution function $\wp(p)$:

$$\wp(p) \equiv \wp(p_{th}) \prod_{k=1}^{N} \wp(p_k). \tag{4.57}$$

Here $\wp(p)$ is product of individual pollution functions $\wp(p_i)$ (which resemble the Fermi distribution function in statistical physics):

$$\wp(p_i) = \frac{\exp[-p_{Ci}/p_{0i}] + 1}{\exp[(p_i - p_{Ci})/p_{0i}] + 1}; \quad i = th, \, k; \quad k = 1, \ldots, N. \tag{4.58}$$

Equations (4.57) and (4.58) model the reaction of the economic system to environmental pollution. As long as all thermal pollution and chemical pollution, p_{th} and p_k, are far below their critical limits p_{Ci}, the pollution function $\wp(p)$ is practically equal to 1, and one has the growth equation (4.24) without thermodynamic limits to growth. If, however, environmental pollution increases and approaches one of the critical limits p_{Ci}, the corresponding function $\wp(p_i)$ and the function $\wp(p)$, which multiply the output elasticities, drop below 1. Then, a given increase of capital, labor, and energy no longer results in the same increase of value added as in the case of $\wp(p) = 1$. This simulates the response of society to the challenge of environmental protection: parts of the production factors are withdrawn from the generation of traditional output and rededicated to the prevention of environmental damage. An example would be the much discussed carbon (dioxide) capture and storage. A power station with (end-of-the-pipe) carbon (dioxide) capture and storage would consume roughly as much primary energy for carbon (dioxide) capture and storage as for electricity generation, require substantial additional equipment, and involve highly qualified people who handle this equipment and monitor the storage of CO_2 in, e.g., aquifers. Since it is unlikely that carbon (dioxide) capture and storage will solve all problems of climate change, countries with large areas only slightly above sea level may also have to invest heavily in dam construction to keep the sea back. Nothing of that contributes to the traditional basket of goods and services.

As we discussed in Chap. 3, and in Sect. 4.2 in the context of the "heat equivalents of noxious substances" (HEONS), it is in principle possible to convert all chemical (and radioactive [17]) pollution into thermal pollution. If one were to do that, because thermal pollution might be considered as least problematical, the pollution function would only consist of the thermal component $\wp(p_{th})$. This can be constructed in the following way for the world economy within the biosphere. (1) thermal pollution p_{th} is proportional to global nonsolar energy "consumption" E_W (about 1.4×10^{13} W in 2009), which finally (more or less) ends up in heat[22];

---

[22]The potential energy stored in buildings is relatively small and turns into heat once the buildings are torn down. Similarly, oxidation of chemical products releases stored chemical energy as heat.

(2) the purification rate p_{0th} is proportional to the infrared power radiated into space by the Earth in radiative equilibrium with the Sun. The absorbed solar radiation is 1.2×10^{17} W; (3) the critical limit of thermal pollution p_{Cth} is proportional to the "heat barrier" of 3×10^{14} W, in whose vicinity and beyond one will have to invest in adaptation to climate change even without the anthropogenic greenhouse effect.[23] Thus, in this simple case, where there is only thermal pollution by nonsolar "energy consumption" E_W, the pollution function becomes

$$\wp(p)_W = \wp(p_{th})_W = \frac{\exp[-3 \times 10^{14}\ \text{W}/1.2 \times 10^{17}\ \text{W}] + 1}{\exp[(E_W - 3 \times 10^{14}\ \text{W})/1.2 \times 10^{17}\ \text{W}] + 1}. \tag{4.59}$$

The growth-limiting effect of finite material resources can be mitigated by recycling, but only for a while. Let us assume that all natural deposits of nonenergetic materials are exhausted. Then, with sufficiently large quantities of capital, labor, and energy, recycling can recover the materials that have been embodied in industrial products. This, however, makes sense only as long as the average service lifetime τ of the industrial products is shorter than the time $1/\nu$ after which they have to be fed into the recycling process in order to provide all the materials needed; ν is the average recycling frequency. One can construct an overall recycling function $\mathscr{R}(\nu, \tau)$ as the product of material-specific recycling functions in a way similar to the construction of the pollution function and include losses of materials during recycling.

Multiplication of α, β, and γ in (4.24) by $\wp(p)\mathscr{R}(\nu, \tau)$ yields the growth equation in the presence of physical limits to growth as

$$\frac{dy}{y} = \left[\alpha \frac{dk}{k} + \beta \frac{dl}{l} + \gamma \frac{de}{e} \right] \wp(p)\mathscr{R}(\nu, \tau) + \delta \frac{dt}{t - t_0}. \tag{4.60}$$

The term $\delta \frac{dt}{t - t_0}$, which takes into account human creativity, is not affected by the physical limits to growth. There is no natural law that forbids us from extending the economic system's boundaries beyond the space of the biosphere, for instance, by space industrialization described in Sect. 2.6.3. This would expand the critical limits in the pollution and recycling functions far beyond the critical limits on Earth.

Equation (4.60) represents the most primitive model that describes the growth-limiting effects of environmental pollution and finite material resources. Even in this simple model the growth of global output y can only be calculated numerically for *scenarios* of energy utilization and its repercussions on the environment and on scarce material resources. Trivially, no quantitative prediction is possible of what people will actually do. Thus, (4.60) serves essentially as a reminder of the fact that the growth of material wealth is definitely limited in a finite world.

---

[23]Whether one may go beyond the heat barrier by increasing the albedo of Earth, e.g., by covering large desert areas with highly reflecting material, is an open question.

The facts that energy conversion in the machines of the capital stock has been the basis of industrial growth and that entropy production, coupled with energy conversion, establishes limits to growth in finite economic systems are summarized by the second law of economics:[24]

Energy conversion and entropy production determine the growth of wealth.

What future growth will be possible on our planet? What might sustainable growth look like? "Dematerialization of growth," "efficiency enhancement by factor $X(\gg 1)$," and "smart, qualitative growth" are popular concepts that offer to save the growth paradigm. Critics respond: "Smart growth is a magic formula working with the principles of efficiency and consistency: Using material and energy resources in an efficient way and closing the cycles of materials by consistent recycling will reconcile economic growth with the environment. But the magic formula of innovation does not work, because plenty of rebound effects overcompensate all efforts on reduction. Innovations add to problematic technologies and practises but seldom replace them. Instead of discontinuing the poison, an antitoxin is applied while the cause of disease remains unaffected. This cause lies in growth, itself which always comes along with growing material and energy flows. The myth of smart growth makes it hard to accept that fact and to find constructive solutions" [98].

Furthermore, the thermodynamic limits to energy conservation, outlined in Sect. 2.5.4, indicate that the energetic efficiency of energy systems in industrial countries cannot be improved by a factor of more than 2 if the demand for energy services is kept constant. In response to that, advocates of dematerialized growth, argue that people ought to change their behavior and should increase their demand for immaterial things. For example, they should go more often to concerts instead of driving around in gas-guzzling heavy limousines, SUVs, and sports cars, or flying to the Caribbean for vacations. Apart from the fact that the transportation energy associated with attending modern concert performances is not negligible, it will not be easy to change the preferences of people. Substituting listening to music for other, more energy intensive activities is not to everybody's taste. Furthermore, great concert and opera performances can be heard and seen at home via radio, television, CDs, and DVDs, but this does not decrease traveling by any means. To the contrary, music enthusiasts feel stimulated to travel long distances to experience the live performance in addition to the "canned" music at home – an example of a rebound effect.

And yet, more and more people agree that we would be better off if we rediscovered the depths of spirituality and reduced the importance of material

---

[24]Needless to say, this does *not* imply an energy theory of value. Human creativity designs the energy-converting machines and labor supervises and services them. Thus, capital, labor, and creativity also contribute to the generation of wealth – although mostly via energy conversion.

things. Certainly, a true mystic experience enriches life more than anything else. But such an experience is a gift that can be only received on one's knees – an attitude we usually do not appreciate too much. Nevertheless, if, for whatever reasons, people in the rich industrial countries were indeed to change their attitude, in the sense that less (material wealth) is more (spiritual well being), why then offer this as a "growth" that could replace the growth of conventional GDP? One should not camouflage the tremendous challenge posed by the need to change the growth paradigm by using the same concept for completely different things.

As mentioned already, the thermodynamic limits to conventional growth do not apply if the system boundary is extended to, say, the fringe of the Sun's planetary system. There is plenty of energy, materials, and space for absorbing emissions. Sure, space industrialization as described in Sect. 2.6.3 would be expensive and risky, at least at the beginning. But physics is not an obstacle. Nevertheless, even people who understand the technological feasibility of space industrialization and see the limits to growth on Earth object to expanding the industrial system beyond the biosphere of Earth. They rather prefer changing human preferences and human nature. The dean of moral theology at the University of Würzburg, the late Prof. Heinz Fleckenstein, made an interesting comment on that. In a discussion on the limits to growth, someone objected to space industrialization, saying, "But then we would not have to change human nature," hoping that a change of human nature would make us better Christians. Fleckenstein replied: "It is a long-standing moral teaching of the Catholic church that a technical solution of a problem is always preferable to any attempt to change human nature."

4.6   Distribution of Wealth

4.6.1   Adam Smith's Concepts and Karl Marx's Error

Adam Smith's *The Wealth of Nations* founded capitalist economics in 1776 – the same year as the first steam engines of James Watt were installed in commercial enterprises. The Industrial Revolution had yet to show its power, and the concept of "energy" was not yet clear at all, as Sect. 2.1.1 recalls. Thus, living within the notions of agrarian society, Adam Smith could not help but consider capital, labor, and land as the only factors of production. In his time capital consisted essentially of tools, agricultural implements, and means of transportation, all powered by human and animal muscle power, partially in combination with wind and water. Labor handled them. And land, bearing photosynthesizing plants, provided food and firewood, without anybody realizing that the two have "energy" in common. Regrettably, even more than 200 years after the Industrial Revolution, the disciples of Adam Smith remain in his conceptual world and ignore the production factor energy.

Furthermore, since the days of Adam Smith, the focus of economic thinkers and economic modeling has been mainly on trade and the workings of the market. It has often been guided by the myth of the "Invisible Hand," which fails its believers whenever markets crash. "The theory of the Invisible Hand states that if each consumer is allowed to choose freely what to buy and each producer is allowed to choose freely what to sell and how to produce it, the market will settle on a product distribution and prices that are beneficial to all the individual members of a community, and hence to the community as a whole. The reason for this is that self-interest will drive actors to beneficial behavior. Efficient methods of production will be adopted in order to maximize profits. Low prices will be charged in order to maximize revenue through gain in market share by undercutting competitors. Investors will invest in those industries that are most urgently needed to maximize returns, and withdraw capital from those that are less efficient in creating value. Students will be guided to prepare for the most needed (and therefore most remunerative) careers. And all these effects will take place dynamically and automatically" [99]. The physical forces that create goods and services have been of minor or no interest to economists.[25]

Thus, the notions of the agrarian society Adam Smith lived in still shape the basic concepts and models of modern economics. A striking example of how little physical aspects matter – in this case the difference between land and labor – was given by Bertola [100] in 1993. In an extensive mathematical analysis with cost-share weighting of factors, he studied the distributional implications of growth-oriented policies and described output by disembodied productivity multiplying a constant-returns-to-scale Cobb–Douglas production function of capital K and a factor L that "might refer to land or (uneducated) labor."

When Karl Marx published the first volume of *Das Kapital* in 1867, he was concerned with the physical origin and the distribution of wealth. He justly worried about the workers' miserable living conditions. The working day had 14–16 h, and people died early. Marx reasoned that all capital and consumer goods are ultimately produced by labor, that therefore they should belong to the working masses, and that private property could only be accumulated by the exploitation of workers. Had he realized the completely new production mechanism that had emerged with the coal-powered steam engine, he would have understood that in the sphere of industrial production value added can also be produced by the exploitation of energy sources without exploiting people. Society would have been spared the theory that capitalism inevitably will collapse because of the pauperization of the masses and the ensuing revolution. And the people in the formerly "socialist" countries would not have sustained the failure of establishing the "dictatorship of the proletariat," but like their luckier contemporaries in the democratic market economies they could have enjoyed the wealth created from the energy sources by human ingenuity. The tragic misunderstanding of the energy-based industrial production process by

---

[25]Occasional discussions of "engineering" production functions have hardly dealt with basic engineering principles of production.

Marxism also shows symbolically in the fact that the hammer and sickle, the tools of the workmen and peasants of the bygone agrarian society, had decorated the national flag of the second most powerful nation on Earth on its way into economic collapse.

4.6.2   Rich and Poor

4.6.2.1   The Global View

Merrill Lynch, a New York-based bank, published 11 World Wealth Reports before it lost US $23.2 billion in 2007 in the wake of the US mortgage crisis. Since January 2009 it has been a subsidiary of the Bank of America. Its World Wealth Reports 2007 and 2009 document impressively the global distribution of wealth.

The 2007 report, published jointly with Capgemini, summarized the "state of the world's wealth" as follows:

- 9.5 million people hold globally more than US $1 million in financial assets, an increase of 8.3% over 2005. (By definition, each one of these people is a High Net Worth individual – HNWI.)
- HNWI wealth totals US$ 37.2 trillion, representing an 11.4% gain since 2005.
- HNWI financial wealth is expected to reach US$ 51.6 trillion by 2011, growing at an annual rate of 6.8%.
- In large measure, the continued growth of HNWI wealth has been driven by the world's wealthiest individuals – the Ultra-HNWIs: those whose financial assets exceed US$ 30 million. In 2006, the number of Ultra-HNWIs grew to 94,970, an 11.3% gain, up from a 10.2% gain in 2005. Total wealth accumulation for this elite group also grew last year, by an impressive 16.8% to US $13.1 trillion – another sign that global wealth is rapidly consolidating among this ultra-wealthy segment.

Merrill Lynch and Capgemini announced that "In this year's Report we also take an in-depth look at HNWIs' investments of passion. We find that although art continues to be a favorite investment of this kind, luxury collectibles account for the largest spend." Luxury collectibles include "vintage yachts, automobiles selling for hundreds of thousands of dollars, and private airplanes." The report also examines Forbes' Cost of Living Extremely Well Index (CLEWI), which was established in 1976. It compares the price inflation of luxury goods against that of everyday consumer items. "In 2006 we found the cost of luxury goods and services rose nearly twice as fast as the cost of everyday consumer products. The cost of luxury items tracked by the CLEWI rose 7.0% while the the cost of consumer goods and services, as monitored by the Consumer Price Index (CPI), rose 4%. This represents a larger increase over 2005, when the CLEWI rose 4% and the CPI by 3.6%. All else kept constant, the higher price increases reported in the most recent CLEWI signals that demand for luxury goods is outpacing demand for everyday consumables."

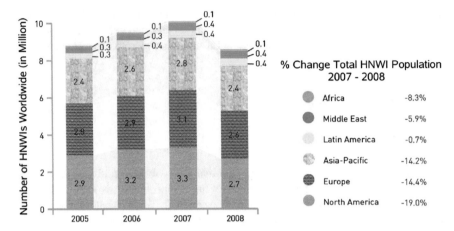

Fig. 4.21   High-net-worth individual (*HNWI*) population 2004–2008 (by region). (Source: Merrill Lynch and Capgemini, World Wealth Report 2009)

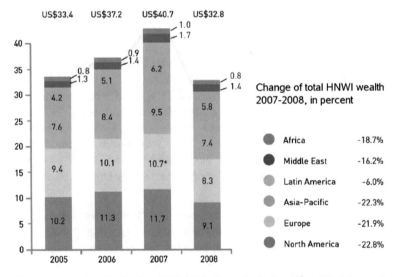

Fig. 4.22   HNWI wealth distribution 2005–2008 (by region); in trillion US dollars. (Source: Merrill Lynch and Capgemini, World Wealth Report 2009)

The regional distribution of the number of HNWIs in Fig. 4.21 shows that most of them live in North America, Europe, and Asia-Pacific. The regional distribution of their wealth in Fig. 4.22 shows that wealth concentrates in the same regions, with one exception: the Latin American HNWIs have a wealth share that is much larger than their HNWI-population share. In 2006 the financial wealth accumulated by all HNWIs was roughly 75% of global GDP. One should add nonfinancial property to get a feeling for the share of global wealth owned by the richest ten million people.

The financial and economic crisis that started with the collapse of the US real estate market in 2007 has deprived HNWIs of some of their wealth. But under the present general framework of markets, they are likely to recover their losses as long as the economy recovers. The World Wealth Report 2009 states:[26]

"At the end of 2008, the world's population of high net worth individuals (HNWIs) was down 14.9% from the year before, while their wealth had dropped 19.5%. The unprecedented declines wiped out two robust years of growth in 2006 and 2007, reducing both the HNWI population and its wealth to below levels seen at the close of 2005. Ultra-HNWIs suffered more extensive losses in financial wealth than the HNWI population as a whole. The Ultra-HNWI population fell 24.6%, as the group's wealth dropped 23.9%, pushing many down into the 'mid-tier millionaire' pool (which consists of HNWI having US $5 million to US $30 million). The global HNWI population is still concentrated, but the ranks are shifting. The USA, Japan and Germany together accounted for 54.0% of the world's HNWI population in 2008, up very slightly from 53.3% in 2007. China's HNWI population surpassed that of the U.K. to become the fourth largest in the world. Hong Kong's HNWI population shrank the most in percentage terms (down 61.3%). HNWI wealth is forecast to start growing again as the global economy recovers. By 2013, we forecast global HNWI financial wealth to recover to $48.5 trillion, after advancing at a sustained annual rate of 8.1%. By 2013, we expect Asia-Pacific to overtake North America as the largest region for HNWI financial wealth."

The 2009 OECD publication "Engaging with High Net Worth Individuals on Tax Compliance" [101] states in its summary that individuals with high net assets represent a significant challenge to taxing authorities. This challenge arises from the complexity of their assets, the multitude of their tax optimization strategies, and the repercussions of their behavior on the integrity of the tax system.

"Poverty Facts and Stats" [102] complements the World Wealth Report. The basic facts are shown by Figs. 4.24 and 4.25. Figure 4.23 shows that in 2005 the wealthiest 20% of world population accounted for 76.6% of global consumption, whereas the poorest 20% accounted for just 1.5%. World population was approximately 6.5 billion people in 2006. One notes from Fig. 4.24 that almost half of them, over three billion people, live on less than US $2.50/day and about 80% of humanity live on less than US $10/day. According to Fig. 4.25, the poorest 10% of the world population had a share of 0.5% and the richest 10% had a share of 59% of world's total private consumption in 2005.

Consequences of poverty are listed in [102]. Some are:

- Nearly one billion people entered the twenty-first century unable to read a book or sign their names.
- In total, 10.6 million children died in 2003 before they reached the age of five. This number equals the child population in France, Germany, Greece, and Italy.

---

[26]http://www.ml.com/media/113831.pdf.

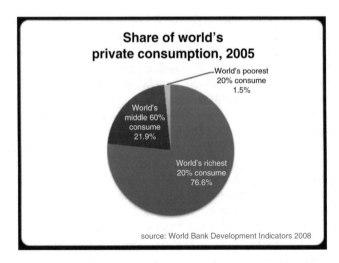

Fig. 4.23 How much rich and poor consume [102]

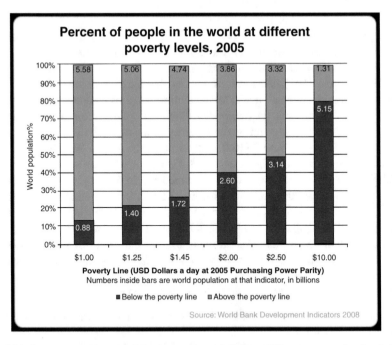

Fig. 4.24 Percentage of people defined as poor according to different poverty levels of daily income. *Numbers inside bars* are the number of billion people [102]

- Approximately half the world's population now live in cities and towns. In 2005, one out of three urban dwellers (approximately one billion people) was living in slum conditions.

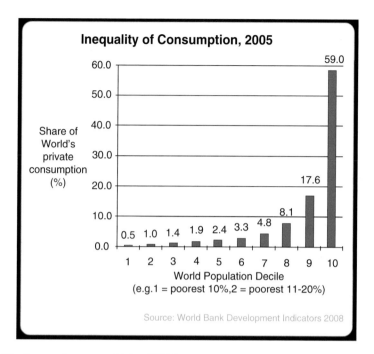

Fig. 4.25 Consumption per wealth level [102]

- In developing countries some 2.5 billion people are forced to rely on biomass – fuelwood, charcoal, and animal dung – to meet their energy needs for cooking. In sub-Saharan Africa, over 80 % of the population depends on traditional biomass for cooking, as do over half of the populations of India and China.

The United Nations considers people who must live on roughly US $1/day at 1993 purchase-power parity as living in *absolute* poverty. The OECD and the European Union consider households whose income is below 60% of national median equivalized household income as living in *relative* poverty. Lack of employment is a main source of poverty. The development of unemployment in the G7 countries is shown in Table 4.6. The average unemployment rate in the G7 countries was 5.9% in 2008 and 8.4% in September 2009.

Of course, unemployed (relatively) poor people in a G7 country will be considered as materially rich by an absolutely poor person in a developing country. On the other hand, there is the danger that the relatively poor will be more and more marginalized in their rich societies. This may breed social tensions and instabilities.

4.6.2.2   Germany

The poverty and wealth reports of the German government published in April 2001 and March 2005 stated that during the time covered by the reports – about

Table 4.6 Unemployment rates in G7 countries, as a percentage. (Source: Institut der deutschen Wirtschaft, Cologne, 1996, and OECD.StatExtracts, December 2009)

Country	1970	1980	1985	1990	1992	1994	2004	2008	2009[a]
Canada	5.7	7.5	10.4	8.1	11.2	10.0	7.3	6.1	8.6
France	2.5	6.3	10.1	9.0	10.0	11.3	9.5	7.9	10.1
Germany[b]	0.6	2.5	7.1	4.8	4.5	6.3	9.9	7.3	7.5
Italy	5.4	7.7	9.6	10.0	10.3	11.9	8.9	6.8	8.0
Japan	1.2	2.0	2.6	2.1	2.2	2.5	4.7	4.0	≈5.3
UK	2.4	6.1	11.4	7.0	10.0	9.0	4.9	5.7	≈7.8
USA	5.0	7.2	7.1	5.5	7.4	5.8	5.5	5.8	10.2

[a]October 2009
[b]Only West Germany until 1994

7 years – the share of total wealth possessed by the richest 10% of all households grew from 42% to 47%, whereas the share in financial and material assets possessed by the poorer 50% of the population shrank from 4.5% to 4%.

The Third Poverty and Wealth Report of the German government was presented in May 2008. It stated that the gap between rich and poor had further widened, that the risk of falling in poverty has increased, and that the earnings of the rich had grown. Poverty risk is highest for people who have been out of work for a long time, and for single mothers and single fathers. A single person with a net monthly income of less than €781 is considered as poor. If net monthly income exceeds €3,418, the person is considered as rich. It is especially alarming that more and more people who have jobs are endangered by poverty. Every fourth German lives below the level of relative poverty. Thirteen percent of Germans are poor, and another 13% of the population are only saved from falling below the poverty line by financial aid from the national welfare system. This includes unemployment benefits, child allowances, and housing subsidies. Gross average real annual earnings of employees dropped between 2003 and 2005 by 4.8%, from €24,873 to €23,684. The middle class, whose income is by definition between 70% and 150% of the median of private net household income, shrank. Until 2001 the number of people with average income had been relatively constant at about 62% of the population. In 2006 only 54% still belonged to the middle class. A number of former middle class households moved up to the wealthier levels, and an even larger number dropped below the level of 70% of the median income. Simultaneously, fewer and fewer poor people have succeeded in entering the middle class.

4.6.3 Taxes and Debts

4.6.3.1 Taxes

Big wealth differences between rich and poor may lead to severe social instabilities. The medieval peasant uprisings, or present-day internal armed conflicts as the one in Colombia, are testimony to that. So far, the industrial democracies have avoided this problem, and even won the "cold war" with the allegedly egalitarian "socialist" camp, because state-established systems of social security mitigate the hardship of poverty by financial aid.

The state-managed transfer of wealth from the rich to the poor was initiated in the nineteenth century by the German chancellor Otto von Bismarck. He had realized the enormous disruptive power of the extreme social discrepancies between the impoverished industrial workers and the German bourgeoisie. He wanted to ease that, not only in order to weaken the socialist movement but also in order to strengthen the ties between the people and the government of the young nation, just unified in 1871. Social legislation introduced compulsory health insurance and accident insurance for workers in 1883 and 1884. Workers had to pay two-thirds of the health insurance fees and their employers one-third. Accident insurance had to be fully paid by the employer. Later, old-age pension insurance was introduced, whose fees were split 50:50 between workers and employers. Bismarck's model was soon adopted by other nations and further developed. Unemployment insurance for all German employees was established in 1927.

Since the Industrial Revolution, the character of the state has changed. Absolute monarchs and feudal nobility represented the state in past agrarian societies, except for some city republics. They collected taxes, tolls, and contributions from peasants, craftsmen, and merchants in order to finance the military and a splendid lifestyle in court. The latter included advancement of the arts and sciences at differing intensity. Jurisdiction was the duty and privilege of the rulers. Maintenance of internal public order was essentially a military obligation, and education was mostly in the hands of religious communities. Infrastructure for traffic, water supply, and sewage was primitive and mostly the business of the people themselves. For the common citizen the state was an entity that took a lot, especially in times of war, and gave little. Paying taxes was a sour duty.

Citizens of modern industrial democracies do not like to pay taxes either. But most of them understand that the state is different from what it was during the agrarian age. Public authorities collect taxes and social insurance contributions for:

- Financing defense against external enemies
- Maintaining internal public order via police forces and a sophisticated judiciary system
- Providing public education, research, and development
- Building and maintaining, at differing degrees, a complex technological infrastructure for traffic, energy supply, water supply, and garbage disposal
- Supervising public health and safety
- Providing benefits from the modern welfare system

The OECD regularly compares the taxes and levies paid in its member countries. The levies represent the taxes and the contributions actually paid to the national systems of social security. Table 4.7 gives recent numbers. They show that especially in the Nordic countries, but also in Austria, Belgium, France, and Italy, the levy quota is relatively high, whereas Japan, the USA, Switzerland, Slovakia, Greece, and Ireland have relatively low levy quotas. High levy quotas usually finance tightly knit social security nets, low quotas indicate preference for private insurance.

Table 4.7 International comparison of tax and levy quotas in the years 2002 and 2007; as a percentage of GDP. (Source: OECD (Ed.), Revenue Statistics 1965-2007, Paris, 2008; and Bundesministerium der Finanzen, Berlin)

Country	Tax Quota 2002	Levy Quota 2002	Tax Quota 2007	Levy Quota 2007
Austria	29.3	44.1	27.8	41.0
Belgium	31.6	46.2	30.7	44.4
Canada	28.4	33.5	28.6	33.3
Czech Republic	21.9	39.2	20.3	36.4
Denmark	47.7	49.4	47.9	48.9
Finland	33.7	45.9	31.1	43.0
France	27.7	44.2	27.4	43.6
Germany	21.7	36.2	23.0	36.2
Greece	23.5	34.8	20.2[a]	31.3[a]
Hungary	26.3	37.7	26.4	39.3
Ireland	23.7	28.0	27.3	33.2
Italy	28.6	41.4	30.2	32.2
Japan	17.0	27.3	18.4	27.9[a]
Luxembourg	30.5	42.3	26.7	36.9
Netherlands	25.4	39.3	24.2	38.0
Norway	33.4	43.1	34.4	43.4
Poland	24.2	34.3	21.4[a]	33.5[a]
Portugal	24.8	34.0	24.9	36.6
Sweden	35.3	50.6	35.6	48.2
Switzerland	23.4	31.3	22.8	29.7
Slovakia	19.2	33.8	17.9	29.8
Spain	23.0	35.6	25.0	37.2
UK	29.8	35.9	29.8	36.6
USA	21.8	28.9	21.6	28.3

[a]Data from 2006

Note that the 2002 tax quotas of Germany (21.7%) and the USA (21.8%) were nearly the same and were only undercut by the quotas of Slovakia (19.2%) and Japan (17.0%). A big difference is in the levy quota, where Germany occupied a middle position and the USA the third lowest, above Ireland and Japan. Things changed somewhat in 2007, but not dramatically. In Japan, companies provide a lot of social security. Therefore, the levies are small and the ties of loyalty between employers and employees are strong. Denmark has the highest tax quota and the minimum difference between tax and levy. Its social security system is essentially financed by taxes, and hardly any European country can match its unemployment rates of a little more than 4%.[27]

---

[27]"Denmark embarked on a 'green tax' reform to shift the tax burden away from income and towards resource use. This system was put into place between 1994 and 1998. Energy taxes were raised progressively, particularly on coal and electricity consumption, leading to an average increase in taxation on heating and power of 30% from 1994 to 1998. The main effect was that households paid lower income taxes and higher environmental taxes" [103].

Table 4.8 Structure of German tax revenues in 2008. (Source: Bundesministerium der Finanzen, Monatsbericht digital, July 2009: Die Steuereinnahmen von Bund, Ländern und Gemeinden im Haushaltsjahr 2008)

Tax	Billion euros	Percentage of total tax
Wage tax	176.8	32
Value added tax	176.0	31
Local business tax	41.5	7
Assessed income tax	32.7	6
"Energy" tax[a]	39.2	7
Electricity tax	6.3	1
Corporation tax	16.7	3
Tobacco tax	13.6	2
Total tax revenue	561.2	100

[a] Of this, 90.7% was from transportation fuels, 5.3% was from natural gas, and 4% was from heating oil, liquid gas, and coal

Table 4.8 shows a number of selected taxes in Germany. Wage taxes and value-added taxes both exceed 30% of total tax revenue, and have the lion's share. Corporation taxes have a 3% share and contribute just one percentage point more than tobacco taxes. The sum of taxes on all fuels and electricity is 8% of the total. In 2002 fuel tax still had a share of 9.6%.

4.6.3.2   Debts

Since the 2008–2009 economic crisis, and the ensuing deficit spending of governments to fight economic recession and bail out the banks whose speculations caused the financial market crash, some politicians have been advocating tax cuts to stimulate economic growth. This would presumably reduce the ballooning state deficits via higher total tax revenues. However, growth rates that could achieve this aim would have to be much larger than the ones any industrially advanced country had in the last decades.

One also resorts to an additional (old) reasoning, which goes like this. If the tax rate is 100%, there are no tax revenues, because nobody works. If the tax rate is zero, there are no revenues either, trivially. Between 0% and 100% there must be a revenue maximum. The advocates of tax cuts assume that the actual tax rate is to the right of the maximum, where the revenue curve rises toward the maximum with decreasing tax rate. The problem is that nobody knows at what tax rate the revenue maximum is actually positioned for a given economy. The Reagan administration, which governed the USA from 1981 to 1989, lowering taxes and enhancing spending, certainly did not know it, because it accumulated more state debts than all US governments before it since George Washington, as Fig. 4.26 shows. In this figure, US public debt, which is also referred to as national debt, is the sum of all securities issued by the US Treasury and held by institutions outside the US government itself. Securities issued by state and local governments are not part of it. Gross debt includes intragovernment obligations such as the Social Security Trust Fund.

Fig. 4.26 US federal debt, nominal in 2009 dollars (*top*) and in percentage of GDP (*bottom*). *Upper curves* gross debt, *lower curves* public debt. (Source: US Government; modification of a file from Wikimedia Commons, http://en.wikipedia.org/wiki/File:USDebt.png)

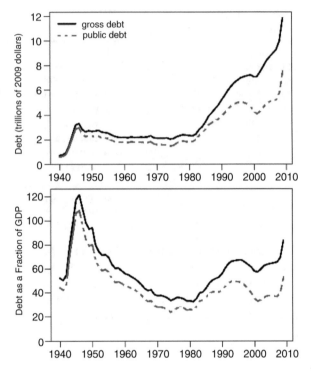

Table 4.9 shows gross debts of G7 countries.[28] The maximum national debt allowed by the European Union to its member states is 60% of GDP. Japan has accumulated its high debt of 192% of GDP by government deficit spending in attempts to overcome stagnation and stimulate economic growth after the bursting of the real-estate bubble in the early 1990s.[29] Germany chose to finance reunification by increasing national debt (see Fig. 4.27), although many Germans would have been happy to pay higher taxes for this unexpected gift from a benevolent fate.

---

[28]Sources: Economy Watch, February 20, 2010, for debt in terms of currency, and http://www.economicshelp.org/blog/economics/list-of-national-debt-by-country/ for debt as a percentage. In 2008, US gross debt of $12,867 billion was estimated at 90.8% of GDP. In 2009, the GDP of the USA was $14,463.4 billion (source: Bureau of Economic Analysis, US Department of Commerce, January 29, 2010). There may be a 5% error in the absolute numbers. For instance, the German *Schuldenuhr* ("debt clock") indicated only a gross debt of €1735 billion on March 10, 2010.

[29]In 1986, during a lunch break at an international conference in Tokyo, I went with a Japanese colleague to the grounds of the Emperor's Palace. "How high do you think is the value of the palace grounds?" my colleague asked me, smilingly. "No idea, you tell me," I replied. "For the value of the palace grounds we can buy all of Canada," he informed me. Later, a German economist told me that he had read an article where the value of the emperor's palace grounds was estimated to be equal to the value of the state of California.

Table 4.9 Estimates of 2009 gross debts of G7 countries in national currency, and as a percentage of GDP for 2008–2009. (Sources are given in the text)

Canada	CAD 1,191.29 billion	73%
France	EUR 1,471.02 billion	80%
Germany	EUR 1,853.87 billion	77%
Italy	EUR 1,761.81 billion	115%
Japan	JPY 1,047,730.45 billion	192%
UK	GBP 962.927 billion	68%
USA	USD 12,093.10 billion	86%

Fig. 4.27 Growth of gross national debt of the Federal Republic of Germany before and after reunification on October 3, 1990; billion euros. (Source: http://www.staatsverschuldung.de/index.html)

Obviously, the conclusion from this and the preceding section is:

The legal framework that determines the allocation of wealth on the global market is such that the gap between the rich high-net-worth individual and the common man is widening more and more, and the debts of nations are growing.

4.6.3.3   World Financial System at the Limit

Analysts of the US-mortgage-induced financial and economic crisis fear that debts will join energy scarcity and pollution in menacing the stability of the global economic system [14, 104]. We close the subject of wealth distribution with a brief summary of their findings.

The present regime of globalization under the rules of market fundamentalism enables certain, economically especially powerful actors to pay small amounts of or no taxes on their earnings [105]. This significantly increases the net return on investments in tangible and financial assets and decreases the revenues of public authorities correspondingly. The facts and consequences are summarized by the economists Wolfgang Eichhorn and Dirk Solte [14] as follows.

Since the 1970s, global earnings from tangible and financial assets grew more rapidly than global GDP, from about 14% of global GDP in 1970 to 40% in 2007. Incomes from tangible and financial assets hold roughly equal shares. Growth of financial assets means growth of debt, because each financial asset is an entitlement to money in the future, which balances a credit, and each credit is debt. One expects interest on each credit. Credits and savings are used to build up tangible assets such as firms and real estate. A return is expected on these tangible assets too. It has to be generated by the sale of goods and services, or by renting and leasing. Extrapolation of the present growth trends into the future yields that as early as 2030 all global GDP would be consumed by payments of interests and returns. Nothing would be left for paying employed persons. This is a definite limit to the world's financial system.[30]

In 2008 new indebtedness of public authorities exceeded total global savings. Meanwhile, the budget deficits of the rich countries are about as high as the sum of unpaid taxes, which the very strong, internationally operating actors, especially in the financial sector, can avoid thanks to globalization and the free movement of financial assets around the globe. This movement occurs at the speed of light on electromagnetic waves after the push of a computer button. Governments finance their budget deficits by borrowing. The big financial players satisfy the growing demand for credit by new "creative" financial instruments, such as "asset-backed securities," notorious for their role in the US mortgage crisis. Thus, a feedback mechanism has been established that more and more inflates the financial sector. In 2007 global GDP was US $_{2007}54.585$ trillion and financial assets totaled US $_{2007}195$ trillion, exceeding global GDP by a factor of 3.6. Here, "financial assets" include the central bank money of 160 currencies, check money, and all other obligations to pay money in the future. "Check money and all other obligations to pay money in the future," also called "leverage money" in [105], exceed central bank money by a factor of more than 50. Thus, if everybody wanted to change their bank deposits into central bank money, which is the only legal tender, there would be a problem.

Can economic growth defuse the bomb that is waiting to go off in the financial sector? Will it be possible to increase production of goods and services so much

---

[30]The year 2030 seems to be the magic year of the future, if one extrapolates trends of the past. In 1972, *The Limits to Growth* commissioned by the Club of Rome saw the trend of economic growth leading to collapse in 2030, because of either resource scarcity or environmental pollution. In 2009, the trend of financial growth indicates that the asset owners would deprive the world population of all income in 2030.

that all payment obligations can be met? Most likely not, according to what we know about energy, entropy, and the economy. Additional skepticism is added by a quantitative integrated analysis of (1) financial markets, (2) energy's importance for economic growth, and (3) declining energy returns on investment (EROIs) [106]; see Table 2.8. Independently from the research presented in this book, the economist Hannes Kunz of the Institute of Integrated Economic Research [104] has looked into the correlations between wealth, energy conversion, and debt, and arrived at the following conclusions:

- Economic activity is impossible without external energy inputs. Energy is a prerequisite for all economic activity, no matter whether it is a product or a service. Many products are highly dependent on significant energy inputs and thus extremely sensitive to the availability and the price of fuels. Despite these facts, traditional economists do not include energy inputs in their macroeconomic production functions.
- GDP per energy input (from external energy sources and human labor) in the economies of industrial and developing nations averages US $\$_{2007} 133/GJ$. Higher-than-average GDP per gigajoule in industrial countries is mostly due to the import of energy-intensive intermediate goods from developing countries, and lower-than-average GDP per gigajoule of developing countries results from the export of raw materials and energy-intensive intermediate goods.
- The price differences between 1 kWh of physical work from humans and 1 kWh from fossil fuels are breathtaking. On the global average, the price ratio is between 300 (oil 2010) and 800 (coal 2010). Fossil energy use transfers wealth from past solar inputs to our economy.
- A significant portion of increased productivity is the result of substituting energy for human labor.
- Over recent decades, our fossil energy sources have become less productive. Independent of the arrival of "Peak Oil," increasing amounts of upfront energy are required to explore the next new units of energy. The concept of EROI [energy return on (energy) investment] describes this as energy units gained from one energy unit invested in energy extraction. EROI ranges are 80–100 for 1930 Texas oil rigs, 8–15 for offshore oil platforms, and 2–10 for Canadian tar sands. Most renewable energy sources operate at EROIs between 2 and 8.
- Significant rises of energy "production" cost destroy the key economic driver of the past.
- Over the past 50 and more years, energy conversion and increasing debt have supported growth. When adjustments are made for contributions from growing debt, remaining growth of GDP is very much in line with growth of energy conversion.
- Owing to globalization, average inflation in industrial countries have stayed low despite ever-growing money supply, because industrial products and low-cost labor from abroad have lowered costs significantly. Increasing amounts of relatively low cost energy consumption in domestic industrial production have also kept costs down.

- Total interest expectations of creditors require profits or growth in asset values. Return on investment expectations (from debts and stocks) require overall GDP growth. Today, return expectations put unrealistic pressure on global growth requirements. In 2009, total debt and stock investments were estimated at approximately 345% of global GDP. Real return expectations of 2% would require 6.9% annual real growth of GDP .
- Growth expectations increasingly conflict with resource constraints. A widening gap is opening between debt and equity service requirements, on the one hand, and possible economic growth, on the other hand. Growth restrictions are due to higher energy extraction costs, expensive renewables, and the increased cost of environmental protection.
- In the mid-term, the Institute of Integrated Economic Research predicts a high likelihood of continued deflationary trends. However, inflation may be observed for certain phases where government interventions offset the speed of debt destruction – at the cost of ever-growing government debt. During those periods of GDP growth, rising energy and commodity prices will relatively quickly undermine the growth pattern, sending economies back into decline and causing deflation. In this case, instability becomes the key characteristic of economic development.

4.6.4   Discounting the Future

Interest and return on investment are expected, because people value future benefits and utility less than present ones. The same is true for future damage, which, therefore, is usually discounted at the interest rate. The magnitude of the discount rate has drastic consequences for the internalization of external environmental costs [107]. It is also decisive for the question of how much should be done against global warming caused by the anthropogenic greenhouse effect.

The problem of discounting the future was clearly and timely illustrated in the 1970s by the climate researchers Chen and Schneider from the National Center for Atmospheric Research, Boulder, Colorado, USA. Consider the example of coastal inundations because of global warming in the following scenario. Within, say, 150 years the West Antarctic Ice Shelf melts and the resulting flooding of Earth's coastal areas causes global damage that amounts to US $_{1971}2$ trillion, at minimum. (If global warming also caused Greenland's ice cap to melt, the damage would be much higher.) The GDP of the USA in 1971 (GDP_0) was US $_{1971}1$ trillion. Discounting the global damage of 2 GDP_0 with the discount rate d yields the sum S one would have to invest today in measures of damage prevention, or for monetary damage compensation in 150 years by a financial investment at annual interest rate d. Table 4.10 indicates this sum S for three discount rates d.

If one takes individual time preferences as a basis, a discount rate of 4% per annum is not too high for a time span of 150 years. Then S is less than 6/1,000 of US GDP in 1971. This is less than the cost of 20 (combined-cycle) gas-steam power

Table 4.10 Compensation S of global warming damage, depending on discount rate d

d	$(1+d)^{150}$	$S = \frac{2}{(1+d)^{150}} GDP_0$
2%	19.5	$0.10\, GDP_0$
4%	359	$5.6 \times 10^{-3} GDP_0$
7%	25580	$8 \times 10^{-5}\, GDP_0$

plants with CO_2 capture and storage according to an estimate by Siemens in 1990 [108]. In this sense it would be economically irrational to invest more in fighting the anthropogenic greenhouse effect – the more so as economic growth in the USA at an annual rate of $w = 1\%$ would yield a GDP equal to $(1 + w)^{150} GDP_0 = 4.45 GDP_0$ in 150 years, and thus overcompensate for the global damage of $2GDP_0$.

Individual time preferences have been known since biblical times. They obey the Esau principle, exposed in the book of Genesis in the Old Testament. Esau, the son of Isaac and grandson of Abraham, was born before his twin brother Jacob. He had primogeniture and was entitled to the inheritance. One day he came home from the hunt and was hungry. Jacob was cooking a lentil dish. Esau said to Jacob: "Let me quickly eat from that meal there, for I am exhausted." Jacob replied: "Very well, but first sell me your primogeniture." Esau answered: "I am wandering around here and there and must die anyway. What use is the primogeniture to me?" And in order to quench his present hunger he sold the promise for the future. Jacob received the blessing of his father, which had been intended for Esau, and became the progenitor of Israel.

But even if one does not discount the future according to individual time preferences, expected future damage is dwarfed by expected economic growth. A striking example is the Stern Review Report on the Economics of Climate Change [6]. It discounts future climate change damage with a discount rate of only 0.1% because of the ethical reasons indicated in the context of intertemporal welfare optimization (see Sect. 4.4.2). This was heavily criticized by Stern's peers [7]. One economist, in an interview aired by the BBC in January 2007, even declared: "If a student of mine were to hand in this report as a masters thesis, perhaps if I were in a good mood I would give him a 'D' for diligence; but more likely I would give him an 'F' for fail. ... There is a whole range of very basic economics mistakes that somebody who claims to be a professor of economics simply should not make." On the other hand, the report was praised by the British prime minister Tony Blair as "The most important report on the future ever published by this government". Environmentalists even heralded it as "the Copernican revolution of climate-change research" because of its message that in the "business as usual" (BAU) scenario losses of welfare (consumption) due to climate change are much bigger than those in the scenario of investing heavily in emission mitigation. However, Stern's BAU economic future looks much brighter if the message of the report is also summarized in the form of equations [109] (which are not printed explicitly in [6]):

1. If per capita consumption at time t *without* climate change damage is

$$c_0(t) = c(t_1) \exp(\alpha t),$$

(4.61)

then the BAU scenario *with* climate change yields per capita consumption at time t as

$$c_{cc}(t) = c(t_1)(1 - x/100)\exp(\alpha t), \quad 5 \le x \le 20. \qquad (4.62)$$

Here α results from assumed growth rates of 1.9 % for global GDP and 0.6% for the world population. The time span considered is from 2001 to 2200. Thus, in Stern's model, even with climate change, per capita consumption would grow exponentially for 200 years. The reported 5–20% reduction in consumption per head is only the small difference between two paths of exponentially growing consumption. Consequently, Stern remarks that "…even with climate change the world will be richer in the future as a result of economic growth" [6].

2. Evaluating a great number of technological options of reducing greenhouse gas emissions and estimating the associated costs, the report obtains a slightly reduced exponential growth of global GDP according to

$$GDP(t) = GDP(t_1)(1 - 1/100)\exp(0.019t). \qquad (4.63)$$

This is meant by "The Review estimates the annual costs of stabilization at 500–550 ppm CO_{2e} to be around 1% of (annual global) GDP by 2050."

The big difference between 5 and 20% (per capita welfare losses), on the one hand, and a 1% loss (of global GDP because of climate stabilization), on the other hand, has convinced many people that their economic interest favors investments in climate stabilization. This is the merit of the Stern Review Report, even if one sees no way how exponential global growth at 1.9% per year can be maintained for 200 years. This growth is given exogenously, and energy as a factor of production is not an issue.

Stern, and most natural scientists who call for action against greenhouse gas emissions, are opposed by people who think that it is cheaper to adapt to climate change than to do something about it. For instance, in the 1990s, the renowned economists *Nordhaus* (Yale), *Beckermann* (Oxford), and *Schelling* (Harvard; Nobel Prize in economics 2005) estimated the economic consequences of the anthropogenic greenhouse effect. Their results were summarized by Daly [110]. These economists assumed that global warming affects only agriculture, which contributed less than 3% to the GDP of the USA in 1992. (The 2009 contribution of US agriculture to GDP was a mere 1.2% of total GDP.) The contribution of agriculture in other industrialized countries has been comparably low (see Table 4.1). Therefore, even a drastic decline of agricultural production should only result in small losses of welfare:

Nordhaus [111], 1991: "…there is no way to get a very large effect on the US economy."
Beckermann [112], 1997: "…even if net output fell by 50% by the end of next century this is only a 1.5% cut in GNP."
Schelling [113], 1997: "If agricultural productivity were drastically reduced by climate change, the cost of living would rise by 1 or 2%, and that a time when per capita income will likely have doubled."

These estimates ignore the fact that famine increases the value and the price of food dramatically, and thus enhances the contribution of agriculture to GDP drastically. In addition, no difference is made between marginal and total utility (in this case of food). This is the more remarkable as modern economics has considered this difference as especially important since the introduction of calculus into economics in the "marginal revolution." Mainstream economists' perspectives on food and energy are the same: what costs little matters little.

4.7 Summary and Discussion

The cost-share theorem of standard economics is not valid in modern industrial economies, where capital, labor, and energy are the main factors of production. Maximization of profit or overall welfare subject to the technological constraints "limits to automation" and "limits to capacity utilization" yields new conditions for economic equilibrium. According to them, output elasticities, which measure the factors' economic weights and indicate their productive powers, are equal not to the factors' cost shares but rather to "shadowed" cost shares, where shadow prices due to the constraints add to factor prices.

Consequently, output elasticities must be computed independently of equilibrium conditions. They are obtained as solutions of a set of partial differential equations that result from the standard requirement that production functions must be twice differentiable; two technology parameters are estimated econometrically. The numbers in Table 4.5 and Figs. 4.10–4.16 and 4.18–4.20 show that cheap energy has a high productive power, whereas expensive labor has a low productive power.

The standard objection to this finding is: "If this were true, money would be lying on the street. One only has to increase the input of cheap energy and decrease that of expensive labor until output elasticities and cost shares are equal." This reasoning of orthodox economists overlooks the insurmountable barriers that block access to that side of the street where the money lies. These barriers are precisely the technological constraints that determine the shadow prices: One can neither increase energy input beyond design capacity of the machines, and decrease labor's handling of the machines correspondingly, nor substitute energy and capital for labor beyond the limit to automation that exists at a given time.

The large discrepancies between productive powers and cost shares of energy and labor explain the pressure to increase automation as quickly as technology permits, substituting cheap energy/capital combinations for expensive labor. (This way one indeed gets bit by bit to a limited street sector where the money lies.) Increasing automation is the main part of what is called "increasing productivity." The imbalance between economic weights and costs of labor and energy also reinforces the trend toward globalization, because goods and services produced in low-wage countries can be cheaply delivered to high-wage countries thanks to cheap energy and increasingly sophisticated, highly computerized transportation systems. Thus, if the disparities between productive powers and cost shares of labor and energy are too pronounced, there is the danger that newly emerging and expanding business sectors cannot generate enough new jobs to compensate for the ones lost to progress in automation and globalization. This, then, will result in the net loss of routine jobs in high-wage countries and increasing unemployment in the less qualified part of the labor force. A slowdown of economic growth, as natural constraints may cause, or economic recessions for whatever reasons, will aggravate the problem of unemployment.

Energy conversion is the basis of life and wealth creation. It is invariably coupled to entropy production, which manifests itself in emissions of heat and particles. These emissions will eventually restrict economic growth in the finite system Earth, when its emission-absorbing and life-supporting capacities are being exhausted. Growth can be maintained if industrialization expands into the space beyond Earth's biosphere. The second law of economics sums this up: Energy conversion and entropy production determine the growth of wealth. It complements the first law of economics: Wealth is allocated on markets, and the legal framework determines the outcome.

The second law of economics exemplifies the importance of thermodynamics for all energy-converting systems, whether they are inanimate, as in the field of physics, or involve people and machines, as in economics. It is a corollary to the first law and the second law of thermodynamics, which represent the "constitution of the universe" [114]. Although the differential equations (4.26) for output elasticities correspond to the Maxwell relations in equilibrium thermodynamics, methods of non-equilibrium thermodynamics may also prove useful in the modeling of economic fluctuations. Spontaneous parameter fluctuations in thermodynamic equilibrium of dissipative physical systems result in Nyquist noise of electric circuits, Brownian motion in fluids, and pressure fluctuations in a gas [115]. Behavioral fluctuations of economic actors, triggered by shocks [31], irrational expectations, pursuit of market dominance irrespective of cost, and other deviations from the idealized behavior of the "homo economicus," perturb macroeconomic equilibrium in the maximum of profit or overall welfare, no matter whether this maximum results from constrained or unconstrained optimization. Adaptation of the statistical methods, used in the derivation of the general fluctuation–dissipation theorem [116], to fluctuations about economic equilibrium may contribute to progress in nonequilibrium economics. This is a subject for future research.

Such research may also help prepare us for economic instabilities financial market analysts expect because of the accumulating global debts. The global economic system may be in for what is known in physics as *critical fluctuations* close to phase transitions. For instance, at the triple point of water where the absolute temperature is 273.16 K the system cannot decide whether to be ice, water, or water vapor. Similarly, the total market credit debt has reached a volume that overly strains the debt-carrying capacity of economies [14, 104]. Additional economic strains may result from resource problems such as peak oil, decreasing EROIs and climate change. Thus, if the pessimistic view prevails, high market volatilities are to be expected as a forerunner to a "phase transition" of the global economy into one of the states of hyperinflation, soft depression, or collapse.

Things may turn out better if the social set of values and the legal framework of the market change. Some thoughts on that from a common citizen's perspective are presented in the epilogue.

Appendix 1: Maximizing Profit

The necessary condition for a maximum of profit $G \equiv Y - \mathbf{p} \cdot \mathbf{X}$, subject to the technological constraints (4.7), is

$$\nabla \left[Y(\mathbf{X}; t) - \sum_{i=1}^{3} p_i(t) X_i(t) + \sum_a \mu_a f_a(\mathbf{X}, t) \right] = 0, \qquad (4.64)$$

where $\nabla \equiv (\partial/\partial X_1, \partial/\partial X_2, \partial/\partial X_3)$ is the gradient in factor space, and the μ_a are Lagrange multipliers. (The sufficient condition for profit maximum involves a sum of second-order derivatives of $Y(\mathbf{X}; t) - \mathbf{p} \cdot \mathbf{X} + \sum_a \mu_a f_a(\mathbf{X}, t)$. One assumes that the extremum of profit at finite X_i is the maximum.) This yields the three equilibrium conditions

$$\frac{\partial Y}{\partial X_i} - p_i + \sum_a \mu_a \frac{\partial f_a}{\partial X_i} = 0, \quad i = 1 \ldots 3. \qquad (4.65)$$

Multiplication of (4.65) with $\frac{X_i}{Y}$, and observing (4.3) brings the equilibrium conditions into the form

$$\varepsilon_i \equiv \frac{X_i}{Y} \frac{\partial Y}{\partial X_i} = \frac{X_i}{Y} \left[p_i - \sum_a \mu_a \frac{\partial f_a}{\partial X_i} \right], \quad i = 1, 2, 3. \qquad (4.66)$$

Combining (4.5) and (4.66) yields

$$Y = \sum_{i=1}^{3} X_i \left[p_i - \sum_a \mu_a \frac{\partial f_a}{\partial X_i} \right]. \qquad (4.67)$$

Inserting this Y into (4.66) results in the equilibrium conditions

$$\varepsilon_i = \frac{X_i \left[p_i - \sum_a \mu_a \frac{\partial f_a}{\partial X_i} \right]}{\sum_{i=1}^{3} X_i \left[p_i - \sum_a \mu_a \frac{\partial f_a}{\partial X_i} \right]} \equiv \frac{X_i [p_i + s_i]}{\sum_{i=1}^{3} X_i [p_i + s_i]}. \qquad (4.68)$$

Here s_i, defined as

$$s_i \equiv -\mu_A \frac{\partial f_A}{\partial X_i} - \mu_B \frac{\partial f_B}{\partial X_i}, \qquad (4.69)$$

is the *shadow price* of the production factor X_i. Comparison of this equation with (4.65) explains the standard definition of "shadow price": In optimization subject to constraints the shadow price measures the change of the objective (here profit) if each constraint is loosened by one unit.

To indicate a framework for the calculation of Lagrange multipliers, shadow prices, and equilibrium values of production factors we define

$$f_{Ai} \equiv \frac{\partial f_A}{\partial X_i}, \quad f_{Bi} \equiv \frac{\partial f_A}{\partial X_i}, \quad i = 1, 2, 3, \tag{4.70}$$

and write the equilibrium conditions, (4.65), as

$$\frac{\partial Y}{\partial X_i} - p_i + \mu_A f_{Ai} + \mu_B f_{Bi} = 0, \quad i = 1, 2, 3. \tag{4.71}$$

Simple algebra eliminates μ_B from these conditions for $i = 1$ and $i = 2$. One obtains

$$\mu_A = \frac{f_{B1}\left(p_2 - \frac{\partial Y}{\partial X_2}\right) - f_{B2}\left(p_1 - \frac{\partial Y}{\partial X_1}\right)}{f_{A2} f_{B1} - f_{A1} f_{B2}}. \tag{4.72}$$

Inserting this μ_A into (4.71) for $i = 1$ yields

$$\mu_B = \frac{p_1 - \frac{\partial Y}{\partial X_1}}{f_{B1}} - \frac{f_{A1}\left[f_{B1}\left(p_2 - \frac{\partial Y}{\partial X_2}\right) - f_{B2}\left(p_1 - \frac{\partial Y}{\partial X_1}\right)\right]}{f_{B1}\left(f_{A2} f_{B1} - f_{A1} f_{B2}\right)}. \tag{4.73}$$

Thus, all quantities entering the shadow prices (4.9) in the equilibrium conditions (4.8) are known in principle if one knows the production function $Y(\mathbf{X}; t)$, the prices per factor unit p_i, and the constraint equations $f_A(\mathbf{X}, t) = 0$ and $f_B(\mathbf{X}, t) = 0$.

Inserting μ_A from (4.72) and μ_B from (4.73) into (4.71) for $i = 3$ yields one equation for the equilibrium vector $\mathbf{X}_{eq}$. Furthermore, there are the two constraint equations. It remains to be seen whether the factor magnitudes that result from these three equations lead in a straightforward manner to the absolute profit maximum for an appropriate set of slack variables, or whether other methods of constrained nonlinear optimization, e.g., the Levenberg–Marquardt method [68], employed in minimizing the sum of squared errors (4.42) subject to the constraints (4.44), are better suited for computing $\mathbf{X}_{eq}$. For our purpose of elucidating the effect of technological constraints on the cost-share theorem, (4.68)–(4.73) are sufficient.

Appendix 2: Maximizing Overall Welfare

Overall welfare,

$$W[s] = \int_{t_0}^{t_1} U[C]dt, \tag{4.74}$$

is a functional of the curve [s] along which the production factors change in time. This curve depends on the variables that enter consumption C. In general, utility may depend on many variables. Here we limit the model to the case that the utility function $U[C]$ is simply output minus capital formation. Output (per unit time) is described by the macroeconomic production function $Y(\mathbf{X}; t)$. Part of Y goes into consumption C and the rest into new capital formation $\dot{X}_1 \equiv \frac{dX_1}{dt}$ plus replacement of depreciated capital. As usual we approximate the annual replacement rate by $\delta^d X_1$, where δ^d is the depreciation rate. (The year is the natural time unit, because the annual cycle of seasons is decisive for agriculture, and is important for construction. It also structures education, vacations – hence tourism and transportation – and some other industrial activities in the moderate climate zones. Thus, for practical purposes $Y(\mathbf{X}; t)$ and $\dot{X}_1 + \delta^d X_1$ are annual output and annual capital formation, respectively.) Then, annual consumption is $C = Y(\mathbf{X}; t) - \dot{X}_1 - \delta^d X_1$. Economic research institutions provide the price of capital utilization p_1 as the sum of net interest, depreciation, and state influences. We use this price later. Since it already includes depreciation, we can omit explicit reference to the depreciation rate, so consumption is given by

$$C = Y(\mathbf{X}; t) - \dot{X}_1. \tag{4.75}$$

(If we kept $\delta^d X_1$ in the optimization procedure, we would have a term proportional to $\delta^d X_1$ added to p_1 everywhere.)

The constraint equations (4.7) and (4.13) are included in the optimization problem with the help of the Lagrange multipliers μ_a and μ. Consequently, we have to maximize

$$W[s] = \int_{t_0}^{t_1} dt \left\{ U[C(\mathbf{X}, \dot{X}_1)] + \mu(c_f(t) - \mathbf{p} \cdot \mathbf{X}) + \sum_a \mu_a f_a(\mathbf{X}, t) \right\}. \tag{4.76}$$

$W[s]$ is a functional of the curve $[s] = \{t, \mathbf{X} : \mathbf{X} = \mathbf{X}(t), \quad t_0 \leq t \leq t_1\}$. Consider another curve $[s, \mathbf{h}] = \{t, \mathbf{X} : \mathbf{X} = \mathbf{X}(t) + \mathbf{h}(t), \quad t_0 \leq t \leq t_1\}$ close to $[s]$ which goes through the same endpoints so that $\mathbf{h}(t_1) = 0$ and $\mathbf{h}(t_0) = 0$. Its functional is

$$W[s, \mathbf{h}] = \int_{t_0}^{t_1} dt \left\{ U[C(\mathbf{X} + \mathbf{h}, \dot{X}_1 + \dot{h}_1)] + \mu(c_f(t) \right.$$

$$\left. -\mathbf{p} \cdot (\mathbf{X} + \mathbf{h})) + \sum_a \mu_a f_a(\mathbf{X} + \mathbf{h}, t) \right\}. \tag{4.77}$$

Since $\mathbf{h}$ is small, the integrand can be approximated by its Taylor expansion up to first order in $\mathbf{h}$ and $\dot{h}_1$. The necessary condition for a maximum of W is that the variation $\delta W \equiv W[s, \mathbf{h}] - W[s]$ vanishes:

$$\delta W = \int_{t_0}^{t_1} dt \left\{ \delta U - \mu \mathbf{p} \cdot \mathbf{h} + \sum_a \mu_a \sum_{i=1}^{3} \frac{\partial f_a}{\partial X_i} h_i \right\} = 0. \tag{4.78}$$

With the chain rule one obtains

$$\delta U \equiv U[C(\mathbf{X} + \mathbf{h}, \dot{X}_1 + \dot{h}_1)] - U[C(\mathbf{X}, \dot{X}_1)]$$

$$= \frac{dU}{dC} dC = \frac{dU}{dC} \left[\sum_{i=1}^{3} \frac{\partial C}{\partial X_i} h_i + \frac{\partial C}{\partial \dot{X}_1} \dot{h}_1 \right]. \qquad (4.79)$$

Partial integration yields

$$\int_{t_0}^{t_1} dt \, \frac{dU}{dC} \frac{\partial C}{\partial \dot{X}_1} \dot{h}_1 = - \int_{t_0}^{t_1} dt \, \frac{d}{dt} \left(\frac{dU}{dC} \frac{\partial C}{\partial \dot{X}_1} \right) h_1(t); \qquad (4.80)$$

$\left[\frac{dU}{dC} \frac{\partial C}{\partial \dot{X}_1} h_1(t) \right]_{t_0}^{t_1}$ is zero, because $h_1(t_1) = 0 = h_1(t_0)$.

Combination of (4.78)–(4.80) results in

$$\delta W = \int_{t_0}^{t_1} dt \left\{ \frac{dU}{dC} \sum_{i=1}^{3} \frac{\partial C}{\partial X_i} h_i - \frac{d}{dt} \left(\frac{dU}{dC} \frac{\partial C}{\partial \dot{X}_1} \right) h_1(t) \right.$$

$$\left. - \mu \sum_{i=1}^{3} p_i h_i + \sum_{a} \mu_a \sum_{i=1}^{3} \frac{\partial f_a}{\partial X_i} h_i \right\} = 0. \qquad (4.81)$$

Since the small h_i are arbitrary for $t_0 < t < t_1$, the integral can only vanish if the coefficients of the h_i vanish in the integrand. This yields the following conditions for $\delta W = 0$:

$$\frac{dU}{dC} \frac{\partial C}{\partial X_1} - \frac{d}{dt} \left(\frac{dU}{dC} \frac{\partial C}{\partial \dot{X}_1} \right) - \mu p_1 + \sum_{a} \mu_a \frac{\partial f_a}{\partial X_1} = 0 \qquad (4.82)$$

and

$$\frac{dU}{dC} \frac{\partial C}{\partial X_i} - \mu p_i + \sum_{a} \mu_a \frac{\partial f_a}{\partial X_i} = 0, \quad i = 2, 3. \qquad (4.83)$$

If one identifies $U[C(\mathbf{X}, \dot{X}_1)]$ with the Lagrangian $L(\mathbf{X}, \dot{X}_1)$, one notes the formal equivalence of these equations with constrained Lagrange equations of motion in classical mechanics.

With C from (4.75), the three equilibrium conditions (4.82) and (4.83) become

$$\frac{dU}{dC} \frac{\partial Y}{\partial X_i} - \mu p_i + \sum_{a} \mu_a \frac{\partial f_a}{\partial X_i} = -\frac{d}{dt} \left(\frac{dU}{dC} \right) \delta_{i,1}, \quad i = 1, 2, 3; \qquad (4.84)$$

the Kronecker delta $\delta_{i,1}$, is 1 for $i = 1$ and 0 otherwise.

Equations (4.84) are the general equilibrium conditions for an economic system subject to cost limits and technological constraints if the behavioral assumption is that society optimizes time-integrated utility.

Note that if Y does not depend explicitly on time t, and if $\mu = 0 = \mu_a$, (4.84) imply the conservation law $U + \frac{dU}{dC}\dot{X}_1 = $ constant. Thus, formally, the conserved Legendre transform of utility, $U + \frac{dU}{dC}\dot{X}_1$, corresponds to the Hamiltonian of classical mechanics.

Dividing (4.84) by dU/dC and multiplying them by X_i/Y changes them to

$$\frac{X_i}{Y}\frac{\partial Y}{\partial X_i} = \frac{\mu X_i}{Y\frac{dU}{dC}}\left[p_i - \sum_a \frac{\mu_a}{\mu}\frac{\partial f_a}{\partial X_i} - \delta_{i,1}\frac{1}{\mu}\frac{d}{dt}\left(\frac{dU}{dC}\right)\right]. \tag{4.85}$$

The left-hand side of this equation is the output elasticity ε_i. Inserting the right-hand side into (4.5) yields

$$\mu = \frac{Y\frac{dU}{dC}}{\sum_{i=1}^{3}X_i\left[p_i - \sum_a \frac{\mu_a}{\mu}\frac{\partial f_a}{\partial X_i} - \delta_{i,1}\frac{1}{\mu}\frac{d}{dt}\left(\frac{dU}{dC}\right)\right]}. \tag{4.86}$$

With that the equilibrium conditions (4.85) become

$$\varepsilon_i = \frac{X_i\left[p_i - \sum_a \frac{\mu_a}{\mu}\frac{\partial f_a}{\partial X_i} - \delta_{i,1}\frac{1}{\mu}\frac{d}{dt}\left(\frac{dU}{dC}\right)\right]}{\sum_{i=1}^{3}X_i\left[p_i - \sum_a \frac{\mu_a}{\mu}\frac{\partial f_a}{\partial X_i} - \delta_{i,1}\frac{1}{\mu}\frac{d}{dt}\left(\frac{dU}{dC}\right)\right]}, \quad i = 1, 2, 3. \tag{4.87}$$

This has the form of (4.8). The shadow prices, however, differ somewhat from those of (4.9):

$$s_i \equiv -\sum_{a=A}^{B}\frac{\mu_a}{\mu}\frac{\partial f_a}{\partial X_i} - \delta_{i,1}\frac{1}{\mu}\frac{d}{dt}\left(\frac{dU}{dC}\right), \quad i = 1, 2, 3. \tag{4.88}$$

To compute the ratios of Lagrange multipliers in the case that (4.15) holds so that $\delta_{i,1}\frac{1}{\mu}\frac{d}{dt}\left(\frac{dU}{dC}\right) = 0$, we define

$$\mu_1 \equiv \frac{\mu_A}{\mu}, \quad \mu_2 \equiv \frac{\mu_B}{\mu}, \tag{4.89}$$

and abbreviate the partial derivatives of the two constraint equations as in (4.70). With that the equilibrium conditions (4.87) become

$$\varepsilon_i = \frac{X_i\left[p_i - \mu_1 f_{Ai} - \mu_2 f_{Bi}\right]}{\sum_{i=1}^{3}X_i\left[p_i - \mu_1 f_{Ai} - \mu_2 f_{Bi}\right]}, \quad i = 1, 2, 3. \tag{4.90}$$

If one resolves the two independent ratios

$$\frac{\varepsilon_1}{\varepsilon_2} = \frac{X_1 [p_1 - \mu_1 f_{A1} - \mu_2 f_{B1}]}{X_2 [p_2 - \mu_1 f_{A2} - \mu_2 f_{B2}]} \tag{4.91}$$

and

$$\frac{\varepsilon_1}{\varepsilon_3} = \frac{X_1 [p_1 - \mu_1 f_{A1} - \mu_2 f_{B1}]}{X_3 [p_3 - \mu_1 f_{A3} - \mu_2 f_{B3}]} \tag{4.92}$$

with respect to μ_1 and μ_2 using the definitions

$$R_{21} \equiv \frac{X_2 \varepsilon_1}{X_1 \varepsilon_2}, \quad R_{31} \equiv \frac{X_3 \varepsilon_1}{X_1 \varepsilon_3}, \tag{4.93}$$

and performs some algebraic manipulations, one obtains

$$\mu_1 = \frac{(p_1 - p_2 R_{21})}{f_{A1} - f_{A2} R_{21}} + \frac{f_{B2} R_{21} - f_{B1}}{f_{A1} - f_{A2} R_{21}} \times \mu_2 \tag{4.94}$$

and

$$\mu_2 = \frac{(p_1 - p_3 R_{31})(f_{A1} - f_{A2} R_{21}) - (p_1 - p_2 R_{21})(f_{A1} - f_{A3} R_{31})}{(f_{B2} R_{21} - f_{B1})(f_{A1} - f_{A3} R_{31}) - (f_{B3} R_{31} - f_{B1})(f_{A1} - f_{A2} R_{21})}. \tag{4.95}$$

The calculation of the equilibrium values of the X_i would be similar to the procedure indicated below (4.73).

Appendix 3: Aggregating Output and Factors in Physical Terms

Macroeconomic production functions are the basis of computing economic equilibrium by optimizing profit and overall welfare. Macroeconomic growth models employ them as well. The concept of the twice-differentiable macroeconomic production function, although widely accepted, has nevertheless been criticized by scholars who have been more concerned than mainstream economists about the physical aspects of production [88–93]. Their criticism centers on three principal objections. The first is the problem of aggregating the heterogeneous goods and services of output into one monetary quantity, represented in the national accounts by the deflated GDP, or parts thereof. The second is the related problem of aggregating the heterogeneous components of the capital stock (machines, structures, etc.) into one monetary quantity "capital," measured by deflated currency in the national accounts. The third problem is the unclear relationship between the micro theory of production in individual firms, for which the concept of the (micro)production function is not questioned, and the macro theory: "The standard practice, para-doxically, is to argue that the production function in the theoretical model is a microproduction function. However, the empirical evidence and examples provided tend to be macro..." [92].

The problem of capital aggregation was at the center of the "Cambridge Contro-versy" between economists from Cambridge, UK, and Cambridge, Massachusetts, USA. That debate was triggered by Joan Robinson [94] asking in what units is "capital" to be measured when used as a factor of production in aggregate production functions. She reasoned that all factors of production should be measured in factor-specific physical terms, whereas the statistics show capital only as a sum of values, expressed in (deflated) monetary terms. Her question was: "How can this be made to correspond to a physical factor of production?" [95]. Later, Usher asked: "Can time series of quantities of capital goods be combined in a single number that may be interpreted as 'the' measure of real capital in the economy as a whole?" [96]. In addition, Pasinetti [91] recently declared: "The problem that arises in the case of capital ...is ...fundamentally the conceptual difficulty of having to treat an aggregate quantity expressed in value terms (capital) in the same way as other quantities (land and labor) which are instead expressed in physical terms. The two types of aggregate quantities do not belong to the same logical class, and thus can neither be placed on the same level nor be inserted symmetrically in the same function .... It becomes a fundamental and indeed abyssal conceptual diversity concerning the 'factors' labor and land on the one hand, and the factor 'capital' on the other."

In view of these concerns, the aggregation of output and factors in physical terms was the starting point for the derivation of the LinEx production function [3, 24], independently from any equilibrium considerations. For the sake of completeness and clarity, the aggregation scheme is outlined here in a somewhat more extended form than in [24].

First is the observation that macroeconomic production functions, both in orthodox theory and in heterodox theory, are *state functions* of the economic system: output depends only on the actual numerical values of the inputs and not on the path along which the system has arrived at these values. This makes them a mathematical tool for quantitative analyses in economics which is as convenient as state functions are in physics. Kinetic energy and potential energy of particles in conservative force fields depend only on the actual magnitudes of the spatial coordinates, the free energy of a thermodynamic system depends unequivocally on temperature and volume, etc. A necessary condition for the existence of state functions in systems subject to the laws of nature is that these laws establish causal relations between the independent state variables and the dependent variable, represented by the function. Thus, if the output of an economic system $Y(\mathbf{X})$ is a state function of its inputs $\mathbf{X}$, there must exist an unequivocal cause-and-effect relation between value added Y and the production factors $\mathbf{X}$. But what can be the origin of such a relation? This is the question that leads to aggregation in physical terms. The answer is provided by the elementary process of production, which is work performance and information processing,[31] as sketched in Fig. 4.1. Since the laws of nature govern (routine) work performance and information processing by machines and humans, they establish the causal relationship between output and inputs that is the necessary condition for the existence of macroeconomic production functions. (Human creativity, tied to the arrow of time, is, of course, above that.)

We postulate that, on average (notwithstanding short-term fluctuations), the monetary value of a good or a service is greater the more physical work has to be performed and the more information has to be processed in the generation of this good or service. We further assume that, although the relative monetary value of work performance and information processing may change for different components of output, the average over all components of output should nevertheless stay approximately constant during short time intervals. (A short time interval is one that is shorter than typical times required for innovation diffusion.) Similarly, the monetary value of a capital good should be related to its capacity to perform work and process information.

Output: Embodied Work Performance and Information Processing

The kilowatt-hour (kWh) is chosen as the unit of work performance in the production of a good or service, and the kilobit (kb) is the unit of the associated information processing.[32] The magnitudes of the units really do not matter if dimensionless, normalized variables are used for output and production factors.

However, there are no statistics of output in terms of these physical units. Therefore, for practical purposes, they have to be related to time series of deflated monetary units, which are provided by the national accounts. We introduce a fictitious unit as a dummy for any real monetary unit and call it "mark."[33]

We link the monetary measure of output to its physical measure in five steps.

1. The monetary output $Y_{mon}(t)$ of the economy at time t, measured in constant mark, is partitioned into $M \gg 1$ pieces $Y_{i,mon}$ that all have the same monetary value, say $\mu \times$ mark: $Y_{mon} = \sum_{i=1}^{M} Y_{i,mon} = M\mu \times$ mark. As $Y_{mon}(t)$ changes with time t, $M(t)$ changes correspondingly; thus, μ is constant and mark $= Y_{mon}(t)/M(t)\mu$.

---

[31] Work performance and information processing is also the elementary process in all biological production systems. Witt [97] discusses production in nature and production in the economy.

[32] One kilobit is one 1,000 "yes–no" decisions as represented, e.g., by the switching on and switching off of energy flows that are carried by electric currents.

[33] In memory of a European currency that vanished with the introduction of the euro.

2. The work performed in the production of $Y_{i,\text{mon}}$ is measured by the required energy input. It includes all energy conversion losses due to frictional work and limits to thermodynamic efficiency. We define the following:

 - W_i is the number of kilowatt-hours of primary energy consumed in the generation of $Y_{i,\text{mon}}$.
 - V_i is the the number of kilobits processed in the generation of $Y_{i,\text{mon}}$.

Machine standards can be established for the measurements of W_i and V_i for all goods and services sold on the market, because these are produced by routine, standardized operations.

3. We define the physical output Y_{phys} of the economy as

$$Y_{\text{phys}} = \sum_1^M Y_{i,\text{phys}} \equiv \sum_1^M W_i \times \text{kWh} \times V_i \times \text{kb}. \tag{4.96}$$

According to the definitions in step 1, $Y_{i,\text{phys}}$ has the monetary value $\mu \times$ mark.

The unit of physical output is given the name ENIN, as shorthand for "energy and information." It is defined as

$$\text{ENIN} \equiv \frac{1}{M} \sum_1^M W_i \times V_i \times \text{kWh} \times \text{kb} = \zeta \times \text{kWh} \times \text{kb}, \tag{4.97}$$

with

$$\zeta \equiv \frac{1}{M} \sum_1^M W_i \times V_i, \tag{4.98}$$

so

$$Y_{\text{phys}} = M \times \text{ENIN}. \tag{4.99}$$

The equivalence factor ζ changes in time t if the monetary value of work performance and information processing changes in such a way that the numbers $W_i \times V_i$, which correspond to constant μ, *and* the right-hand side of (4.98) become time-dependent.

4. It follows from the above equations that the relation between the monetary and the physical output is

$$\frac{Y_{\text{mon}}(t)}{Y_{\text{phys}}(t)} = \frac{M\mu \times \text{mark}}{M \times \text{ENIN}} = \frac{\mu}{\zeta} \times \frac{\text{mark}}{\text{kWh} \times \text{kb}}. \tag{4.100}$$

For constant ζ there is proportionality between Y_{phys} and Y_{mon}. Then the time series of output in constant currency is proportional to the time series of the technological output, aggregated in the physical ENINs.

5. When working with dimensionless variables that are normalized to their magnitudes in a base year t_0, dimensionless output at time t is

$$y(t) \equiv Y(t)/Y_0, \qquad (4.101)$$

where $Y_0 = Y(t_0)$. Thus, the dimensionless physical output at time t is

$$y_{\text{phys}}(t) \equiv \frac{Y_{\text{phys}}(t)}{Y_{\text{phys}}(t_0)}. \qquad (4.102)$$

By virtue of (4.100) the monetary output in the base year t_0 is given by

$$Y_{\text{mon}}(t_0) = \frac{\mu}{\zeta_0} \times \frac{\text{mark}}{\text{kWh} \times \text{kb}} Y_{\text{phys}}(t_0), \qquad \zeta_0 \equiv \zeta(t_0). \qquad (4.103)$$

The combination of (4.100), (4.102), and (4.103) shows that the dimensionless time series of monetary output is equal to the dimensionless, physically aggregated time series of output multiplied by $\frac{\zeta_0}{\zeta}$:

$$y_{\text{mon}}(t) \equiv \frac{Y_{\text{mon}}(t)}{Y_{\text{mon}}(t_0)} = \frac{\zeta_0}{\zeta} \frac{Y_{\text{phys}}(t)}{Y_{\text{phys}}(t_0)} \equiv \frac{\zeta_0}{\zeta} y_{\text{phys}}(t). \qquad (4.104)$$

As long as the equivalence factor ζ can be considered as time-independent and equal to ζ_0, the dimensionless monetary time series of output is equal to the dimensionless physical time series. In this sense, monetary aggregation and physical aggregation of inhomogeneous output are the same. A time dependence of the equivalence factor ζ causes a time dependence of the parameter y_0 in the LinEx function (4.35).

Capital: Capacity to Perform Work and Process Information

The capital stock of industrialized countries consists of all energy conversion devices and information processors and all buildings and installations necessary for their protection and operation.[34] This is what is meant by "machines." With respect to machines, Samuelson and Solow [64] state: "Even though there is no such thing as a single abstract capital substance that transmutes itself from one machine form to another like a restless reincarnating soul, the rigorous investigation of a heterogeneous capital-goods model shows that over extended periods of time an

---

[34]We ignore hammers, sickles, and other tools operated by muscle power.

economic society can in a perfectly straightforward way reconstruct the composition of its diverse capital goods ...." In other words, Samuelson and Solow reject the notion of an abstract capital substance. However, we think there *is* something like that. It is the *capacity* to perform physical work and process information, thus providing the services that flow from the capital stock per unit time. What transmutes itself from one machine form to another is the monetary value of this capacity. The value varies with the materials, mass, and complexity of the capital goods. Therefore, aggregation in the terms of the work-performance and information-processing capacity has to average over the total capital stock in a way similar to output aggregation. This provides a physical concept of capital that is directly related to the monetary concept of capital "over extended periods of time." Time series of capital in deflated monetary units are provided by the national accounts.

As above, we proceed in five steps.

1. The gross capital stock of the total economy, K_{mon}, measured in deflated "mark," is partitioned into $N \gg 1$ pieces $K_{i,mon}$, which all have the same monetary value, say $\nu \times$ mark: $K_{mon} = \sum_{i=1}^{N} K_{i,mon} = N\nu \times$ mark. As $K_{mon}(t)$ changes with time t, $N(t)$ changes correspondingly; thus, ν is constant and mark$= K_{mon}/N\nu$.

2. The capacity to perform work per unit time is measured in kilowatts, (kW), and the capacity to process information per unit time is measured in kilobits per second (kb/s). We define the following:

   - S_i as number of kilowatts performed by the fully employed capital good i of monetary value $K_{i,mon}$.
   - T_i as number of kilobits per second processed by the fully employed capital good i of monetary value $K_{i,mon}$.

   The S_i can be read off machine specifications, and the T_i may be measured by the number of switching processes per unit time that pass along or shut down energy flows in the fully employed machines.

3. We define the physical capital stock K_{phys} of the economy as

$$K_{phys} = \sum_{1}^{N} K_{i,phys} \equiv \sum_{1}^{N} S_i \times kW \times T_i \times kb/s. \qquad (4.105)$$

According to the definitions in step 1, $K_{i,phys}$ has the monetary value $\nu \times$ mark.

The unit of physical capital is given the name ATON, as shorthand for automation. It is defined as

$$ATON \equiv \frac{1}{N} \sum_{1}^{N} S_i \times T_i \times kW \times kb/s = \kappa \times kW \times kb/s, \qquad (4.106)$$

with

$$\kappa \equiv \frac{1}{N} \sum_{1}^{N} S_i \times T_i, \qquad (4.107)$$

so

$$K_{\text{phys}} = N \times \text{ATON}. \tag{4.108}$$

The equivalence factor κ changes in time t if the monetary values of the capacities to perform work and process information change in a such way that the numbers $S_i \times T_i$, which correspond to constant ν, *and* the right-hand side of (4.107) become time-dependent.

4. The relation between monetary and physical capital follows from the above equations as

$$\frac{K_{\text{mon}}(t)}{K_{\text{phys}}(t)} = \frac{N\nu \times \text{mark}}{N \times \text{ATON}} = \frac{\nu}{\kappa} \times \frac{\text{mark}}{\text{kW} \times \text{kb/s}}. \tag{4.109}$$

For constant κ there is proportionality between K_{phys} and K_{mon}. Then the time series of capital in constant currency is proportional to the time series of technological capital, aggregated in the physical ATONs.

5. When working with dimensionless variables, the physical capital stock normalized to its magnitude at time t_0 is

$$k_{\text{phys}}(t) \equiv \frac{K_{\text{phys}}(t)}{K_{\text{phys}}(t_0)}. \tag{4.110}$$

By virtue of (4.109), the monetary capital stock in the base year t_0 is

$$K_{\text{mon}}(t_0) = \frac{\nu}{\kappa_0} \times \frac{\text{mark}}{\text{kW} \times \text{kb/s}} K_{\text{phys}}(t_0), \qquad \kappa_0 \equiv \kappa(t_0). \tag{4.111}$$

The combination of (4.109)–(4.111) shows that the dimensionless time series of monetarily aggregated capital is equal to the dimensionless, physically aggregated time series of capital multiplied by $\frac{\kappa_0}{\kappa}$:

$$k_{\text{mon}}(t) \equiv \frac{K_{\text{mon}}(t)}{K_{\text{mon}}(t_0)} = \frac{\kappa_0}{\kappa} \frac{K_{\text{phys}}(t)}{K_{\text{phys}}(t_0)} \equiv \frac{\kappa_0}{\kappa} k_{\text{phys}}(t). \tag{4.112}$$

As long as the equivalence factor κ can be considered as time-independent and equal to κ_0, the dimensionless monetary time series of capital is equal to the dimensionless physical time series. In this sense monetary aggregation and physical aggregation of inhomogeneous capital are the same. A time dependence of κ contributes to the time dependence of the technology parameter a in the LinEx function (4.35).

Labor and Energy

Aggregation of output and capital in the physical terms of work performance and information processing is linked to the observation that energy activates capital, and labor handles both. Energy conversion in the machines of the capital stock provides all mechanical, chemical, and electrical work required for production. Routine labor, supervising and manipulating the capital stock, steers the energy flows.

Economic activities of humans are modeled by the two components "routine labor" and "creativity." Machines can meanwhile perform most of the routine activities of the human brain and hand, as progress in automation shows. The national labor statistics measure and aggregate routine labor in hours worked per year. (One could also measure routine labor in multiples of the daily energy requirement of an average employee – about 2 kWh per person per day.) When one works with dimensionless variables, the unit of routine labor does not matter. In quantitative economic studies "labor" means "routine labor." The units used by us are "man-hours worked per year.". The impact of "creativity" is measured ex post facto by the time dependence of the LinEx technology parameters.

Discriminating between routine labor and creativity when considering human contributions to production and growth dissolves Felipe and Fisher's objection to the aggregation of labor in hours worked per year. They ask: "Was one woman-hour of labor by Joan Robinson really the same as one of Queen Elizabeth II or one of Britney Spears in terms of productivity?" [92]. Questioning the creative contributions of the three women to wealth production would be more to the point. Even so, the answer "No" would not be an argument against the aggregation of labor in man-hours and woman-hours, because we are dealing with averages, not individuals. Moreover, the economic system is largely based on activities of firms, and transactions between structured collections of individuals, called firms, that do not behave like individuals. We recognize that the quality of labor increases over time, thanks to learning and education that, in turn, reflects the impact of creativity in the past.[35]

The national energy balances aggregate energy in enthalpy units such as peta-joules, tons of oil or coal equivalents, or quads. (Fortunately, these enthalpy units measure also exergy, because all primary energy carriers are practically 100% exergy.) Quite familiar to many consumers is the kilowatt-hour, because it appears on the bills of electricity and gas suppliers. Any one of these measuring units of energy is appropriate. When one works with dimensionless variables, the energy unit does not matter anyway.

In summary, aggregation of output, capital, routine labor, and energy in physical terms is possible. These physical terms represent work performance and information

---

[35] Some economists in recent years have preferred to incorporate the quality aspect as something called knowledge, which can be regarded as a component of capital. This may have a certain theoretical justification, but it introduces a measurement problem. Knowledge is undoubtedly created to some extent by formal research and development, and it is conveyed to some extent by years of education, but those two proxies are by no means sufficient. We rather prefer to consider the measurable factor energy also as a proxy variable for the knowledge embodied in physical capital, and for the professional qualification of labor as well. This is based on the observation that, as a rule, energy-intensive economies have more complex capital stocks and more highly trained labor forces than economies with less energy input per GDP.

processing – the basic elements of economic production. They are the response to the conceptual criticism of Joan Robinson and others. And since they are related by cause and effect according to the laws of nature, they also satisfy the necessary condition for state functions. (The change with time of technology parameters by creativity does not affect the relation between the other factors at a given time.) Thus, there is some engineering justification for the standard assumption that twice-differentiable macroeconomic production functions do make sense in the real economy.

Appendix 4: Explicit Constraint Equations

The capital stock $k_m(y)$, which enters the constraint equation (4.19) as the instrumental capital that is required for the maximally automated production of output y by the factors k_m, l_m, and e_m, can be calculated in the LinEx approximation, (4.35) by demanding that the LinEx function in the state of maximum automation, $y_{L1}[k_m, l_m, e_m; t]$, is equal to the LinEx production function $y_{L1}[k, l, e; t]$ of the actual inputs:

$$y_{L1}[k_m, l_m, e_m = ck_m; t] = y_{L1}[k, l, e; t]. \tag{4.113}$$

The technology parameters in both LinEx functions are the ones at time t, because one is interested in the factors that are required for the maximally automated production of a given output y in a given state of technology. If one assumes that the routine labor l_m that remains in the state of maximum automation is much smaller than k_m, so that one can disregard $l_m/k_m \ll 1$, (4.113) becomes

$$y_0 c(t) k_m(y) \exp[2a(t)(1 - c(t))]$$
$$= y_0 e \exp\left[a(t)\left(2 - \frac{l+e}{k}\right) + a(t)c(t)\left(\frac{l}{e} - 1\right)\right]. \tag{4.114}$$

This yields the capital stock for the maximally automated production of an output y that at time t is produced by the factors $k(t)$, $l(t)$, and $e(t)$:

$$k_m(y) = \frac{e(t)}{c(t)} \exp\left[a(t)c(t)\left(1 + \frac{l(t)}{e(t)}\right) - a(t)\frac{l(t) + e(t)}{k(t)}\right]. \tag{4.115}$$

Inserting $k_m(y)$ into (4.19), where the technical limit to automation $\rho_T(t)$ and the slack variable k_ρ model the technological constraint, we obtain

$$f_A(K, L, E, t) \equiv \frac{(k + k_\rho)}{k_m(y)} - \rho_T(t)$$
$$= (k + k_\rho)\frac{c}{e} \exp\left[-ac\left(1 + \frac{l}{e}\right) + a\frac{l+e}{k}\right]$$
$$- \rho_T(t) = 0. \tag{4.116}$$

Here, and in the following, we drop the time arguments of factors and parameters for the sake of simplicity.

The equation for the constraint on capacity utilization results from (4.20) and (4.21) as

$$f_B(K, L, E, t) \equiv \eta_0 \left(\frac{1 + l_\eta(t)}{k}\right)^\lambda \left(\frac{e + e_\eta(t)}{k}\right)^\nu - 1 = 0. \tag{4.117}$$

Equations (4.116) and (4.117) yield the slack-variable relations

$$k + k_\rho = k_m(y)\rho_T(t) \tag{4.118}$$

and

$$e + e_\eta = \frac{k}{\eta_0^{1/\nu} \left(\frac{1+l_\eta}{k}\right)^{\lambda/\nu}}. \tag{4.119}$$

The derivatives of f_A and f_B are calculated by observing (4.16) and the chain rule so that $\partial f_A / \partial K = (1/K_0)(\partial f_A / \partial k)$, etc. From (4.116)–(4.119) we obtain

$$\frac{\partial f_A}{\partial k} = \frac{1}{k_m(y)} - a\frac{l + e}{k^2}\rho_T, \tag{4.120}$$

$$\frac{\partial f_B}{\partial k} = -\frac{\lambda + \nu}{k}, \tag{4.121}$$

$$\frac{\partial f_A}{\partial l} = -a\left(\frac{c}{e} - \frac{1}{k}\right)\rho_T, \tag{4.122}$$

$$\frac{\partial f_B}{\partial l} = \frac{\lambda}{l + l_\eta}, \tag{4.123}$$

$$\frac{\partial f_A}{\partial e} = \left(\frac{a}{k} + \frac{acl}{e^2} - \frac{1}{e}\right)\rho_T, \tag{4.124}$$

$$\frac{\partial f_B}{\partial e} = \frac{\nu}{e + e_\eta} = \frac{\nu}{k}\eta_0^{1/\nu}\left(\frac{l + l_\eta}{k}\right)^{\lambda/\nu}. \tag{4.125}$$

Inserting these equations into the shadow price equations (4.22) and (4.23), one gets the explicit equations for all shadow prices.

To compute the shadow prices from the general theoretical framework for an existing economic system, one has to take the following steps.

1. The technology parameters a and c have to be determined econometrically for the system.

2. In a rough approximation one may assume proportionality between the slack
   variables in the constraint on capacity utilization:

$$e_\eta(t) = d(t) \times l_\eta(t); \qquad (4.126)$$

here $d(t)$ is the second constraint parameter besides $\rho_T(t)$. We call it the "labor–energy-coupling parameter for full capacity." Ideally, one should be able to determine it from measurements of the energy and labor increases required in order to go from any degree η of capacity utilization to 1. With that (4.119) becomes the relation between l_η (or e_η) and k, l, and e.

3. The multiplier η_0 and the exponents λ and ν may be obtained by fitting the phenomenological η of (4.21) to empirical time series of η, which are available from economic research institutions.

4. The technical limit $\rho_T(t)$ to the degree of automation can be any number between 0 and 1. General business inquiries should give clues to it. Alternatively, one has to compute the time series of the shadow prices for a number of scenarios for $\rho_T(t)$.

Then (4.8) could be used, in principle, to determine the production factors that satisfy the equilibrium conditions in the LinEx approximation: One substitutes the α, β, and γ from (4.34) for the ε_i on the left-hand side of (4.8) and proceeds as indicated below (4.73). Comparison of the results, i.e., the $\mathbf{X}_{eq}$, with the empirical inputs (K, L, E) would allow one to judge the validity of the behavioral assumptions of profit or overall welfare maximization, which lead to the equilibrium conditions.

Appendix 5: Empirical Data of Output and Inputs

The German empirical data for the systems "FRG, total economy" and "FRG, industries" (Tables 4.11 and 4.12) were obtained from the German system of national accounts (SNA), labor statistics, and energy balances. Help from O. Schmalwasser (Statistisches Bundesamt) and H.-U. Bach (Institut für Arbeitsmarkt und Berufsforschung) is gratefully acknowledged. The data are those for West Germany from 1960 to 1990 and of reunified Germany from 1991 to 2000. "FRG, Industries" is "goods-producing industries," which consists of the SNA subsectors "mining and quarrying", "manufacturing", "electricity, gas, and water," and "construction".

Residential fixed capital and the energy demand of households are excluded in the German, Japanese, and US data.

The empirical data for "Japan, industries" (Table 4.13), which produces about 90% of Japanese GDP, were obtained with the kind help of Shigeru Yasukawa and Osamu Sato from the Energy System Assessment Laboratory, Tokai Mura, and Kokichi Itoh from the The Institute of Energy Economics, Tokyo, during personal

Table 4.11 Empirical data of output and inputs in "FRG, total economy" (see the text for an explanation), 1960–2000; $y(t) = \frac{Y(t)}{Y(1960)}$, $k(t) = \frac{K(t)}{K(1960)}$, $l(t) = \frac{L(t)}{L(1960)}$, $e(t) = \frac{E(t)}{E(1960)}$. $Y(1960) = DM_{1991}852.8 \times 10^9$, $K(1960) = DM_{1991}1517.4 \times 10^9$, $L(1960) = 56341 \times 10^6$ h, $E(1960) = 4458.8$ PJ

Year	$y(t)$	$k(t)$	$l(t)$	$e(t)$	Year	$y(t)$	$k(t)$	$l(t)$	$e(t)$
1960	1.000	1.000	1.000	1.000	1981	2.042	2.923	0.829	1.594
1961	1.048	1.075	1.002	1.009	1982	2.020	3.016	0.819	1.546
1962	1.096	1.157	0.988	1.035	1983	2.053	3.095	0.805	1.545
1963	1.128	1.240	0.976	1.066	1984	2.114	3.175	0.803	1.582
1964	1.205	1.323	0.982	1.128	1985	2.164	3.251	0.797	1.608
1965	1.271	1.415	0.981	1.156	1986	2.221	3.330	0.808	1.597
1966	1.308	1.513	0.966	1.143	1987	2.245	3.416	0.805	1.621
1967	1.305	1.607	0.917	1.129	1988	2.332	3.506	0.810	1.666
1968	1.382	1.684	0.912	1.214	1989	2.419	3.604	0.809	1.680
1969	1.490	1.761	0.917	1.315	1990	2.555	3.711	0.814	1.709
1970	1.562	1.856	0.921	1.394	1991	3.031	4.800	1.065	2.491
1971	1.609	1.965	0.911	1.397	1992	3.099	4.968	1.060	2.445
1972	1.680	2.078	0.902	1.461	1993	3.070	5.115	1.032	2.411
1973	1.763	2.186	0.896	1.551	1994	3.138	5.237	1.030	2.379
1974	1.765	2.291	0.871	1.526	1995	3.205	5.348	1.019	2.378
1975	1.733	2.379	0.828	1.448	1996	3.240	5.449	1.005	2.438
1976	1.827	2.461	0.841	1.531	1997	3.301	5.546	0.999	2.403
1977	1.882	2.544	0.828	1.526	1998	3.377	5.646	1.006	2.392
1978	1.941	2.630	0.824	1.580	1999	3.445	5.753	1.011	2.350
1979	2.027	2.720	0.831	1.673	2000	3.570	5.846	1.018	2.357
1980	2.040	2.820	0.838	1.649					

visits of the author to Japan. They are taken from the Japanese SNA, publications of the Economic Planning Agency, the Ministry of Labor, and the Energy Balances of Japan.

"USA, total economy" (Table 4.14) comprises "private industries" (with its subsectors "agriculture, forestry, and fishery", "mining", "construction", "manufacturing", "transportation and public utilities", "wholesale trade", "retail trade", "finance, insurance, and real estate", "services", and "government"; the latter produced about 13% of US GDP in 1996. Data sources were various editions of *Statistical Abstracts of the United States*, *Survey of Current Business*, *Handbook of Labor Statistics*, *EnergyBalances of OECD Countries*, and *Monthly Energy Review*. Dale W. Jorgenson of Harvard University provided valuable advice for the selection of the data sources. They were evaluated during visits to Harvard University, the Institute of Energy Analysis in Oak Ridge, and the University of California at Berkeley.

The empirical time series of output and inputs are presented in Tables 4.11–4.14.

Table 4.12 Empirical data of output and inputs in "FRG, industries" (see the text for an explanation), 1960–1999; $y(t) = \frac{Y(t)}{Y(1960)}$, $k(t) = \frac{K(t)}{K(1960)}$, $l(t) = \frac{L(t)}{L(1960)}$, $e(t) = \frac{E(t)}{E(1960)}$. $Y(1960) = DM_{1991}453.5 \times 10^9$, $K(1960) = DM_{1991}692.7 \times 10^9$, $L(1960) = 26{,}132 \times 10^6$ h, $E(1960) = 3{,}798$ PJ

Year	$y(t)$	$k(t)$	$l(t)$	$e(t)$	Year	$y(t)$	$k(t)$	$l(t)$	$e(t)$
1960	1.000	1.000	1.000	1.000	1980	1.896	2.657	0.771	1.497
1961	1.051	1.090	1.008	0.997	1981	1.864	2.714	0.747	1.449
1962	1.099	1.184	0.992	1.016	1982	1.801	2.758	0.729	1.389
1963	1.122	1.275	0.972	1.041	1983	1.828	2.793	0.705	1.379
1964	1.221	1.367	0.988	1.100	1984	1.874	2.823	0.701	1.409
1965	1.294	1.464	0.991	1.125	1985	1.914	2.853	0.690	1.437
1966	1.321	1.559	0.973	1.096	1986	1.941	2.895	0.693	1.399
1967	1.282	1.642	0.893	1.079	1987	1.915	2.944	0.687	1.410
1968	1.380	1.716	0.903	1.165	1988	1.968	2.995	0.688	1.443
1969	1.513	1.799	0.929	1.253	1989	2.039	3.050	0.691	1.449
1970	1.593	1.904	0.943	1.332	1990	2.135	3.115	0.697	1.456
1971	1.620	2.021	0.925	1.334	1991	2.370	3.596	0.838	1.869
1972	1.677	2.133	0.905	1.372	1992	2.359	3.712	0.823	1.814
1973	1.767	2.230	0.897	1.469	1993	2.229	3.787	0.769	1.758
1974	1.735	2.308	0.854	1.454	1994	2.301	3.819	0.753	1.748
1975	1.648	2.372	0.783	1.345	1995	2.299	3.838	0.739	1.691
1976	1.766	2.431	0.784	1.423	1996	2.240	3.850	0.708	1.719
1977	1.799	2.488	0.774	1.395	1997	2.274	3.862	0.694	1.701
1978	1.832	2.540	0.762	1.437	1998	2.294	3.876	0.691	1.665
1979	1.921	2.595	0.768	1.531	1999	2.268	3.893	0.680	1.636

Table 4.13 Empirical data of output and inputs in "Japan, industries" (see the text for an explanation), 1965–1992; $y(t) = \frac{Y(t)}{Y(1965)}$, $k(t) = \frac{K(t)}{K(1965)}$, $l(t) = \frac{L(t)}{L(1965)}$, $e(t) = \frac{E(t)}{E(1965)}$. $Y(1965) = JPY_{1985}97{,}751 \times 10^9$, $K(1965) = JPY_{1985}85{,}084 \times 10^9$, $L(1965) = 72 \times 10^9$ h, $E(1965) = 6{,}354$ PJ

Year	$y(t)$	$k(t)$	$l(t)$	$e(t)$	Year	$y(t)$	$k(t)$	$l(t)$	$e(t)$
1965	1.000	1.000	1.000	1.000	1979	2.625	3.791	1.083	2.556
1966	1.116	1.092	1.023	1.110	1980	2.746	4.049	1.096	2.535
1967	1.241	1.212	1.043	1.268	1981	2.850	4.314	1.106	2.464
1968	1.396	1.368	1.053	1.434	1982	2.941	4.574	1.118	2.411
1969	1.568	1.558	1.054	1.645	1983	2.980	4.872	1.175	2.479
1970	1.718	1.785	1.030	1.926	1984	3.116	5.175	1.237	2.543
1971	1.811	2.015	1.040	1.998	1985	3.276	5.729	1.256	2.593
1972	1.972	2.262	1.046	2.137	1986	3.329	6.155	1.273	2.571
1973	2.110	2.511	1.059	2.345	1987	3.491	6.618	1.273	2.699
1974	2.107	2.744	1.026	2.347	1988	3.714	7.053	1.302	2.847
1975	2.160	2.956	1.010	2.283	1989	3.903	7.550	1.326	2.950
1976	2.256	3.160	1.031	2.385	1990	4.105	8.123	1.348	3.074
1977	2.359	3.357	1.047	2.415	1991	4.252	8.833	1.369	3.161
1978	2.475	3.560	1.059	2.471	1992	4.266	9.437	1.390	3.292

Table 4.14 Empirical data of output and inputs in "USA, total economy" (see the text for an explanation), 1960–1996; $y(t) = \frac{Y(t)}{Y(1960)}$, $k(t) = \frac{K(t)}{K(1960)}$, $l(t) = \frac{L(t)}{L(1960)}$, $e(t) = \frac{E(t)}{E(1960)}$. $Y(1960) = \$_{1992}2263 \times 10^9$, $K(1960) = \$_{1992}2685 \times 10^9$, $L(1960) = 118 \times 10^9$ h, $E(1960) = 39051$ PJ

Year	$y(t)$	$k(t)$	$l(t)$	$e(t)$	Year	$y(t)$	$k(t)$	$l(t)$	$e(t)$
1960	1.000	1.000	1.000	1.000	1979	2.046	1.861	1.457	1.851
1961	1.023	1.027	1.018	1.009	1980	2.039	1.920	1.454	1.778
1962	1.085	1.057	1.021	1.059	1981	2.086	1.978	1.466	1.740
1963	1.131	1.089	1.040	1.096	1982	2.042	2.027	1.425	1.653
1964	1.197	1.127	1.064	1.161	1983	2.123	2.074	1.451	1.649
1965	1.273	1.172	1.101	1.215	1984	2.271	2.128	1.523	1.737
1966	1.356	1.220	1.144	1.289	1985	2.353	2.197	1.558	1.735
1967	1.391	1.268	1.154	1.312	1986	2.425	2.256	1.587	1.744
1968	1.456	1.316	1.182	1.388	1987	2.497	2.365	1.631	1.808
1969	1.500	1.350	1.216	1.479	1988	2.592	2.387	1.684	1.881
1970	1.502	1.397	1.202	1.542	1989	2.679	2.450	1.722	1.908
1971	1.551	1.458	1.201	1.588	1990	2.711	2.505	1.738	1.919
1972	1.636	1.504	1.245	1.659	1991	2.686	2.555	1.711	1.908
1973	1.730	1.557	1.296	1.723	1992	2.759	2.566	1.722	1.933
1974	1.719	1.609	1.304	1.679	1993	2.824	2.662	1.756	1.968
1975	1.712	1.653	1.267	1.635	1994	2.921	2.730	1.819	2.015
1976	1.804	1.696	1.307	1.723	1995	2.979	2.819	1.855	2.051
1977	1.889	1.744	1.351	1.778	1996	3.061	2.894	1.892	2.114
1978	1.990	1.801	1.415	1.825					

Appendix 6: Determining Technology Parameters

The nonlinear optimization calculus for computing the technology parameters of the LinEx functions that yield Figs. 4.3–4.6 was performed by Jörg Schmid [71]. His notation for output is that used in prior publications (and in Samuelson's textbook *Economics*), where the symbols Q and q represent output and dimensionless output, respectively. In this appendix (and in Figs. 4.3–4.6), Schmid's notation is used, and the following identities hold: $q_{Lt} \equiv y_{L1}$, $q_{empirical} \equiv y_{empirical}$, and $q_0 \equiv y_0$. The time series of $y_{empirical}$ are shown in Tables 4.11–4.14 as $y(t)$.

According to the theory of partial differential equations, the most general solutions of equations (4.27) are any differentiable functions of l/k and e/k. One could determine them uniquely – and thus compute the exact production function for any given economic system – if one knew α on some boundary curve in k, l, e-space and β at all k, l, e points of a boundary surface [3] that has not more than one point in common with every characteristic basis curve of the third of the differential equations (4.27). A parameter representation of the characteristic basis curves is $k(\theta) = k_0 e^\theta$, $l(\theta) = l_0 e^\theta$, $e(\theta) = e_0 e^\theta$ [3]. Then one would also know the exact output elasticity of energy, $\gamma = 1 - \alpha - \beta$. However, the boundary surface for β and the boundary curve for α that would result from the method of the characteristics are unknown even near $\beta \to 0$ and $l/k \to 0$, $e/k \to 0$, where the asymptotic boundary conditions (4.33) hold. Therefore, one has to rely on the approximate output elasticities (4.34) and determine the technology parameters a and c (and q_0, if necessary) by fitting the LinEx function q_{Lt}, with k, l, and e from the empirical time series for the inputs, to the empirical time series $q_{empirical}$ of the output, subject to the constraints of nonnegative output elasticities, (4.43).

Prior to nonlinear optimization, various methods of fitting had been used. The simplest one, fitting q_{Lt} to $q_{empirical}$ in three subsequent years, was done for the West German total economy and its industrial sector "goods-producing industries" from 1960 to 1978 [24]. The economic recession caused by the first energy crisis and the subsequent recovery were reproduced, and the residuals were small. This method did not work for the sector "industries" of the USA (1960–1978) Because of the nearly parallel rise of k and l in the USA between 1960 and 1973, not observed in other countries such as Germany and Japan, the fit equations for a and c involved quotients of small differences of large numbers, resulting in a wide spectrum for a and c within the error margins of k, l, and e. This well-known problem of collinearity in the USA makes fitting for the USA always difficult and leads some people to question fitting in principle. But, fortunately, collinearity is not such a problem in general. The problem was dealt with (not very elegantly, however) by playing around with the constants a and c until the residuals and the reproduction of the energy crisis were comparable with the results for Germany [24]. Still, in all systems considered there were systematic deviations of q_{Lt} from $q_{empirical}$.

Subsequently, the technology parameters were determined by minimizing the sum of squared errors (SSE),

$$\text{SSE} = \sum_i \left[q_{empirical}(t_i) - q_{Lt}(t_i) \right]^2, \tag{4.127}$$

subject to the constraints (4.43) of nonnegative output elasticities. The sum goes over all years t_i between the initial and the final observation time.

The a and c obtained this way reduced the systematic deviations of the theoretical growth curves from the empirical growth curves, first for West Germany, 1960–1981), and the USA, 1960–1978, and then for West Germany. 1960–1989, and Japan, 1965–1992. Recalibrating the technology parameters in 1978 by increasing the capital effectiveness parameter a and decreasing the energy demand parameter c in the form of step functions improved things further for West Germany, 1960–1989, Japan, 1965–1992, and the USA, 1960–1993, without changing the time-averaged LinEx output elasticities $\bar{\alpha}$, $\bar{\beta}$, and $\bar{\gamma}$ significantly: the elasticity $\bar{\beta}$ of labor does not exceed the order of 0.2, that of energy $\bar{\gamma}$ is above 0.4, and only capital's elasticity $\bar{\alpha}$, with magnitudes between 0.34 and 0.45, is roughly in equilibrium with capital's cost share [20, 28, 29, 37, 38]. Energy-dependent Cobb–Douglas functions (CDEs) with output elasticities close to the time-averaged LinEx elasticities also reproduce the observed economic growth not too badly.

The recalibrated a and c after 1978 are consistent with observed net efficiency improvements and energy-demand reductions of the capital stock in response to the first and the second oil price shocks. The associated abrupt change with time may be interpreted as the result of a "creativity pulse" between 1977 and 1978. However, such a 1-year pulse is a crude approximation of the working of "creativity" at best. Therefore, modeling the technology parameters by functions that change continuously in time tries to improve that, providing also the means of calculating creativity's output elasticity δ. This allows one to see how much of the Solow residual is removed by energy and its non-cost-share weighting, and how much remains unexplained by the factors capital, labor, and energy. Therefore, we have modeled $a(t)$ and $c(t)$ by the logistic functions (4.40) and Taylor series expansions in terms of $t - t_0$ as described in Sect. 4.5.4. In all cases the free coefficients $a_1 \ldots a_s$ and $c_1 \ldots c_r$ are determined by nonlinear regression analysis observing the constraints (4.44). The Levenberg–Marquardt method [68] for the SSE minimization (4.127) and the SAS software package were used. The proper starting values for the numerical iteration (with up to 32,000 iteration steps) are crucial for convergence to the true minimum. Two methods were employed to obtain them:

1. The "brute force" method. Here a lattice is projected into the multidimensional space spanned by the free coefficients, and the SSE is computed for each lattice point. The "coordinates" of the lattice point with the smallest SSE are used as starting values. This method was employed for the total economy of reunified Germany, and as a control of the results for the other systems as well.
2. A new iteration method developed by Julian Henn and employed first in [70]. It is illustrated in Fig. 4.28 and involves the definitions $x_i \equiv 2 - (l_i + e_i)/k_i$ and $y_i \equiv l_i/e_i - 1$, so the LinEx function (4.35) at time t_i can be written as

$$y_{\text{LI}} \equiv q_{\text{LI}} = q_0 \cdot e_i \cdot \exp[a(t_i)(x_i + c(t_i)y_i)] \equiv q_i. \qquad (4.128)$$

In the iteration scheme in Fig. 4.28, $c_{\min}(t)$ and $a_{\max}(t)$ are given by (4.44). The search for the proper starting values according to this scheme proceeds as follows.

(a) Make an educated guess for q_0 and $a(t)$. One option is $q_0 = 1$ and for $a(t)$ the choice of a step function that corresponds to a LinEx fit with piecewise constant a and c and recalibration, e.g., in 1978.
(b) With these q_0 and $a(t_i)$ compute $c(t_i)$ for each point in time t_i from $c(t_i) = \ln(q_i/q_0e_i)/a(t_i)y_i - x_i/y_i$; this is (4.128), resolved with respect to $c(t_i)$.
(c) If many of the $\{c(t_i)\}$ are smaller than the $\{c_{\min}(t_i)\}$ from (4.44), decrease q_0 and repeat step b.
(d) If most of the $\{c(t_i)\}$ are larger than $\{c_{\min}(t_i)\}$, estimate the free coefficients $c_1 \ldots c_r$ of the logistic function or Taylor expansion for $c(t)$, $f(c_1 \ldots c_r)$ so that this $f(c_1 \ldots c_r)$ fits the $\{c(t_i)\}$ satisfactorily.
(e) Compute $a(t_i)$ for each point in time t_i from $a(t_i) = \ln(q_i/q_0e_i)/[x_i + c(t_i)y_i]$; this is (4.128), resolved with respect to $a(t)$. Insert $c(t) = f(c_1 \ldots c_r)$ and q_0 from step d into this equation.
(f) If many of the $\{a(t_i)\}$ are outside the range of the $a(t_i)$ values that are allowed by the first of equations (4.44), decrease q_0 and repeat step e.

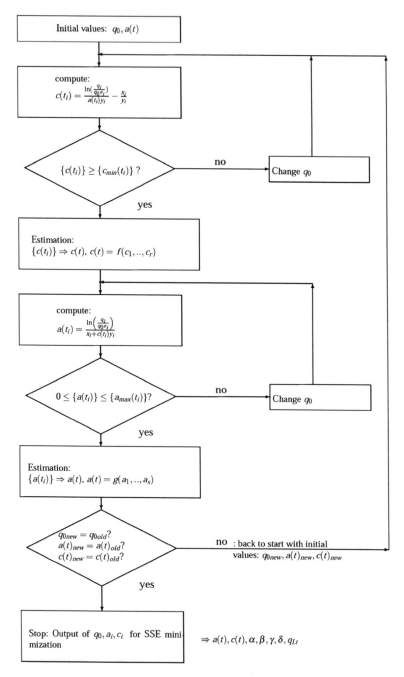

Fig. 4.28 Iteration scheme for the determination of the starting values of the free coefficients in $a(t)$ and $c(t)$ that are to be used for minimization of the sum of squared errors [71]

(g) If most of the $\{a(t_i)\}$ are within the allowed range, estimate the free coefficients $a_1 \ldots a_s$ of the logistic function or Taylor expansion for $a(t)$, $g(a_1 \ldots a_s)$ so that this $g(a_1 \ldots a_s)$ fits the $\{a(t_i)\}$ satisfactorily.

(h) Repeat steps b–g until the q_0, $a(t)$, and $c(t)$ do not change anymore. The corresponding $a_1 \ldots a_s$, $c_1 \ldots c_r$, and q_0 are the starting values for SSE minimization.

References

1. Leontief, W. (Nobel Laureate in Economics): Academic Economics. Science **217**, 104–107 (1982)
2. Gurland, A.R.L.: Wirtschaft und Gesellschaft im Übergang zum Zeitalter der Industrie. In: Mann, G. (ed.) Propyläen Weltgeschichte, Vol. 8, pp. 280–336. Propyläen, Berlin–Frankfurt (1991)
3. Kümmel, R.: Growth Dynamics of the Energy-Dependent Economy. Mathematical Systems in Economics **54**, Eichhorn, W., Henn, R. (eds.). Oelgeschlager, Gunn & Hain, Cambridge (1980)
4. Kümmel, R., Ayres, R.U., Lindenberger, D.: Thermodynamic laws, economic methods and the productive power of energy. J. Non-Equilib. Thermodyn. **35**, 145–179 (2010); doi:10.15.15/JNETDY.2010.009
5. Samuelson, P.A.: Economics, Tenth Edition, International Student Edition. MacGraw-Hill Kogagusha, Tokyo (1976)
6. Stern Review Report on the Economics of Climate Change, ISBN number: 0-521-70080-9, Cambridge University Press (http://www.cambridge.org/9780521700801) 2007.
7. Stern, D.I.: The Economics of Climate Change. Amer. Econ. Rev. **98(2)**, 1–37 (2008)
8. Pack, H.: Endogeneous Growth Theory: Intellectual Appeal and Empirical Shortcomings. J. Econ. Perspect. **8**, 55–72 (1994)
9. Barbier, E.B.: Endogeneous Growth and Natural Resource Scarcity. Environ. Resource Econ. **14**, 51–74 (1999)
10. Welsch, H., Eisenack, K.: Energy Costs, Endogeneous Innovation, and Long-run Growth. Jahr. Nationalökon. Statist. **222/4**, 490–499 (2002)
11. Meadows, D.H., Meadows, D.L., Randers, J., Behrens III, W.W.: The Limits to Growth. Universe Books, New York (1972)
12. Laherrère, J. H.: The reliability of oil and gas reserves data. In: Tolba, M.K. (ed.) Our Fragile World (Forerunner to the Encyclopedia of Life Support Systems) pp. 427-451. UNESCO and Eolss Publishers, Oxford, (2001)
13. Strahan, D.: The Last Oil Shock, John Murray, London (2007)
14. Eichhorn, W., Solte, D.: Das Kartenhaus Weltfinanzsystem. Fischer Taschenbuch Verlag, Frankfurt (2009)
15. Kammer, H.-W., Schwabe, K.: Thermodynamik irreversibler Prozesse. Physik-Verlag, Weinheim (1986)
16. Kluge, G., Neugebauer, G.: Grundlagen der Thermodynamik. Spektrum Fachverlag, Heidelberg (1993)
17. Kümmel, R., Schüssler, U.: Heat equivalents of noxious substances: a pollution indicator for environmental accounting. Ecol. Econ. **3**, pp. 139–156 (1991).
18. Deutscher Bundestag: Dritter Bericht der Enquete Kommission Vorsorge zum Schutz der Erdatmosphäre, Drucksache 11/8030. Bonn (1990)
19. von Buttlar, H.: Umweltprobleme. Phys. Blätter **31**, 145–155 (1975)
20. Hall, C., Lindenberger, D., Kümmel, R., Kroeger, T., Eichhorn, W.: The need to reintegrate the natural sciences with economics. Bioscience **51**, 663–673 (2001)

21. Tryon, F. G.: An index of consumption of fuels and water power. J. Amer. Statistical Assoc. **22**, 271–282 (1927)

22. Binswanger, H.C., Ledergerber, E.: Bremsung des Energiezuwachses als Mittel der Wachstumskontrolle. In: Wolf, J. (ed.) Wirtschaftspolitik in der Umweltkrise, pp. 103–125. dva, Stuttgart (1974)

23. Georgescu-Roegen, N.: The Entropy Law and the Economic Process. Harvard University Press, Cambridge (1971)

24. Kümmel, R.: The impact of energy on industrial growth. Energy—Intntl. J. **7**, 189–203 (1982)

25. Berry, R.S., Salamon, P., Heal, G.: On a relation between thermodynamic and economic optima. Resources Energy **1**, 125–137 (1978)

26. Berry, R.S., Andresen, P.: Thermodynamic constraints in economic analysis. In: Schieve, W.C., Allen, P.M. (eds.) Self-organization and Structures. Application to the physical and economic sciences, Chap. 20. University of Texas Press, Austin (1982)

27. Ayres, R. U., Nair, I.: Thermodynamics and Economics. Phys. Today **37**, 62–71 (1984)

28. Kümmel, R., Strassl, W., Gossner, A., Eichhorn, W.: Technical progress and energy-dependent production functions. J. Econ. (Z. Nationalökon.) **45**, 285–311 (1985).

29. Kümmel, R., Lindenberger, D., Eichhorn, W.: Energie, Wirtschaftswachstum und technischer Fortschritt. Phys. Blätter **53**, 869–875 (1997)

30. van Gool, W., Bruggink, J.J.C. (eds.): Energy and Time in the Economic and Physical Sciences. North-Holland, Amsterdam (1985)

31. Salamon, P., Komlos, J., Andresen, B., Nulton, J.D.: A geometric view of welfare gains with non–instantaneous adjustment. Math. Soc. Sci., **13/2**, 153–163 (1987)

32. Faber, M., Niemes, H., Stephan, G.: Entropy, Environment, and Resources. Springer, Berlin (1987)

33. Faber, M., Proops, J.: Evolution, Time, Production, and the Environment, 2nd ed. Springer, Berlin (1994)

34. Ayres, R.U.: Information, Entropy, and Progress. AIP Press, New York (1994)

35. Daly, H. E.: On Nicholas Georgescu-Roegen's contributions to economics: an obituary essay. Ecol. Econ. **13**, 149–154 (1995)

36. Söllner, F.: Thermodynamik und Umweltökonomie. Physica, Heidelberg (1996)

37. Kümmel, R.: Energy as a factor of production and entropy as a pollution indicator in macroeconomic modelling. Ecol. Econ. **1**, 161–180 (1989).

38. Kümmel, R., Lindenberger, D., Eichhorn, W.: The Productive Power of Energy and Economic Evolution. Indian J. Appl. Econ. **8**, 231–262 (2000). (Special Issue Essays in Honor of Professor Paul A. Samuelson.)

39. Tsirlin, A. M.: Extremal principles and the limiting capabilities of open thermodynamic and economic macrosystems. J. Autom. Remote Ctrl. **66**, 449–464 (2005)

40. Tsirlin, A. M.: Mathematical models and equilibrium in irreversible economics. Mathem. Mod. **21/11**, 47-56 (2009)

41. Solow, R.M.: The Economics of Resources and the Resources of Economics. Amer. Econ. Rev. **64**, 1–14 (1974).

42. Glaser, P.E.: Solar Power from Satellites, Phys. Today, February 1977, 30–38 (1977)

43. Lior, N.: Power from Space. Energy Convers. Manage. **42**, 1789–1805 (2001), and references therein.

44. National Security Space Office: Space-Based Solar Power as an Opportunity for Strategic Security. Report to the Director, National Security Space Office, Interim Assessment, Release 0.1, 10 October 2007.

45. Hudson, E. H., Jorgenson, D.W.: U.S. energy policy and economic growth, 1975–2000. Bell J. Econ. Manag. Sc. **5**, 461–514 (1974)

46. Tintner, G., Deutsch, E., Rieder, R.: A Production Function for Austria Emphasizing Energy. In: Altman, F.L., Kýn, O., Wagner, H.-J. (eds.) On the Measurement of Factor Productivities, pp. 151-164. Vandenhoek & Ruprecht, Göttingen (1974)

47. Griffin, J.M., Gregory, P.R.: An intercountry translog model of energy substitution responses. Amer. Econ. Rev. **66,** 845–857 (1976)

48. Berndt, E.R., Jorgenson, D.W.: How energy and its cost enter the productivity equation. IEEE Spectr. **15**, 50–52 (1978)
49. Berndt, E.R., Wood, D.O.: Engineering and econometric interpretations of energy–capital complementarity. Amer. Econ. Rev. **69**, 342–354 (1979)
50. Jorgenson, D.W.: The role of energy in the U.S. economy. Nat. Tax J. **31**, 209–220 (1978)
51. Allen, E.L.: Energy and economic growth in the United States. MIT Press, Cambridge (1979)
52. Jorgenson, D.W.: The role of energy in productivity growth. Amer. Econ. Rev. **74/2**, 26-30 (1984)
53. Denison, E.F.: Explanation of declining productivity growth. Surv. Curr. Bus. **59/8**, Part II , 1–24 (1979)
54. Kümmel, R.: Energie und Wirtschaftswachstum. Konjunkturpolitik **23**, 152–173 (1977)
55. Groscurth, H.-M., Kümmel, R., van Gool, W.: Thermodynamic limits to energy optimization. Energy—Intntl. J. **14**, 241–258 (1989)
56. Kunkel, A., Schwab, H., Bruckner, T., Kümmel, R.: Kraft–Wärme–Kopplung und innovative Energiespeicherkonzepte. Brennst.–Wärme–Kraft **48**, 54–60 (1996)
57. Lindenberger, D., Bruckner, T., Morrison, R., Groscurth, H.-M., Kümmel, R.: Modernization of local energy systems, Energy—Intntl. J. **29**, 245–256 (2004)
58. Deutsche Bundesbank: Makroökonomisches Mehr-Länder-Modell. Frankfurt, 1996.
59. Solow, R.M.: Perspectives on growth theory. J. Econ. Perspect. **8**, 45–54 (1994)
60. Goldman Sachs: The GCC Dream: Between the BRICs and the Developed World. Global Economics Paper 155, April 17, 2007. https://portal.gs.com
61. Nordhaus, W.: A Question of Balance. Weighting the Options on Global Warming Policies. Yale University Press, New Haven & London (2008) (http://nordhaus.econ.yale.edu/~nordhaus/homepage/Balance_2nd_proofs.pdf.)
62. Hoedl, E.: Socio–ecological market economy in Europe. In: Unger, F. (ed.) European Academy of Sciences & Arts, Activities 2009, pp. 82–88. Salzburg 2010
63. Atkinson, A.B., Rainwater, L., Smeeding, T.M.: Income Distribution in OECD Countries—Evidence from the Luxembourg Income Study. OECD, Paris (1995)
64. Samuelson, P.A., Solow, R.M.: A complete capital model involving heterogeneous capital goods. Quart. J. Econ. **70**, 537–562 (1956)
65. Ramsey, F.P.: A Mathematical Theory of Saving. Econ. J. **38(152)**, 543–559 (1928)
66. Arrow, K.J.: Some ordinalist-utilitarian notes on Rawl's "Theory of Justice". Journal of Philosophy **70(9)**, 245–263 (1973)
67. Hellwig, K., Speckbacher, G., Wentges, P.: Utility maximization under capital growth constraints. J. Math. Econ. **33**, 1–12 (2000)
68. Press, W.H., Teukolsky, S.A., Vetterlin, W.T., Flannery, B.P.: Numerical Recipes in C. Cambridge University Press, Cambridge (1992)
69. Lindenberger, D.: Wachstumsdynamik industrieller Volkswirtschaften: Energieabhängige Produktionsfunktionen und ein faktorpreis-gesteuertes Optimierungsmodell. Metropolis, Marburg (2000)
70. Kümmel, R., Henn, J., Lindenberger, D.: Capital, labor, energy and creativity: modeling innovation diffusion. Struct. Change Econ. Dynam. **13**, 415–433 (2002)
71. Schmid, J.: Diploma Thesis, Fakultät für Physik und Astronomie, Universität Würzburg (2002)
72. Grahl, J., Kümmel, R.: Das Loch im Fass. Energiesklaven, Arbeitsplätze und die Milderung des Wachstumszwangs. Wissenschaft & Umwelt Interdisz. **13**, 195–212 (2009)
73. Shulman, B.: Working and Poor in the USA. The Nation, February 9, 2004; see also http://www.thenation.com/doc/20040209/shulman
74. Lindenberger, D., Eichhorn, W., Kümmel, R.: Energie, Innovation und Wirtschaftswachstum. Z. Energiewirtschaft **25**, 273–282 (2001)
75. Lindenberger, D.: Service production functions. J. Econ. (Z. Nationalökon.), **80**, 127–142 (2003)

76. Stresing, R., Lindenberger, D., Kümmel, R.: Cointegration of output, capital, labor, and energy. Eur. Phys. J. B **66**, 279–287 (2008); see also http://www.ewi.uni-koeln.de/fileadmin/user/WPs/ewiwp0804.pdf

77. Stresing, R.: Energie und Wirtschaftswachstum: Produktionsfunktionen und Kointegrationsanalysen für Deutschland, Japan und die USA. Diploma Thesis, Fakultät für Physik und Astronomie, Universität Würzburg (2005)

78. Engle, R.F., Granger, C.W.J.: Cointegration and Error Correction: Representation, Estimation, and Testing. Econometrica **55**, 251–276 (1987)

79. Dickey, D.A., Fuller, W.A.: Distribution of the Estimators for Autoregressive Time Series with a Unit Root. J. Amer. Statistical Assoc. **74**, 427–431 (1979)

80. Dickey, D.A., Fuller, W.A.: Likelihood Ratio Statistics for Autoregressive Time Series with a Unit Root. Econometrica **49**, 1057–1071 (1981)

81. Hamilton, J.D.: Oil and the Macroeconomy since World War II. J. Polit. Economy **91**, No.2, 224–248 (1993)

82. Hamilton, J.D.: Time Series Analysis. Princeton University Press, Princeton (1994)

83. MacKinnon, J.G.: Critical Values for Cointegration Tests. In: Engle, R.F., Granger, C.W. (eds.) Long-run Economic Relationships: Readings in Cointegration Oxford (1991)

84. Ayres, R.U., Ayres, L.W., Warr, B.: Exergy, power and work in the US economy, 1900- 1998 Energy—Intntl. J. **28**, 219–273 (2003)

85. Ayres, R.U., Warr, B.: Accounting for growth: the role of physical work. Struct. Change Econ. Dynam. **16**, 181–209 (2005)

86. Ayres, R.U., Warr, B.: The Economic Growth Engine. Edgar Elgar, Cheltenham, 2009

87. Ayres, R.U., Warr, B.: Accounting for growth: the role of physical work. In: Max-Planck-Institute for Research into Economic Systems (ed.) Proceedings of the workshop Reappraising Production Theory. Jena (2001)

88. Fisher, F.M.: Aggregation. Aggregate Production Functions and Related Topics. MIT Press, Cambridge, (1993)

89. Kurz H.D., Salvadory, N.: Theory of Production. Cambridge University Press, Cambridge (1995)

90. Kurz H.D.: Wicksell and the problem of the "missing equation" Hist. Polit. Economy **32**, 765–788 (2000)

91. Pasinetti, L.: Critique of the neoclassical theory of growth and distribution. Moneta Credito (Banca Nazionale del Lavoro Quarterly Review) **210**, 187–232 (2000)

92. Felipe, J., Fisher F.M.: Aggregation in production functions: what applied economists should know. Metroeconomica **54**, 208–262 (2003)

93. Silverberg, G.: Private communication (2007)

94. Robinson, J.: The production function and the theory of capital. Rev. Econ. Stud. **21**, 81–106 (1953-54)

95. Robinson, J.: The measure of capital: the end of the controversy. Econ. J. **81**, 597–602 (1971)

96. Usher, D.: Introduction. In: Usher D. (ed.) The Measurement of Capital, pp. 1–21. University of Chicago Press, Chicago(1980)

97. Witt, U.: "Production" in Nature and Production in the Economy—Second Thoughts About Some Basic Economic Concepts. Struct. Change Econ. Dynam. **16(2)**, 165–179 (2005)

98. Paech, N.: Wachstum "light"—Qualitatives Wachstum ist eine Utopie. Wissenschaft & Umwelt **13**, 84–91 (2005)

99. Wikipedia, The Free Encyclopedia: "Invisible Hand".

100. Bertola, G.: Factor Shares and Savings in Endogenous Growth. Amer. Econ. Rev. **83(5)**, 1184–1198 (1993)

101. OECD Berlin Centre, OECD-Publikationen, Neuerscheinungen Januar 2010, ISBN 978-92-64-06883-4

102. Shah, A.: Poverty Facts and Stats. Global Issues http://www.globalissues.org/article/26/poverty-facts-and-stats

103. OECD/IEA 2009, http://iea.org/textbase/pm/?mode=pm&id=1573&action=detail

104. Kunz, H.: A primer on current economic conditions. Institute for Integrated Economic Research, Meilen (2009); http://www.iier.ch

105. Solte, D.: Weltfinanzsystem am Limit—Einblicke in den "Heiligen Gral" der Globalisierung. Terra Media Verlag, Berlin (2007)

106. Hall, C., Powers, R., Schoenberg, W.: Peak oil, EROI, investments and the economy in an uncertain future. In: Pimentel, D. (ed.) Biofuels, Solar and Wind as Renewable Energy Systems, pp. 113–136. Elsevier, London (2008)

107. Hohmeyer, O., Ottinger, R.L. (eds.): External Environmental Costs of Electric Power, Springer, Berlin (1991)

108. The Siemens cost estimate was presented during a hearing in the German Federal Ministry of Research and Technology in 1990.

109. Kümmel, R.: Ökonomische Bewertungen der Klimawandel-Folgen. In: Keilhacker, M. (ed.) Weltklima und zukünftige Energieversorgung, pp. 73–89. Deutsche Physikalische Gesellschaft, Bad Honnef (2007)

110. Daly, H.: When smart people make dumb mistakes. Ecol. Econ. **34**, 1–3 (2000).

111. Nordhaus, W.: Science, Sept. 1991, 1206 (1991)

112. Beckermann, W.: Small is Stupid. Duckworth, London (1997)

113. T.C. Schelling, T.C.: The Cost of Combating Global Warming. Foreign Affairs, November/December 1997, 9 (1997)

114. Knizia, K.: Kreativität, Energie und Entropie. Econ, Düsseldorf (1992)

115. Callen, H.B., Welton, Th.A.: Irreversiblity and Generalized Noise. Phys. Rev. **83**, 34–39 (1951)

116. Reif, F.: Fundamentals of Statistical and Thermal Physics, Chap. 15, pp. 594–600. McGraw-Hill, New York (1965)

Chapter 5
Epilogue: Decisions Under Uncertainty

5.1 Ethics

Economic growth at the rates of the past will become increasingly difficult if restrictions on energy conversion tighten, for whatever reasons. Sure, there is still plenty of coal, and renewable energies are nearly inexhaustible. But the burning of fossil fuels threatens climate stability, and shifting climate zones will stimulate large-scale migrations. Energy conservation may provide unchanged energy services with declining primary energy inputs until its thermodynamic limits have been reached. Chances are that nuclear fission and fusion, geothermal and wind power, terrestrial solar farms, and solar power satellites will open up abundant energy sources. But whether any or a combination of these options will allow us to maintain sufficient growth so that hunger, disease, and pollution can be fought successfully, and social breakdowns and major wars can be avoided in a world of nuclear weapons, growing population, and threatening financial collapses is an open question. To find out what can be done, intensive research and development is necessary. This requires heavy investments. Can the people in the rich industrial countries be motivated to spend their money on that rather than continuing with their lavish consumption?

This is a question of morals – the set of binding values accepted by a person – and of ethics – the standards of behavior within the social system in which the person lives. Philosophy and religion deal with both. This is not the place to go into this in more detail. Instead, it is sufficient to quote a statement from *John Bardeen*,[1] when he was once surprised by a journalist with a question concerning religion. He said: "I am not a religious person, and I do not think about it very much. ... I feel that science cannot provide an answer to the ultimate questions about the meaning and purpose of life. With religion, one can get answers of faith. Most scientists leave

---

[1] Bardeen was awarded the Nobel Prize in physics twice: in 1956 for the invention of the transistor and in 1972 for the theory of superconductivity.

R. Kümmel, *The Second Law of Economics: Energy, Entropy, and the Origins of Wealth,* 275
The Frontiers Collection, DOI 10.1007/978-1-4419-9365-6_5,
© Springer Science+Business Media, LLC 2011

them open and perhaps unanswerable, but do abide by a code of moral values. For civilized society to succeed, there must be a common consensus on moral values and moral behavior, with due regard to the welfare of our fellow man. There are likely many sets of moral values compatible with successful civilized society. It is when they conflict that difficulties arise" [1].

Due regard to the welfare of our fellow humans is *the* ethical command. But opinions differ on how to implement it in practice. The problems of appropriate energy utilization and consumption of material goods reveal conflicting morals and ethics when we have to answer the question: "Who are our fellow humans?" Thinking of our contemporaries, we have to choose from the local, the national, and the global level. Thinking of our descendants, we must decide for how many generations we should feel responsible.

Some examples may illustrate the ethical problems that are interwoven with the economic and physical problems:

- Discounting the future, i.e., assigning less value to future benefits and losses than to present ones, is a controversial issue in business ethics. There are economic thinkers and philosophers who reject it [2–4], whereas most practitioners accept it, because of the well-known individual time preference for the present, demonstrated, e.g., by smokers. If one believes that the social optimum is obtained by the "Invisible Hand" when everybody follows his self-interest in competitive markets with a minimum of regulations, and if one extends this belief to the future, one may favor high discount rates and reject precautionary measures against future damage that might slow down economic growth. If, on the other hand, one shares the view of many natural scientists that we should preserve the ecological stability of "Our Fragile World" [5], so that our descendants can enjoy the beauty of the Blue Planet as we do, one prefers small or zero discount rates, even at the cost of economic growth. What will benefit present and future generations more? Can models of intertemporal welfare optimization, as sketched in Chap. 4, provide answers?
- There are people who evaluate the risk of nuclear waste disposal for future generations higher than the risk of forgoing the present economic benefits of electricity from nuclear power plants. They feel morally and ethically obliged to oppose atomic energy at the ballot box and in the streets. And there are people with the opposite risk assessment. The first group is quite influential in Germany, the second group dominates in France. This establishes conflicting energy ethics even among those who agree that one can build nuclear power plants with vanishing probability of a meltdown.
- Hypothetically, 4,030 TWh, the equivalent of German primary energy consumption in 2005, could be produced annually in Germany by solar energy if one would cover about 40,000 km^2, a little more than 11% of German territory, with photovoltaic cells. More readily available is 800 km^2 of roof area. But the price of electricity from photovoltaics is two–five times average residential electricity tariffs. Although the price is expected to decrease with technical progress in photovoltaic module production, it is not clear whether the people who oppose

nuclear and fossil energy will be willing to pay for the high photovoltaic-generating cost once it dominates electricity bills. Even now, supporters of photovoltaics are pleading for a reduction of feed-in tariffs so that the profits of firms manufacturing photovoltaic modules no longer increase at the expense of electricity consumers. This is countered by reasoning that these profits stimulate further expansion of photovoltaics and the cost should be considered as an insurance premium against energy scarcity at the end of the fossil-fuel era. Further controversies are to be expected about where to put the photovoltaic modules once the readily available roof area has been covered. The alternative of building solar thermal power plants in the Sahara to supply electricity to North African and European countries requires the solution of many political questions and investments of about €400 billion, as estimated by the DESERTEC consortium. And how seriously should one take the difficulties that may arise from social, cultural, and religious differences between the countries involved in the DESERTEC project?

- The very issue of economic growth. Quite a few academics feel obliged to advocate an end of economic growth in the rich industrial countries in order to conserve natural resources, protect the environment, and leave room for some growth in the emerging and developing economies. They point to wasteful consumption and production such as driving around in cities in big, gas-guzzling cars, flying thousands of miles for a shopping spree, shipping potatoes, just to wash them, from one country to another, and flying air freight from one city to its close-by neighbor via a very distant central-distribution terminal; also huge amounts of energy and materials are expended on products that are only bought because of the status they are supposed to confer, high-quality oil and gas are burned just to heat poorly insulated homes, and so on. Disagreeing with Say's theorem – "Each supply creates its own demand" – they expect an end to economic growth anyway, because of market saturation, and reason that we ought to stop growth of consumption before we vomit. On the other hand, under the present legal framework of the market, no one knows how to avoid rising unemployment and state indebtedness without economic growth. And, perhaps most importantly in the long run, historic experience with the no- or slow-growth agrarian economies before the Industrial Revolution raises the concern that future stationary societies can be stable only if they restrict personal freedom and social mobility by unbending rules of class, guild, and caste. Examples are medieval Europe, the Aztec and Inca empires in Central America and South America, decapitated by the Spanish conquistadors, and India before the invasion of conquerors from Central Asia and Europe. A certain social mobility existed in China within the Mandarin class, since at least theoretically the examinations on knowledge of Confucian texts, which provided access to and promotion within this class of privileged and wealthy civil servants, were open to even the most humble born. Nevertheless, on the whole, China changed very little under the administrative system of Confucian scholar-bureaucrats. Which risks are ethically more acceptable: those of stationary societies or those of attempts to maintain growth?

- Advocates of space industrialization, as outlined by O'Neill's *The High Frontier* and the proceedings of the Princeton University Conferences on Space Manufacturing Facilities [6], proclaim the ethical command: "If you love the Earth, leave it!" Doubting that stationary societies are compatible with freedom and progress, they favor investments in research and development that open up the resources of the Sun and space. One can expect more than sufficient volunteers who accept the personal risks involved in building solar power satellites and space habitats from extraterrestrial materials and solar energy. But what is ethically preferable: employ resources to surmount the limits to growth by going into space, or employ them to improve the living conditions on Earth within the limits drawn by the first law and the second law of thermodynamics?

Since ethics does not provide clear answers to the new questions raised by modern rapid technological development and social change, we must feel our way along a path of economic evolution that avoids a Dark Age of battles for diminishing resources. The first decision to be taken under uncertainty is how to adjust the legal framework of the market to the challenges of a future ruled more clearly than ever by energy and entropy. It should be guided by socially enlightened self-interest and reason.

5.2 Reason

The peaceful end of the Cold War gives hope in reason. True, the threat from the atomic mushroom constrained military adventurers in East and West. But still, the temptation must have been strong for the supreme command of the Warsaw Pact troops to use their tremendous superiority in conventional arms to grab the meat pots of western Europe in a 5-day dash from East Germany to the Bay of Biscay. This might have delayed the economic decline of the Soviet empire for some time if NATO had chosen not to turn Germany and Poland into a nuclear battlefield.

The danger of World War III being reduced, the task is now to stabilize the global social and ecological system. We must find answers to the question of how the gifts of nature and of human creativity ought to be shared among the people of the present and the future, under due observation of human dignity and responsibility. Our foresight and moral powers are only too limited. But we can strengthen and empower the international institutions created in the spirit of the Declaration of Human Rights after the devastation and moral catastrophes of World War II. These institutions should promote economic instruments that stimulate progress in energy technologies of large potential and small environmental impact, and stop the trend toward growing wealth for the few and diminishing resources for the many. The supranational institutions will need the authority and power to make and enforce internationally binding laws – and democratic control of that power. The task of establishing such institutions of global governance is tremendous. It may be accomplished within the framework of the United Nations. Hopefully, the evolution of the European Union will provide an encouraging example.

After the ratification of the treaty of Lisbon, the European Union acquired supranational powers beyond matters of trade, for instance, in foreign policy. Taxes, however, have remained exclusively in the hands of the individual member states. On the other hand, a strong need to harmonize the international tax system is one of the conclusions drawn by analysts of the first big financial and economic crisis of the twenty-first century [7].

Blaise Pascal (1623–1662), the French mathematician, physicist, philosopher, and religious thinker, muses in his thoughts about human nature [8]: "We run unconcernedly into the abyss after we have put something in front of us that prevents us from seeing it." Despite this sobering view of what we usually do, let us envision the following political scenario.

Shocked by the financial and economic crisis that was triggered by the bursting of the US mortgage bubble, the governments of Europe understand that they must closely cooperate in economic matters and improve the legal framework of the market. Simultaneously, the peoples of Europe realize that their voting decisions should be more determined by their long-term interest in environmental and social stability than by the short-term interest in enjoying the services of cheap energy slaves for a little more time. They agree to transfer taxation power to a European Parliament to which they elect wise, independent, and courageous men and women. These deputies have learned that financial and tangible assets have grown more rapidly than global GDP, and that the share of lowly taxed income from property has increased relative to other, higher-taxed earnings, with a relative decrease of public revenues [7]. They are also aware of the tremendous social, political, and institutional obstacles to changes of long-established tax systems [9]. Nevertheless, complying with their promise of legal action against unemployment, growing state indebtedness, and pollution, they introduce legislation for a sweeping social–ecological tax reform, with the following preamble:

"Considering that

- there are abundant opportunities for jobs that can best be done by human hands, hearts, and brains,
- these jobs are in social services – intensive training and education for the socially or physically handicapped and care for children, the sick, and old people; in handicraft services – maintenance and repair instead of running down, throwing away, and purchase anew; in general education, higher learning, and research; and in all fields that require and promote human creativity,
- individuals and society cannot afford these jobs in sufficient quantities because of the high taxes and levies on labor,
- the contributions of a decreasing number of employees, and of their employers, to the system of social security are insufficient to finance the social benefits to which an increasing number of retired persons are entitled,
- the most powerful laws of nature, which couple wealth creation to energy conversion and emissions, call for innovative investments in energy conservation and in those non-fossil-fuel technologies that emit little of the most bothersome pollutants,

- with low energy prices the return on these investments is usually smaller than the return on the conventional combustion of fossil fuels,
- it is easy to conceal financial assets and evade their taxation,
- it is hard to conceal energy flows and easy to measure them,

be it resolved that

- to fight increasing unemployment and state deficits, and stimulate energy conservation and emission mitigation, the differences between the productive powers and cost shares of labor and energy are to be reduced by a substantial decrease of the taxes and levies on labor and a corresponding increase of the taxes on energy, where gradually increasing energy taxes will not result in economic recessions as after the oil price shocks, because the money will be redistributed only within the European Union countries and will not be transferred to the foreign owners of energy sources,
- the tax per energy (exergy) unit shall increase at the same rate as it succeeds in stimulating energy conservation [10], where the tax basis will never shrink to zero, thanks to the laws of thermodynamics, as long as boring, hard, and dangerous work is performed by energy-powered machines,
- to minimize problems in the competitiveness of energy-intensive industries and minimize energy leakage, negotiations shall be initiated by the Commission of the European Union with all trading partners to introduce such energy taxes in an internationally harmonized and gradual manner [11],
- alternatively, border tax adjustments shall be introduced, refunding the energy tax to exporters and imposing duties on imported goods and services according to the energy consumed in their production and transportation [12],
- social hardships are to be avoided by refunding the energy tax to low-income consumers via existing transfer channels."

The vision of such a legislative initiative is not completely unrealistic, because in the 1990s the Commission of the European Union proposed several energy taxation schemes to its member states [13]. They were initially well received by governments, but then successfully torpedoed by powerful special interest groups. Nevertheless, the idea of the ecological tax reform is alive. And the proposal of shifting financial burdens from the production factor labor to the production factor energy is supported not only by the most fundamental laws of physics and their consequences for ecological stability, but also by production theory and its consequences for social and fiscal stability.

A BBC World Service poll [14], in cooperation with the Program on International Policy Attitudes, asked over 22,000 citizens all over the world whether they would support higher energy taxes on the most harmful types of energy so that individuals/industry use less.[2] The proportion of Chinese favoring higher taxes was

---

[2] A total of 22,182 citizens in Australia, Brazil, Canada, Chile, China, Egypt, France, Germany, India, Indonesia, Italy, Kenya, Mexico, Nigeria, the Philippines, Russia, South Korea, Spain, Turkey, the UK, and the USA were interviewed face to face or by telephone between May 29

24 points greater than the next largest majorities in Australia and Chile (61% in both). This was followed by Germans (59%), Canadians (57%), Indonesians (56%), Britons (54%), and Nigerians (52%). In countries such as France and India, opinion was mixed, and in seven countries, among them the USA, a majority was initially opposed to higher energy taxes. The overall percentage in favor was 50% when the tax revenue was not explained. However, if the energy tax revenue were dedicated to clean/efficient energy or to reducing other taxes so that the total tax bill would stay the same, about 75% would favor higher energy taxes. People seem to be ready for a change from wasteful to thoughtful energy use. The question is whether reason will also prevail in the masters of the energy slaves.

Energy taxation is a necessary but not sufficient condition for system stabilization. In the face of uncertainties, it should be just the first step to be taken in a cautious strategy of steering the economy away from the path of self-destruction. Additional legal instruments must be developed and applied to domesticate the financial markets and establish a general framework that guides the economy toward sustainable development. The design of such a general framework is a challenge to knowledge and reason.

Hope in reason, from Socrates to the Age of Enlightenment, has produced mixed results. Will ethics and reason suffice to move us to actions that help prevent a Dark Age? Or will people only learn by suffering, as so often in history?

The historian and philosopher Arnold J. Toynbee [15] points to the principle of challenge and response in the evolution of civilizations: "Man achieves civilization as a response to a challenge in a situation of special difficulty which rouses him to make a hitherto unprecedented effort." Ours is one global civilization. If it fails, there is no other one to take over. But the spark of human creative power is still alive. It can ignite the successful response to the challenges from energy and entropy if the intellectual and spiritual conditions of society are right. This implies that the complex system "economy and nature" must be thoroughly understood. And the moral code of individuals, and global ethics as well, should be shaped by the Golden Rule "Treat others as you would like to be treated," which has been handed down to our times from the ancient civilizations of Babylon, China, Egypt, Greece, India, and Judea.

References

1. Hoddeson, L., Daitch, V.: True Genius. The Life and Science of John Bardeen. Joseph Henry Press, Washington (2002), p. 169
2. Ramsey, F.P.: A Mathematical Theory of Saving. Econ. J. **38(152)**, 543–559 (1928)

---

and July 26, 2007. Polling was conducted for the BBC World Service by the international polling firm GlobeScan and its research partners in each country. In eight of the 21 countries, the sample was limited to major urban areas. The margin of error per country ranged from 2.4% to 3.5%.

3. Arrow, K.J.: Some ordinalist-utilitarian notes on Rawl's "Theory of Justice". Journal of Philosophy **70(9)**, 245–263 (1973)
4. Birnbacher, D.: Intergenerationelle Verantwortung oder: Dürfen wir die Zukunft der Menschheit diskontieren? In: Klawitter, J., Kümmel, R. (eds.) Umweltschutz und Marktwirtschaft, pp. 101–115. Königshausen & Neumann, Würzburg (1989)
5. Tolba, M.K. (ed.): Our Fragile World—Vols. I and II of the Forerunner to the Encyclopedia of Life Support Systems. UNESCO and Eolss Publishers, Oxford (2001)
6. Grey, J., Hamdan, L.A., Faughanan, B., Maryniak, G. et al. (eds): Space Manufacturing, Vols. 1–10. American Institute of Aeronautics and Astronautics, Reston (1975–1995)
7. Eichhorn, W., Solte, D.: Das Kartenhaus Weltfinanzsystem. Fischer Taschenbuch Verlag, Frankfurt (2009)
8. Pascal, B.: Pensées, No. 187. Translated by Wolfgang Rüttenauer, Deutsche Buch-Gemeinschaft, Berlin (1964); p. 83
9. Armingeon, K.: Energiepolitik in Europa: Hindernisse umweltpolitischer Reformen. In: Pfister, Ch. (ed.) Das 1950er Syndrom, pp. 377–389. Haupt, Bern (1995)
10. Verbruggen, A.: Electricity intensity backstop level to meet sustainable backstop supply technologies. Energy Policy **34**, 1310–1317 (2006)
11. von Weizsäcker, U., Jesinghus, J.: Ecological Tax Reform: A Policy Proposal for Sustainable Development. Zed Books, London (1992)
12. Baron, R.: Competitive Issues Related to Carbon/Energy Taxation. Annex I Expert Group on the UN FCCC. Working paper 14. Paris, ECON-Energy (1997)
13. Kümmel, R.: Energie und Kreativität, pp. 94–110. B.G. Teubner, Stuttgart (1998)
14. BBC World Service Poll (2007): http://www.bbcworldservice.com; contact with the participating pollsters via: dough.miller@globescan.com and skull@pipa.org .
15. Toynbee, A.J.: A Study of History—Abridgement of Vols. I-VI by D.C. Sommervell. Oxford University Press, Oxford (1946)

Glossary

Aggregation represents different things by one single quantity. Output and fixed capital are aggregated monetarily in deflated dollars, euros, and other currencies. Alternatively, they can be aggregated physically in terms of work performance and information processing. Equivalence factors relate one aggregation scheme to the other.

Cost-share theorem says that the output elasticity of a production factor is equal to the ratio of the cost of that factor to the total cost of all factors. The theorem follows from the optimization of profit or time-integrated utility if one disregards technological constraints on factor combinations. Inclusion of such constraints in the optimization calculus destroys the cost share–theorem via shadow prices.

Energy is the capacity to cause changes in the world. It is stored in matter and force fields. Energy, including the energy equivalent of mass, is a conserved quantity, which can be neither created nor destroyed.

Energy return on investment (EROI) is the ratio of the energy supplied by a process to the energy used directly and indirectly in that process.

Energy slave is the energy service performed by an energy conversion device that is energetically (in enthalpy terms) equivalent to the daily human work – calorie requirement of 2,500 kcal (equivalent to 2.9 kWh) for a very heavy work load.

Enthalpy measures energy quantity, for instance, in joules (legal unit), kilowatt-hours (appear on consumer energy bills), and tons of coal/oil equivalents, or Board of Trade Units (shown in energy balances).

Entropy is the physical measure of disorder. Energy conversion processes produce entropy. Entropy production is associated with the emissions of heat and particles.

Equilibrium is the state of a system in which an appropriate objective function assumes an extremum.

R. Kümmel, *The Second Law of Economics: Energy, Entropy, and the Origins of Wealth*, The Frontiers Collection, DOI 10.1007/978-1-4419-9365-6, © Springer Science+Business Media, LLC 2011

Exergy is the valuable part of energy which can be converted into any form of useful physical work. The fossil and nuclear fuels, and solar radiation as well, are practically 100% exergy. The quality of an energy quantity is given by the ratio "exergy/enthalpy." Entropy production decreases exergy and increases anergy, which is the useless part of energy. Heat at the temperature of the environment is practically 100% anergy.

Gross domestic product (GDP) is the market value of all goods and services produced within a country in 1 year. It is value added, generated by work performance and information processing within the country. It does not include the value of imported intermediate goods and services.

Objective function (objective) describes a quantity to be maximized or minimized in optimization. Examples of objective functions in physics are the Gibbs free energy for a many-particle system in contact with a reservoir at constant temperature and pressure, and the time integral of the Lagrange function for many particles in force fields. Examples of objective functions in economics are profit and time-integrated utility (overall welfare).

Optimization determines extremal values of objective functions.

Output of an economic system is the value added of all goods and services produced by the system within a given time span. The output of a country in 1 year is the gross domestic product.

Output elasticity of a production factor gives, roughly speaking, the percentage of output change when the factor changes by 1% while the other factors stay constant.

Peak oil means that oil production, as a function of time, will increase and then decline inevitably in the form of a roughly bell shaped curve. The maximum of oil production is "Peak Oil."

Production factors (inputs) produce the output of an economic system.

Production functions describe the dependence of output on the production factors.

Productive power is synonymous with "output elasticity."

Reserves of energy carriers are the occurrences of energy carriers that have been identified and measured and that are known to be technically and economically recoverable.

Resources of energy carriers are all occurrences of energy carriers with less certain geological assurance and/or doubtful economic feasibility.

Shadow prices of production factors that are subject to constraints in optimization translate these constraints into monetary terms. They are given by the Lagrange multipliers and the gradients of the constraint functions in the extremum. The derivation of the optimum shows that the shadow price measures the change of the objective if each constraint is loosened by one unit.

Slack variables change inequality constraints into equality constraints, which can be taken into account by the method of Lagrange multipliers in optimization calculus.

Systems consist of components, constraints, and boundaries. Components are animate or inanimate parts of the energy–matter world, or both. Constraints restrict the motions, actions, and interactions of the components. Boundaries – material or immaterial ones – separate systems from one another.

Technological constraints on the combinations of capital, labor, and energy are limits to automation and limits to capacity utilization.

Index

Printed in the United States
by Baker & Taylor Publisher Services